Microwave Assisted Organic Synthesis

This book is dedicated to the living memory of

Robert Charles Lazell
High Throughput Chemistry, GSK Harlow

Born 14th January 1976
Died 5th August 2004

Microwave Assisted Organic Synthesis

Edited by

Jason P. Tierney
GlaxoSmithKline
Harlow, UK

Pelle Lidström
Biotage AB
Uppsala
Sweden

Blackwell
Publishing

CRC Press

Editorial offices:
Blackwell Publishing Ltd, 9600 Garsington Road, Oxford OX4 2DQ, UK
   Tel: +44 (0)1865 776868
Blackwell Publishing Asia Pty Ltd, 550 Swanston
   Street, Carlton, Victoria 3053, Australia
   Tel: +61 (0)3 8359 1011

ISBN 978-1-4051-1560-5

Published in the USA and Canada (only) by CRC Press LLC, 2000 Corporate Blvd., N.W., Boca Raton,
FL 33431,   USA
Orders from the USA and Canada (only) to CRC Press LLC

USA and Canada only:
ISBN 0-8493-2371-1

First published 2005

4   2007

Library of Congress Cataloging-in-Publication Data:
A catalog record for this title is available from the Library of Congress

British Library Cataloguing-in-Publication Data:
A catalogue record for this title is available from the British Library

Set in 10.5/12 pt Minion
by TechBooks
Printed and bound in India
by Replika Press Pvt. Ltd, Kundli

For further information on Blackwell Publishing, visit our website:
www.blackwellpublishing.com

Contents

Contributors

D. Michael P. Mingos Chemistry Department and St Edmund Hall, University of Oxford, Queens Lane, Oxford OX1 4AR, UK. michael.mingos@seh.ox.ac.uk

Kristofer Olofsson Biolipox, Box 6280, SE-10234, Stockholm, Sweden. kristofer.olofsson@biolipox.com

Mats Larhed Department of Medicinal Chemistry, Organic Pharmaceutical Chemistry, BMC, Uppsala University, Box 574, SE-75123 Uppsala, Sweden. mats@orgfarm.uu.se

Thierry Besson Laboratoire de Biotechnologie et Chimie Bioorganique associé au CNRS, UFR Sciences Fondamentales et Sciences pour l'Ingénieur, Bâtiment Marie Curie, Université de La Rochelle, 17042 La Rochelle cedex, France. thierry.besson@univ-lr.fr

Christopher T. Brain Novartis Institute for Medical Sciences, 5 Gower Place, London WC1E 6BS, UK. christopher.brain@pharma.novartis.com

Timothy N. Danks The Oratory School, Woodcote, Reading, Berkshire RG8 0PJ, UK. T.Danks@oratory.co.uk

Gabriele Wagner Department of Chemistry, University of Surrey, Guildford, Surrey GU2 7XH, UK. G.Wagner@surrey.ac.uk

Jacob Westman Associate Professor, MedChemCon, Ålands-Västerby, SE-74020 Vänge, Sweden. jacob.westman@medchemcon.com

Ian R. Baxendale RSC Wolfson Senior Research Scientist, Department of Chemistry, University of Cambridge, Lensfield Road, Cambridge CB2 1EW, UK. irb21@cam.ac.uk

Dr. Ai-Lan Lee Department of Chemistry, University of Cambridge, Lensfield Road, Cambridge CB2 1EW, UK. leevz@bc.edu

Prof. Steven V. Ley B01702 Prof. of Organic Chemistry, Department of Chemistry, University of Cambridge, Lensfield Road, Cambridge CB2 1EW, UK. svl1000@com.ac.uk

Alexander Stadler Karl-Franzens-University Graz, Institute of Chemistry, Heinrichstrasse 28, A-8010 Graz, Austria. Alexander_stadler@gmx.at

C. Oliver Kappe Karl-Franzens-University Graz; Institute of Chemistry, Heinrichstrasse 28, A-8010 Graz, Austria. oliver.kappe@uni-graz.at

Christopher R. Sarko Department of Medicinal Chemistry, Boehringer Ingelheim Pharmaceuticals, Inc., Research Development Center, 900 Ridgebury Rd, 06877-0368, Ridgefield, CT, USA. csarko@rdg.boehringer-ingelheim.com

Christopher R. Strauss Centre for Green Chemistry, Monash University, Clayton, Victoria, 3800, Australia. chris.strauss@sci.monash.edu.au

Brett A. Roberts CSIRO Molecular Science, Private Bag 10, Clayton South 3169, Victoria, Australia. Brett.Roberts@csiro.au

Preface

Microwave-assisted organic chemistry has during the last years moved from being an obscurity in the laboratory environment to be an invaluable tool within chemistry research. Although the first reports on microwave-assisted organic synthesis dates back as far as 1986, the breakthrough of the technique as a routine tool in synthesis has been slow. The main reason has been difficulties in conquering the forces of the flame, i.e. there has been a lack of dedicated equipment available to perform chemistry using microwave irradiation. This lack of dedicated equipment led to the use of domestic appliances, leading to very unpredictable and sometimes devastating results. It also gave the technique an aura of black art. However, with the introduction of dedicated equipment, novel, interesting, reproducible chemistry has been and is continuously performed.

In this book we have tried to assemble a selection of authors to shine light on the underlying principles of microwave dielectric heating, how this dielectric heating has been used in chemistry to give us microwave-assisted organic synthesis applied on a wide variety of reaction types as well as on how microwave-assisted organic synthesis has impacted the chemistry research within industry.

These chapters have been written by some of the most prominent researchers of modern microwave-assisted organic synthesis and we hope that you will find it both interesting and enlightening.

Jason P. Tierney
Pelle Lidström

1 Theoretical aspects of microwave dielectric heating

D. MICHAEL P. MINGOS

1.1. Introduction

The growth of chemistry has been closely associated with the discovery of new reagents and new modes of introducing energy into chemical reactions. In the days of alchemy, energy could only be generated thermally by means of fire, but the discovery of the lens, which could focus sunlight onto a chemical reaction vessel, showed that the energy necessary to drive a chemical reaction to completion could be derived from other sources. Humphrey Davy's isolation of sodium, potassium, calcium, strontium and barium as pure metals, using electrolytic methods in the early nineteenth century, provided an excellent illustration of how an alternative energy source could open up whole new areas of chemistry. Since those days chemists have maintained a watching brief on new energy sources in the hope that these might similarly lead to innovative synthetic possibilities.

The first reliable device for generating fixed frequency microwave radiation was designed by Randall and Booth at the University of Birmingham during World War II. The magnetron was produced in large numbers during the war because it formed the basis of radar transmitters, particularly in aircraft and anti-aircraft batteries. At that time it was widely recognised that infrared radiation and visible light were capable of stimulating chemical reactions and therefore it did not come as a major surprise when it was first observed that microwave radiation was able to heat foodstuffs. Indeed there are several apocryphal tales recounting how bars of chocolate, eggs or popcorn reacted dramatically when exposed accidentally to microwave sources. The first patent for microwave dielectric heating was filed by the Raytheon Company in 1946 and a prototype oven was installed in a Boston restaurant. Commercial microwave ovens became available in 1947, at a cost of $5000; however, they were nearly 2 m high and weighed more than 350 kg. Notwithstanding these clumsy origins, by 1976 60% of US households had microwave ovens[1]. This widespread use of microwave ovens in the homes can be attributed to the effective technology transfer, which was achieved by Japanese electronics companies and their professional marketing of the products all over the world. It owed much to the Japanese genius for producing electronic goods that were cheap, reliable and consumer friendly.

From these early days, it was recognised that the rapid heating of foodstuffs in a microwave cavity arose because of their high water content and the consequent efficient conversion of microwave energy into thermal energy by water molecules at microwave frequencies. Therefore, in addition to the development of large-scale applicators for the food processing industry, the advantages of microwave dielectric heating for drying was recognised in the 1950s and 1960s; for example, DuPont built large-scale facilities for drying nylon based on microwave technology. Since the microwave radiation interacted solely with the water molecules and not the nylon, this technique had the advantage that the drying process would halt once all the water molecules had evaporated and the substrate was not scorched[2]. The advantages of microwave dielectric heating for analytical processes were also recognised and specially designed microwave applicators for chemical analyses were soon available in the marketplace. These applications utilised the interactions between microwave energy with a frequency of 2.45 GHz and water either on the surface of the analytical samples or in the acidic solutions and therefore techniques were developed initially for drying of wet analytical samples and eventually for the acid digestion of samples. The rate-determining step in a complete analytical procedure is often the dissolution of the sample and therefore microwave-accelerated acid digestion made an important contribution to improving the overall efficiency of the analytical process[3]. There were also reports in the materials literature during the 1970s of microwave dielectric heating being used for ceramic processing and calcining[4]. Examples of encapsulating radioactive samples in inorganic glasses by using microwave dielectric heating were also patented. During this time, the application of microwave dielectric heating in chemical laboratories remained very limited. The perception that the microwave dielectric heating phenomenon arose exclusively from specific interactions between microwave radiation and water molecules contributed to this lack of interest. Early measurements by von Hippel and his co-workers at MIT[5] established that this property was not limited to water and there were other materials capable of coupling effectively with microwaves, but these results did not percolate into the chemical consciousness. The realisation that it was possible to use microwave energy to accelerate reactions in organic solvents resulted from empirical observations rather than theoretical considerations, but nonetheless represented an important development for synthetic chemistry. Following on from the important observations by Gedye[6], Majetich[7] and their co-workers in 1986 that a range of organic reactions could be accelerated under microwave conditions, the use of microwave dielectric heating in organic, inorganic and organometallic chemistry[8,9] has expanded very rapidly and now there are more than 2000 papers describing the application of this technique for the synthesis of new compounds. The early papers in this field used either domestic microwave ovens, adapted versions of them or cavities initially designed for analytical science, and their use raised many questions and speculations concerning the mechanism by which chemical reactions were accelerated. In particular, it was necessary to solve a range of technical problems associated with the safe containment of flammable organic liquids, sometimes under pressurised conditions, and required the development of accurate and reliable temperature measurement before the scientific fundamentals of microwave-accelerated chemical reactions were defined fully. It also proved necessary to return to the basics and define more clearly the basis of microwave dielectric heating and the measurement of the appropriate dielectric parameters responsible for this phenomenon. The purpose of this chapter is to provide a chemically intelligible account of microwave dielectric heating and to clearly distinguish the similarities and differences between conventional and microwave heating of reactions.

1.1.1. *Microwave radiation – frequencies available for dielectric heating*

The microwave region of the electromagnetic spectrum corresponds to wavelengths of 1 cm to 1 m (30 GHz to 300 MHz). The wavelengths of 1–25 cm are also used extensively for radar and telecommunications, and by international agreement only the frequencies 2.45 GHz (12.2 cm) and 900 MHz (33.3 cm) are available for dielectric heating, unless rigorous precautions are taken to limit the leakage of stray microwave radiation. Later in this chapter the question of whether the availability of limited frequencies for microwave dielectric heating represents a serious limitation is addressed in more detail. Domestic and commercially available microwave applicators for chemical purposes generally operate at 2.45 GHz (12.2 cm) and the vast majority of the literature pertaining to chemical applications of microwave dielectric heating are based on experiments conducted at this frequency[2,9].

The interactions of microwave radiation with chemicals may be interpreted in a manner which chemists will be more familiar with or more traditionally from a physics standpoint. Both interpretations are summarised below[10]:

In the gas phase, small molecules with permanent dipole moments interact with microwave energy and they display a well-defined spectrum that may be used to define the moment of inertia of the molecule. In this phase the rotation of molecules is quantised and the transitions between the energy levels may be observed as sharp lines in the microwave spectrum, as long as the molecule has a permanent dipole moment. The spectral lines are very sharp and lines as close as 1 MHz apart may be distinguished. At low pressures, the mean free paths of the molecules are significant and consequently the lifetimes of the excited states are long. However, as the pressure is increased the lifetimes of the excited states are reduced and this leads progressively to broader bands. In liquids the continuum of rotational states generated for a very large ensemble of frequently interacting molecules means that the phenomenon loses its identity as a quantum mechanical description and the rotational motions become less distinguishable from translational processes. For example, if one imagines a molecule to be like a paddle, then the stimulation of its rotation to a higher frequency can influence the translational properties of neighbouring molecules in the liquid by a 'batting' type of process, whereby the rotational energy in the first molecule is lost in the increase in translational energy of the second. Even in the absence of such an obvious physical model, the rotations at one centre will influence the translational and rotational motions of neighbouring molecules *via* intermolecular interactions. The intermolecular perturbations will be greatest when the intermolecular forces arise from hydrogen bonding and strong dipole–dipole interactions. Such a process would not occur in the gas phase at low pressures because of the large mean free paths. The only feature common to the gas phase and the solution phase interactions of molecules with the microwave radiation is that it is essential that the molecules have a permanent dipole moment. Therefore, in the solution phase, the physical phenomenon associated with microwave dielectric heating is not appropriately considered as a quantum mechanical phenomenon and may be interpreted using classical electromagnetic theory. The interaction between microwave radiation and solutions of polar molecules may be adequately described using classical models, which may be derived from Maxwell's equations.

In summary, it is important to emphasise that microwave dielectric heating is not a quantum mechanical phenomenon localised at one molecular centre, but is a collective property that occurs in a semi-classical manner and involves aggregates of molecules. Energy transfer is rapid between these molecules and this limits the extent of localisation of the heating.

1.2. Theoretical basis of dielectric heating

1.2.1. *Relaxation times of solvents*

As indicated above, the solution-phase behaviour of molecules blurs the distinction between rotational and translational energy and the characteristic rotational frequencies for molecules in the gas phase are replaced by a broad band of rotational and translational phenomena. Therefore, an important parameter in deciding the extent of interaction between microwave radiation and a particular solvent is the average rotational frequency or its inverse, the *average relaxation time*. Debye first proposed that this relaxation time depends on the size of the molecule and the nature of the intermolecular forces. The average relaxation time of a molecule in a liquid, therefore, depends on the average time taken for a collection of molecules to take up a random orientation when an electric field is switched off. This of course depends on the molecular size and the extent of the intermolecular forces. In the Debye interpretation for a spherical molecule with a radius r rotating in a viscous continuum, the relaxation time, τ, is defined as

$$\tau = 4\pi r^3 \eta / kT$$

In this expression, the size of the molecule is identifiable as the volume element and the intermolecular forces are represented by the viscosity, η. This is a somewhat hybrid equation since the size of the molecule is represented in molecular terms, but the intermolecular forces are represented by a macro-physical property, that is, the viscosity. This leads to a very simple equation, but not one that is readily amenable to molecular interpretation by chemists. Clearly the viscosity depends on the strength of intermolecular forces, but not in a simple transparent way. This duality results because it is problematic to provide a simple expression for the intermolecular forces, which may have contributions from dipole–dipole, induced dipole–dipole and hydrogen bonding effects. The relaxation time is temperature dependent since the molecules rotate faster as the temperature is increased. If we compare a series of molecules with similar hydrogen-bonding capabilities, then there is likely to be reasonable correlation between molecular size and the average relaxation time; however, if the viscosity varies significantly then the Debye expression is likely to be less useful. Figure 1.1 provides a plot of relaxation times *versus* V (V = molecular volume) for a series of closely related alcohols and demonstrates the overall validity of the Debye expression. The variation in viscosities is not too large in this homologous series and therefore the correlation proves to be satisfactory; however, for glycols that have several OH groups, the viscosities can vary by more than 10^3 and this can lead to large increases in the relaxation times. Tables 1.1 and 1.2 summarise the relaxation times, dielectric properties and viscosities for alcohols and glycols and present these generalisations on a more quantitative basis.

For the great majority of molecules, only one broad relaxation peak is observed and therefore it represents a continuum of the rotations of the molecule and the envelope cannot be delineated into rotations associated with specific bond groupings. There are some rare cases, however, where a more localised rotation is observed. For example, benzyl alcohol, which would be expected to have a relaxation time of more than 1000 ps on the basis of its volume, actually has a relaxation time of 188 ps and this has been attributed to a more localised rotation of the OH group relative to the larger benzyl

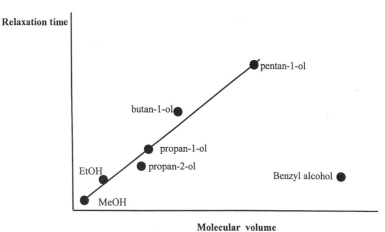

Figure 1.1 A plot of relaxation times *versus* molecular volume for a series of closely related alcohols demonstrating the overall validity of the Debye equation. The outlying point refers to benzyl alcohol (see text for further discussion).

Table 1.1 Relaxation times (20°C) and dielectric properties of some common organic solvents[10]

Solvent	Relaxation time τ (ps)	Dipole moment (debye)	Loss tangent at 2.45 GHz
H_2O	9.04	1.54	0.123
MeOH	51.5	1.70	0.659
EtOH	170	1.69	0.941
Propan-1-ol	332	1.68	0.757
Me_2SO	20.5		0.825
$HCONMe_2$	13.05		0.161
$MeNO_2$	4.51		0.064
THF	3.49		0.047
CH_2Cl_2	3.12		0.042
$CHCl_3$	8.94		0.091
MeCOMe	3.54		0.054
$MeCO_2Et$	4.41		0.059
HCO_2H	76.7 (25°C)		0.722
$MeCO_2H$	177.4 (25°C)		0.174
MeCN	4.47		0.062
PhCN	33.5		0.459
CH_2OHCH_2OH	113 (25°C)		1.35

Table 1.2 Relaxation times and viscosities of some solvents[10]

Solvent	Relaxation time τ (ps)	Viscosity (mP)	Loss tangent at 2.45 GHz
H_2O	9.04	10.1	0.123
MeOH	51.5	5.5	0.659
EtOH	170	10.8	0.941
Propan-1-ol	332	20	0.757
CH_2OHCH_2OH	113 (25°C)	1.4	1.35
Glycerol	1216	9450	0.651

moiety (see Fig. 1.1). More generally, the presence of an OH or NH$_2$ fragment attached to a large molecule behaves as though it were anchored to an immobile raft and the more localised rotations dominate the microwave spectrum. For example, triphenylhydroxy methane has a maximum at 10 ps, which has been attributed to a localised OH group rotation, and for tri-*tert*-butylphenol two peak maxima were observed and attributed to localised and whole molecule rotations[10].

For water and simple aliphatic alcohols (Fig. 1.1), the relaxation time varies from 9 to ~1000 ps and the variation may be attributed primarily to the increase in volume of the molecules. For an individual alcohol, the relaxation time represents a spread of ~10^{10} Hz and the width of the relaxation peak at half-height is ~10^9 Hz and therefore the coupling of microwaves is neither very specific nor very frequency dependent. For example, although the relaxation time for water is 9.04 ps, the operating frequency for dielectric heating corresponds to a maximum relaxation time of 65 ps and if anything that corresponds more closely to the relaxation for MeOH. Therefore, it transpires that a whole range of organic molecules have relaxation time profiles that incorporate a relaxation time of 65 ps and therefore are able to couple effectively with microwave energy with a frequency of 2.45 GHz. These data support the empirical observation of many synthetic chemists who have noted that a wide range of polar organic solvents heat rapidly in a microwave cavity operating at 2.45 GHz[9,10].

The average relaxation time is of course temperature dependent and may be related to a rate constant k for the relaxation process of the molecules in solution. Table 1.3 gives some representative data for EtOH and illustrates the extent to which the relaxation time decreases with temperature. It is noteworthy that the relaxation time decreases from 270 to 49 ps as the temperature rises from 10 to 70°C, and therefore, as the temperature increases the alcohol couples more effectively with the microwave source at 2.45 GHz. Such a situation is ripe for superheating the solvent, since the extent of conversion increases as the temperature rises. It also follows that some organic solvents with very long relaxation times at room temperature may appear to be unsuitable candidates for dielectric heating, but since the match becomes more favourable with temperature then they may behave as effective couplers as the temperature rises, that is, after a slow start they may very well heat very rapidly.

An Arrhenius-type analysis of temperature dependence can be used to calculate the enthalpy and entropy of activation for the relaxation process. For liquid water, the enthalpy of activation is 19 kJ mol^{-1}, which corresponds approximately to the energy required to break one hydrogen bond. For ice, the equivalent enthalpy is 54 kJ mol^{-1},

Table 1.3 Relaxation times and dielectric properties as a function of temperature for EtOH

T (°C)	Relaxation time τ (ps)	Relaxation frequency (GHz)	ε'	ε''
10	270	0.59	7.05	5.81
20	199	0.80	7.49	6.46
30	134	1.19	8.05	7.06
40	104	1.53	8.95	7.39
50	78.4	2.03	10.11	7.28
60	55.2	2.88	11.15	6.76
70	48.8	3.26	11.71	6.35

thereby suggesting that three hydrogen bonds are broken in the relaxation process in this more rigid structure. Since all the water molecules in the ice structure form four hydrogen bonds to neighbouring molecules, this observation is consistent with a concerted rotation of the water molecule, which involves retention of one of the hydrogen bonds. It is well documented that the water molecules in the liquid state clearly retain some of the structure of ice and it is possible to think of water molecules forming parts of clusters of molecules that interact with each other or as a dynamic assembly of molecules, where the number of hydrogen bonds varies with time. Either way, the rotation of a water molecule in the solution phase requires the breakage of fewer hydrogen bonds and the relaxation time decreases accordingly.

The importance of hydrogen bonding is apparent from Table 1.1, which indicates that solvents with weak hydrogen-bonding capabilities have much shorter relaxation times. In addition, their lower viscosities reduce the interaction of molecules in the rotational process, which is induced by the microwave coupling and therefore will generally not heat as effectively in a microwave cavity.

The formation of strong hydrogen bonds in glycerol leads to a very high viscosity and also a long relaxation time (see Table 1.2); however, a comparison with the data for alcohols with comparable volumes suggests that the relationship between viscosity and relaxation time is not the simple linear one implied by the Debye expression.

1.2.2. *Loss tangents*

The more formal theoretical basis of microwave dielectric heating can be associated with the following names: Debye, Frohlich, Daniel, Cole and Cole, Hill and Hasted and the reader should refer to standard texts for a more detailed account of the basic physics[11]. The classic work of von Hippel and his co-workers at MIT in the early 1950s provided the basic dielectric parameters that govern the extent of interaction between microwaves and solid and solution samples[5]. His basic data provided an important basis for the development of microwave dielectric heating of foodstuffs and the drying of plastics, Woods' paper and agricultural products. The dielectric analysis of the interactions of microwaves with biological materials has increased in importance in recent years and Craig has provided an excellent summary of the relevant data[12].

A dielectric material is one that contains permanent or induced dipoles, which when placed between two electrodes acts as a capacitor, that is, the material allows charge to be stored and no d.c. conductivity is observed if the electrodes are connected by a circuit. The polarisation of dielectrics results from the finite displacement of charges or rotation of dipoles in an electric field and must not be confused with conduction, which results from the translational motion of electric charges when a field is applied. At the molecular level, polarisation may be associated either with a distortion of the distribution of the electron cloud of the molecule or the physical rotation of dipoles. The phenomenon of dielectric heating in the microwave region is strongly connected with the latter since, as we mentioned earlier, the rotational relaxation time occurs in the microwave region for many polar organic molecules. The permittivity of a material, ε', is the property that defines the charge-storing ability of that material irrespective of the sample's dimensions. The dielectric constant or relative permittivity of a material is the permittivity of the material relative to free space. Table 1.4 gives some typical dielectric constants and it

Table 1.4 Dielectric constants (relative permittivities) at 20°C for some common solvents

Solvent	Dielectric constant (ε_s)
Water	80.4
MeOH	33.7
Me_2CO	21.4
C_6H_6	2.3

is noteworthy that water has a particularly high dielectric constant, whereas benzene, which has no permanent dipole moment, has a low dielectric constant. Unlike a compass magnet, a collection of molecules in a liquid is not static, but the molecules are rotating all the time and in the absence of an electric field have random orientations. In solutions, molecules are able to respond to electric field fluctuations, which occur at up to 10^6–10^9 s^{-1} or higher, because they rotate naturally at these frequencies.

If the electric field is reversed more rapidly, for example, at 10^{12} s^{-1}, then the smallest molecules are no longer able to respond sufficiently rapidly before the electric field is reversed and therefore the permittivity falls, that is, the solution is no longer able to store the energy as a capacitor. For a polar liquid, the permittivity is frequency dependent in the range of 10^6 (radio frequencies) to 10^{12} s^{-1} (infrared frequencies).

In terms of classical electromagnetic theory, the re-orientation of dipoles and the displacement of charge is equivalent to an electric current, known as the Maxwell current. For an ideal dielectric, there is no lag between the orientation of the molecules and the variations of the alternating voltages. The displacement current is 90° out-of-phase with the oscillating electric field as shown in Fig. 1.2. The relevant phase diagram in Fig. 1.2(a) shows that for a dielectric material where the molecules can keep pace with the field changes no heating occurs. There is no component of the current in-phase with the electric field, that is, the product $E \times I$ is zero because of the 90° phase lag between the field and the current. If the frequency of the electromagnetic radiation is pushed up into the microwave region (~10^9 Hz), the rotations of the polar molecules in the liquid begin to lag behind the electric field oscillations. The resulting displacement δ, shown in Fig. 1.2(b), requires a component $I \times \sin \delta$ in-phase with the electric field and so resistive heating occurs in the dielectric medium – this is described as the dielectric loss and causes energy to be absorbed from the electric field. Since the dipoles are unable to follow the higher frequency electric field oscillations, the permittivity falls at the higher frequencies and the substance behaves increasingly like a non-polar material. Therefore, the dielectric loss spectra show a bell-shaped graph, which reaches a maximum at the relaxation time of the solvent and drops off again as the relaxation time and the microwave frequency progressively mismatch. Typical bell-shaped graphs for some alcohols are illustrated in Fig. 1.3. It is apparent that they reach a maximum at a frequency that corresponds to half way down the declining graph of the permittivity[10].

At frequencies for which the loss tangent (δ) differs significantly from 90°, the liquid has a dual role. It functions both as a dielectric and a conductor. Since $\sin \delta$ is an in-phase

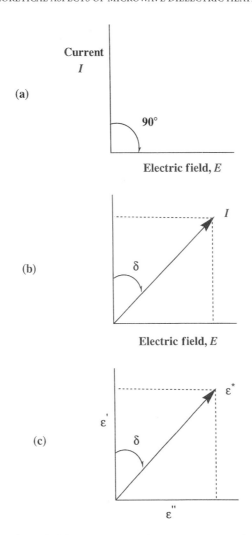

Figure 1.2   Phase diagrams for an ideal dielectric (a) where the energy is transmitted without loss and (b) where there is a phase displacement δ and the current acquires a component $I \times \sin\delta$ in-phase with the voltage and consequently there is a dissipation of thermal energy. (c) The relationship between ε^*, ε' and ε'' is illustrated.

current component, it gives the total relative permittivity a complex character:

$$\varepsilon^* = \varepsilon' - j\varepsilon''$$

where ε' is the real part of the relative permittivity (the dielectric constant) and ε'' is the loss factor that reflects the conductance of the material. In the phase diagram in Fig. 1.2(c), $\varepsilon'/\varepsilon'' = \tan\delta$ and $\tan\delta$ is described as the *energy dissipation factor* or *loss tangent*. Tan δ provides a convenient parameter for defining and comparing the abilities of different materials to convert microwave energy into thermal energy at that frequency. The relationship between the permittivities and the relaxation times discussed in the

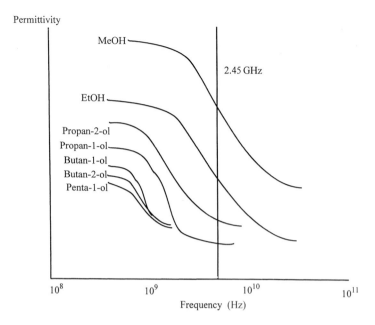

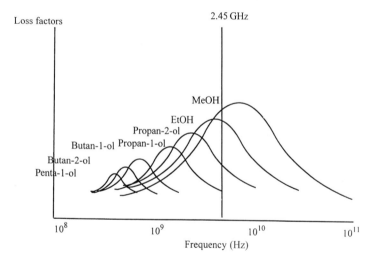

Figure 1.3 Dielectric spectra for range of alcohols in the frequency range of 10^7–10^{11} Hz. The absolute permittivities at low frequencies fall as the size of the alcohol increases and they began to respond to the microwave fields at lower frequencies because their relaxation times become longer. The loss factors that control the efficiency of conversion of microwave into thermal energies also reach their maxima at lower frequencies. The loss tangent is the ratio of the loss factor and permittivity at that frequency. (Idealised from the raw data illustrated in Ref. 10.)

previous section is apparent from the following equations. A polar liquid with a single relaxation time has complex permittivity. Separating the real and imaginary parts of the permittivity leads to the following expressions[2]:

$$\varepsilon' = \varepsilon_\infty + \frac{\varepsilon_s - \varepsilon_\infty}{1 + \omega^2\tau^2}$$

$$\varepsilon'' = \frac{(\varepsilon_s - \varepsilon_\infty)\,\omega\tau}{1 + \omega^2\tau^2}$$

The frequency dependence of ε' and ε'' and their magnitudes control the extent to which a substance is able to couple with the microwave radiation and therefore are fundamental parameters for interpreting the dielectric heating phenomenon. Although tan δ is a helpful parameter for comparing the heating rates of a series of dielectrics with similar physical and chemical characteristics, for more complex mixtures expressions, which take into account the complexity of the electric field pattern, the heat capacity of the compound and the density, have been proposed.

The maximum value of ε'' occurs when ε^1 reaches half of its declining value between ε_s (the permittivity at frequencies much greater than the inverse of t) and ε_∞. Figure 1.3 illustrates the variation in permittivities and loss factors for a range of alcohols and illustrates the manner in which the loss factor comes to a maximum at a frequency that corresponds to the average relaxation time of the solvent. The figures also indicate the point where the frequency is 2.45 GHz, corresponding to the most widely used frequency for microwave dielectric heating. It is also noteworthy that the maximum in the loss factor occurs at a frequency that corresponds to the midpoint in the part of the permittivity curve where it drops rapidly. In absolute terms, the permittivity and the loss factor decrease in magnitude as the chain length of the alcohol increases.

Tables 1.1 and 1.2 summarise the loss tangents of water and some common organic solvents at 2.45 GHz. It is clear that many organic solvents, and especially those that have strong hydrogen bonding capabilities, have loss tangents that are significantly greater than those of water. The high loss tangents of alcohols and poly-ols are particularly noteworthy[2] in the context of microwave heating.

The efficiency of conversion of microwave energy into thermal energy depends both on the dielectric and thermal properties of the materials. The fundamental relationship is[2]

$$P = \sigma|E|^2 = (\omega\varepsilon_0\varepsilon'')|E|^2$$

$$= (\omega\varepsilon_0\varepsilon'\tan\delta)|E|^2$$

where P represents the power dissipation per unit volume in a material of conductivity σ, the electric field in the sample is E and ω is the angular frequency. In terms of the Debye parameters this becomes

$$P = \frac{\varepsilon_0\,(\varepsilon_s - \varepsilon_\infty)\,\omega^2\tau}{1 + \omega^2\tau^2}\,|E|^2$$

Table 1.5   Effect of microwave exposure (600 W) for 1 min (Taken from Ref. 13)

Solvent	Temperature after 1 min exposure	Boiling point (°C)	Dipole moment (debye)
H_2O	81	100	5.9
EtOH	78	78	5.8
$MeCO_2H$	110	119	5.6
$HCONMe_2$	131	153	10.8
n-hexane	25	98	0.0
CCl_4	28	77	0.0

Assuming negligible heat losses and diffusion, the rate of heating or temperature rise ΔT in a time interval t can be expressed as follows:

$$\frac{\Delta T}{t} = \frac{\sigma\,|E|^2}{\rho C}$$

$$\frac{\Delta T}{t} = \frac{\omega\varepsilon_0\varepsilon''\,|E|^2}{\rho C} = \frac{\omega\varepsilon_0\varepsilon'\,\tan\delta\,|E|^2}{\rho C}$$

In practice, the electric field is governed by the characteristic shape of the sample and dimensions of the cavity and it is neither a constant nor easy to define. It is also strongly dependent on the dielectric properties of the material.

Table 1.5 provides some data for organic solvents and compares the temperatures achieved after 1 min of microwave exposure with the dipole moment of the solvent[13]. Table 1.6 summarises heating rates for some typical organic solvents and also the final temperatures achieved when the solvents are exposed to a source with constant power[13]. Interestingly, despite the wide range of relaxation times and dielectric loss tangents at 20°C for these polar organic solvents, the heating rates are remarkably similar and lie between 2 and 2.5°C s^{-1}. Clearly, there are compensatory factors that lead to an equalisation of the observed heating rates. Since the majority of solvents have longer relaxation times than that corresponding to 2.45 GHz, they have loss tangents that increase with temperature. The accelerating heating rates provide suitable conditions for superheating the solvents. The boiling of a solvent is a kinetic as well as a thermodynamic

Table 1.6   Heating rates and nucleation limited boiling points (NLBP) for solvents under microwave conditions

Solvent	Heating rate (°C s^{-1})	Conventional bp (°C)	NLBP^a (°C)	Difference (°C)
H_2O	1.01	100	104	4
EtOH	2.06	79	103	24
MeOH	2.11	65	84	19
CH_2Cl_2	2.16	40	55	15
thf	2.04	66	81	15
MeCN	2.36	82	100	18
Me_2CO	2.23	56	81	25
diglyme	2.17	162	175	13

^{a} NLBP: nucleation limited boiling points.

phenomenon and depends critically on the availability of nucleation sites where bubble formation can be initiated as a prelude to boiling[14]. The heating rates exceed the rate of generation of nucleation sites at the conventional boiling point. Consequently organic solvents boil at temperatures considerably above their thermodynamic boiling points in a microwave cavity. Table 1.6 shows that these nucleation limited boiling points can lie 15–25°C above the conventional boiling point. The difference may be reduced either by increasing the number of nucleation sites in the vessel by adding boiling chips, etc., or by decreasing the rate of power input.

The elevation of boiling point can be enhanced by allowing the pressure to rise in the vessel, and when the pressure is allowed to rise to ~10 atm, the solvent boils at ~100° above the conventional boiling point[9]. Translated into the Arrhenius expression this means that the rate constant for a reaction may be increased by ~10^3. In this pressure range, it is not necessary to use thick-walled steel vessels so as to contain the pressure and thick-walled glass or Kevlar-based containers are sufficient. The great majority of the rate enhancements of organic reactions under microwave conditions may be attributed to the superheating effects described above. Indeed when the temperature of a reaction, which is studied in a microwave cavity, is carefully and accurately measured then the rates are found to correspond closely to those found for the same reaction under conventional heating conditions. There is little evidence of specific or quantum mechanically based phenomenon for solution reactions that have been studied carefully under microwave conditions. However, the very different temperature profiles associated with microwave heating makes it a very attractive and convenient method for carrying out organic, inorganic and organometallic reactions under flash heating conditions, that is, reactions where the energy is generated very rapidly for a short period of time allowing the attainment of a high temperature very quickly followed by rapid cooling at the end of the reaction (see Fig. 1.4). These conditions can result in higher conversion rates for reactions with large activation energies, lower yields of decomposition products and alternative isomer distributions.

It is instructive to examine whether it is worth using alternative frequencies for dielectric heating. Table 1.7 summarises the variation in the dielectric constants and loss tangents for a series of organic solvents for microwave frequencies from 14 MHz to 2.45 GHz. For low-molecular-weight polar solvents, the decrease in frequency leads to a decrease in the loss tangents and therefore there is not an obvious gain in changing the frequency[10]. For very viscous solvents such as glycerol, the loss tangent shows an increase if the frequency is reduced to 444 MHz, but since the loss tangent is high even at 2.45 GHz it is unlikely to lead to a significant increase in heating rates. Therefore, there appears to be little practical advantage for organic reactions to be undertaken at different microwave frequencies.

In summary, microwave dielectric heating appears to be a very effective technique for undertaking organic and inorganic reactions that have high-activation energies. It means that the total time for such reactions may be reduced from approximately 1 day of conventional refluxing to a matter of minutes. Many inorganic syntheses utilise hydrothermal techniques, whereby the reactants in aqueous solutions are heated in a sealed pressured steel bomb at temperatures in excess of 150°C and several atmospheres for several hours. Many of these reactions may also be successfully carried out in microwave cavities[8].

Table 1.7   Loss tangents as a function of frequency for some common solvents[10]

Frequency	ε'	ε''	$\tan \delta$
Water			
14 MHz	78.3	0.10	0.001
444 MHz	779	1.70	0.022
900 MHz	78.6	3.51	0.045
2.45 GHz	77.4	9.48	0.122
Hexan-1-ol			
14 MHz	8.0	0.70	0.088
444 MHz	5.2	3.6	0.702
900 MHz	4.0	2.3	0.568
2.45 GHz	3.4	1.2	0.341
Nitrobenzene			
14 MHz	35.1	0.20	0.006
444 MHz	35.3	4.0	0.113
900 MHz	33.7	7.7	0.229
2.45 GHz	25.2	14.7	0.584
Glycerol			
14 MHz	42.5	3.70	0.087
444 MHz	11.4	9.9	0.866
900 MHz	8.41	6.40	0.759
2.45 GHz	6.33	3.42	0.540

1.3. Dielectric properties of solids

The dielectric constants and dielectric loss factors of some solid materials were presented in von Hippel's compilations of the 1950s[5]. In addition to those reported for foodstuffs, it is possible to find some data for conventional inorganic elements and compounds. Unfortunately, a more recent critical review of the relevant data does not exist in the same way as that for organic and inorganic solvents. Since solid-state reactions occur over a much wider temperature range than organic reactions, data that gives the dielectric constants and loss factors over the temperature range of 20–1500°C would be most useful[15,16]. In the absence of such data, it is necessary to rely on empirical data such as the increase in the temperature for a given weight of compound in a microwave cavity at a specified power and given length of exposure. The limited data, which are at hand, do enable one to propose some tentative, but hopefully useful generalisations.

    The molecules in a covalent solid have very restricted rotations even in the absence of hydrogen-bonding effects and consequently their relaxation times fall well outside the microwave range. In infinite solids, molecular rotations have no meaning and therefore the dipolar rotation mechanism (discussed earlier in this chapter) for organic solvents no longer provides a mechanism for dielectric heating. Indeed, the predominant mechanism for the dielectric heating of solids results from conductivity and therefore is more akin to resistive heating. It follows that insulators and wide band gap semiconductors are generally transparent to microwaves, that is, they have low loss factors and loss tangents and do not heat very rapidly in a microwave cavity. These materials do, however, take on a very important role as insulators and containment materials. For example, silica glasses, pyrex and aluminium oxides may be used as suitable materials for the constructions of containment vessels and crucibles. Similarly, non-polar organic polymers may be used as containment vessels particularly if they have high-melting points. Therefore, teflon and Kevlar-based containment vessels are commonly used for acid digestion and chemical reactions up to $250°C$[17].

Since the predominant contribution to the loss factor is conduction, insulators and wide band semiconductors, which show increased conduction with temperature, become less useful as insulators and containment materials at high temperatures. Indeed, one limitation of microwave dielectric heating is that it is increasingly difficult to find suitable containment materials above $1500°C$ and almost impossible above $2000°C$. At higher temperatures some solids undergo a phase transition, which enables ionic mobility to make a significant contribution to conduction. Therefore, certain glasses that contain quite mobile sodium or potassium ions may be used for glass vessels at low temperatures, but at higher temperatures they begin to absorb the microwave energy by virtue of ion mobility conduction and consequently melt catastrophically. The integrity of insulators is also very sensitive to impurities spread on their surface. If these impurites absorb microwaves very strongly they can create local hot spots, which cause the insulator to achieve high temperatures locally. At these higher temperatures, the insulator has a sufficiently high loss tangent that it begins to absorb the microwaves strongly. The subsequent spread of these hot spots may degrade the insulator and actually convert it into an effective microwave absorber.

Solids that are either semiconductors or exhibit significant conduction *via* ion mobility even at low temperatures do exhibit significant loss tangents and may heat up in a microwave cavity. The conductivity of these materials is very sensitive to temperature, impurities and defects and consequently the loss tangent may increase dramatically as a function of temperature. It follows that diamond and sulphur are transparent to microwaves, but graphite, silicon carbide and boron have high loss tangents and convert microwave energy into thermal energy very efficiently. Similarly, many transition metal oxides and sulphides couple efficiently with microwaves at 2.45 GHz in a conventional microwave cavity and the temperature/time data given in Table 1.8 indicate that very high temperatures may be achieved in relatively short periods of time. Indeed, the heating rates for such solids are at least an order of magnitude greater than those for organic solvents. In part, this may be attributed to a phenomenon described as *thermal runaway*[4,15–17]. The time/temperature profiles of some solids exhibiting thermal runaway are illustrated in Fig. 1.5. SiO_2 behaves like a typical insulator and the temperature does not rise very rapidly with time in the microwave cavity. In contrast, NiO and Cr_2O_3 suddenly exhibit very dramatic rises in their heating rates and this persists until the solid melts. This ability has been widely used for accelerating solid-state syntheses; for example, a solid that exhibits thermal runaway may be combined with other components that are transparent to microwaves in order to synthesise mixed metal oxides[18]. Reaction times are much shorter than those required conventionally to produce mixed metal oxides and the resulting samples have X-ray powder diffraction patterns of good quality suggesting high crystallinity, despite the samples having been produced in a melt and quench procedure[17]. Several groups have prepared samples of superconducting $YBa_2Cu_3O_{7-x}$, utilising the coupling of CuO with microwave radiation[19]. It was found that the shorter reaction times under microwave temperatures obviated the need to have a subsequent annealing phase under an oxygen atmosphere, in order to obtain the superconducting orthorhombic phase[18,19]. Similarly the high loss

Table 1.8 Effect of exposure of microwave radiation at 2.45 GHz on solid samples

Compounds	Final T (°C)	t (min)
Insulators		
NaCl	83	7
SnCl$_4$	49	8
CaO	83	30
SnO	102	30
TiO$_2$	122	30
Semiconductors		
Carbon	1283	1
NiO	1305	6.25
CuO	701	0.5
V$_2$O$_5$	701	9
WO$_3$	532	0.5
Metal powders		
Al	577	6
Ni	384	1
Magnetic materials		
Fe$_3$O$_4$	510	2
Co$_2$O$_3$	1290	3

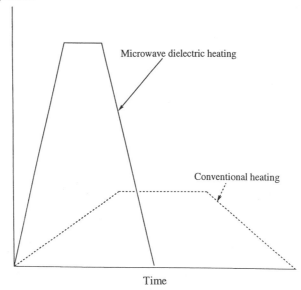

Temperature

Microwave dielectric heating

Conventional heating

Time

Figure 1.4 Differences in the temperature–time profiles for conventional and microwave dielectric heating. Particularly noteworthy are the far higher heating and cooling rates and the greater reaction temperatures achieved. Since at these higher temperatures the rate of reaction is much larger, it is not necessary to hold the reaction at this temperature for an extended period of time. Since the rate of reaction depends exponentially on temperature the translation of these profiles into product yields as a function of time will magnify these effects.

tangents of elemental boron and carbon have been utilised in the syntheses of carbides and borides and the carbon has also been used as a reducing agent in high temperature transformations[20].

The overall time for a chemical reaction is governed mainly by the speed of the slowest step. For many solid-state reactions, this slowest step is the diffusion of the reactants towards one another through an unreactive medium. Any mechanism that can increase the rate of diffusion may lead to a dramatic enhancement of the reaction rate. Much of the work on ceramic processing suggests that the enhanced transport properties occur under microwave conditions. The calculation of the activation energy for diffusion, using Arrhenius plots of the logarithm of diffusion coefficient *versus* temperature under microwave and conventional thermal conditions, show a significant reduction in the activation energy under microwave conditions.

Microwave energy has been widely used to process and sinter ceramics and the following advantages compared to conventional heating have been quoted[4]:

(1) In microwave processing a material which has a high loss tangent, the heat is generated inside the material, whereas in conductive heating the outside is preferentially heated and this can lead to thermal stress and cracking. Hence microwave dielectric heating has proved to be particularly effective for increasing the densities of ceramics.

(2) The mechanical properties of ceramics may be increased by minimising impurity segregation, decreasing grain size and increasing sintered density.

(3) The rapid heating by microwaves limits the extent of non-isothermal processes, such as segregation of the impurities to grain boundaries. Since the sintering time is often reduced, the possibility of secondary recrystallisation, which leads to exaggerated grain growth, may be reduced.

(4) Contamination of the sample being formed in a solid-state reaction may occur from the walls of the containment vessel, but is reduced under microwave conditions because of the lower temperatures at the interfaces between the sample and the crucible.

(5) The remote nature of the interaction between microwaves and the sample means that when the power is switched off the sample rapidly cools. Consequently, it is possible to use the microwave applicator in a more flexible fashion than a normal heating furnace. This becomes an important consideration in the economics of ceramics processing, since the ability to use the furnace flexibly for several processes combined with the shorter reaction times can lead to a more economic use of plant than the conventional thermal processing.

Although this analysis has concentrated on the interaction between the oscillating component of the electromagnetic field of the microwave radiation, it is also possible to get interactions between the oscillating magnetic field and magnetic dipoles in the sample. The high-heating rates observed for Fe_3O_4 may be attributed to such interactions[21].

It is common knowledge that metal objects should not be placed in a microwave cavity, because major sparking and arcing results. However, it is possible to get effective and safe dielectric heating using metal samples as long as they are in a powder form. This property has been widely examined and utilised effectively for synthetic procedures. The methodology has proved to be particular useful using fluidised bed techniques.

The metal particles are suspended in a flow of the gas, which is also a reactant. Nothing happens until the microwave power is switched on and the metal particles begin to heat up very rapidly – at times giving rise to an incandescent glow – causing a rapid reaction between the metal and the gas flowing over it. The solid compound that forms sublimes to the edge of the glass tube that defines the outer surface of the fluidised bed. This has proved to be a very effective way of making oxides, halides and nitrides of the transition metals[22–24]. Even in the absence of a fluidised bed, this methodology may be applied for the synthesis of metal sulphides, selenides, etc.[23].

The dominant problem for microwave-induced syntheses in metal powder organic solvent systems is the selection of conditions that exhibit the minimum degree of arcing[24]. In an ideal reaction system, no arcing would be observed although this may be extremely difficult to achieve, particularly in the case of metal powders in aromatic solvents. Mingos and Whittaker[24] have defined the important parameters that control the onset of arcing. The model they proposed suggests that the ability of metal particles to heat the solvent locally so as to create a vapour pathway through which electric discharges occur is very important. In general, the following conditions limit the extent of arcing: low-microwave powers, high-boiling-point solvents, high pressures and non-aromatic solvents (indeed polar saturated solvents) and small well-dispersed metal particles. The synthetic utility of this methodology has been illustrated for the synthesis of a range of sandwich organometallic compounds of the transition metals.

1.4. Comparison of microwave and conventional heating

It is clear from the preceding discussion that microwave dielectric heating is a non-quantum mechanical effect and it leads to volumetric heating of the samples. Therefore, it is necessary to question whether it has any significant advantages compared to thermal heating of chemical reactants. There are significant differences in the mode of interaction and these may confer advantages for dielectric heating; however, the effects require a greater understanding of the temperature profile and the nature of the interaction.

(1)   The introduction of microwave energy into a chemical reaction that has at least one component, which is capable of coupling with the microwaves can lead to much higher heating rates than those that can be achieved conventionally. Using very cheap and readily available microwave cavities, heating rates of $2–4°C\,s^{-1}$ may be readily achieved even for common organic solvents. Such heating rates are more difficult to achieve using conventional heating, although of course dropping sealed tubes into heated sand furnaces at $>1000°C$ could result in comparable heating rates.

(2)   The microwave energy is introduced into the reactor remotely and therefore there is no direct contact between the energy source and the sample undergoing heating. This combined with point (1) may lead to very different temperature–time profiles for the reaction and as a consequence may lead to an alternative distribution of chemical products in the reaction. Figure 1.4 depicts the differences in temperature–time profiles for thermal and microwave heating. The remote nature of the interaction results in much higher heating and

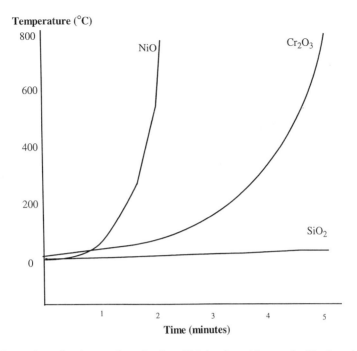

Figure 1.5   Comparison of heating rates for an insulator (SiO_2) and transition metal oxides that exhibit thermal runaway (see Ref. 20 for more details).

cooling rates for the microwave technique and also in a reaction temperature that is higher than that achieved conventionally. Therefore, microwave dielectric heating resembles a flash heating process, whereby the energy is generated much more rapidly and the sample cools more rapidly at the end of the reaction. The different profiles may therefore lead to significantly different products, particularly if the reaction product distribution is controlled by complex and temperature-dependent kinetic profiles.

(3) Chemicals and the containment materials for chemical reaction do not interact equally with the commonly used microwave frequencies for dielectric heating and consequently selective heating may be achieved. Specifically, it is possible to cool the outside of the vessel with a coolant that is transparent to microwaves (solid CO_2 or liquid N_2) and thereby have cold walls that still allow the microwave energy to penetrate and heat the reactants, which are microwave active, in the vessel. Also for solid-state reactions contamination from the crucible walls may be minimised.

These differential heating effects may also be used in solid-state reactions involving metal powders that are made into a fluidised bed by a counter stream of gas. The metal particles interact very strongly with the microwaves and rapidly heat, whereas the gases are transparent to microwaves; therefore, the reaction is induced by a very selective interaction between the metal particles and microwaves.

(4)   The degree of selective heating should not be exaggerated for solvent mixtures. For example, if a mixture of MeOH (2%), which has a high loss factor, and benzene (98%), which is transparent to microwaves, is exposed to a microwave field, the whole mixture heats up very rapidly. The microwave process involves translation and rotation, and although the effect may have its origins in the vicinity of methanol molecules the rate of energy transport is so fast that benzene molecules are also heated rapidly. Therefore, it is not possible to store the microwave energy selectively either within parts of a molecule or in active molecules in two component mixtures.

(5)   The boiling phenomenon is a kinetic as well as a thermodynamic process and therefore solvents heated under microwave conditions often boil at elevated temperatures even though they remain contained under 1 atm pressure. The precise elevation of this nucleation-limited boiling point depends on the power input, the occurrence of effective stirring and the limitation of the number of nucleation sites, for example, by having smooth surfaces and no boiling chips. This effect can be enhanced by allowing the pressure to rise above atmospheric level. The data presented above, for example, has shown that it is possible to increase the temperature of a reaction in a common organic solvent by up to 100°C above the conventional boiling point. Ethanol, which has a conventional boiling point of 79°C, when microwave dielectrically heated boils at 164°C at a pressure of 12 atm. This temperature rise when translated into the Arrhenius expression leads to an enhancement of $\sim 10^3$ in the reaction rate.

(6)   In solid samples, the rate of energy transport is less and consequently the development of hot spots is more significant. A careful analysis of heterogeneous catalysis suggests that hot spot formation around the catalyst not only enhances the reaction rate but may also contribute to shifts in the equilibrium constant[25]. Loupy and co-workers[13] have utilised this difference to study reactions of organic compounds on solid supports such as alumina and silica. Such solvent-free reactions have proved to have widespread applications and have been proposed as an effective means of doing 'green' synthetic organic chemistry.

(7)   In a microwave cavity a standing wave pattern is generated, which depends of course on a multiple of the wavelength of the radiation (12.5 cm at 2.45 GHz), and therefore depending on the dielectric properties and size of the sample one can get considerable variations in temperature. The penetration depth, D_p, of the radiation is clearly an important consideration.

$$D_P \propto \lambda_0 (\varepsilon' / \varepsilon'')^{1/2}$$

Therefore, for large samples it is possible to develop quite large temperature gradients and it is necessary to introduce electronic and mechanical perturbations to even out the field pattern. There are well-determined methods for evening out the field pattern in this way. However, the procedures are complicated because the sample itself perturbs the field pattern and therefore there may not be a universal solution to the problem.

In summary, it is apparent that the application of dielectric heating to chemical reactions may result in different reaction rates and product distributions, and chemists can take advantage of the remote nature of the interaction. To maximise the difference between microwave and conventional heating, one generally needs to use relatively high-power levels, develop a temperature profile that ramps up and cools off more quickly than the conventionally one and work under conditions where higher pressures may develop in the reaction vessel. Therefore, one is working outside the bounds of conventional equilibrium conditions. For example, a single-temperature measurement, which under equilibrium conditions is representative of the bulk, may hide a wide variation in temperatures across the sample under microwave conditions where the system is not at equilibrium. A first-order treatment of this would modify the conventional thermodynamic and kinetic expressions in order to model such a variation, and a second-order treatment may have to establish whether the original equations are still valid under such conditions. It is, therefore, not surprising that there have been a number of suggestions in the literature that account for the differences between conventional and microwave heating in terms of specific microwave effects. Loupy et al.[13] have suggested that these differences may originate from

(1) changes in the pre-exponential factor of the Arrhenius expression, because of preferential orientation of molecules in the transition state because of the electric field component of the microwave radiation,
(2) decrease in the entropy of activation of the reaction because of the ordering of the solvent in the microwave field and
(3) intervention of microscopic high temperatures (i.e. 'hot spots') as a result of the microwave heating.

However, definitive proof for these proposals are difficult to obtain because of the difficulties in temperature measurement alluded to above and that the reaction might be occurring under non-equilibrium conditions, where the conventional rate expressions and transition state theory assumptions may not be valid.

1.5. Acknowledgement

The Alexander von Humboldt Foundation is thanked for their financial support.

1.6. References

1. Hillman, D. and Gibbs, D., *Century Makers*, Rigel Publications, London, 1998, p. 120.
2. Metaxas, A.C. and Meredith, R.J., *Industrial Microwave Heating*, Peter Perigrinis, London, 1983.
3. Kingston, H.M. and Haswell, S.J., *Microwave Enhanced Chemistry*, American Chemical Society, Washington, DC, 1997.
4. Sutton, W.H., Microwave processing of materials, *Am. Ceram. Soc. Bull.*, 1989, **68**, 1601.
5. Von Hippel, A.R., *Dielectric Materials and Applications*, MIT Press, Cambridge, MA, 1954.
6. Gedye, R., Smith, F., Westaway, K., Ali, H., Balderisa, L., Laberge, L. and Roasell, J., The use of microwave ovens for rapid organic syntheses, *Tetrahedron Lett.*,1986, **27**, 279.
7. Giguere, R.J., Bray, T.L., Duncan, S.N. and Majetich, G., Application of commercial microwave ovens to organic syntheses, *Tetrahedron Lett.*, 1986, **28**, 4945.
8. Baghurst, D.J., Mingos, D.M.P. and Watson, M.J., Application of microwave dielectric loss heating effects for the rapid and convenient synthesis of organometallic compounds, *J. Organomet. Chem.*, 1989, **368**, C43.

9. Mingos, D.M.P. and Baghurst, D.R., Applications of microwave dielectric heating effects to synthetic problems in chemistry, *Chem. Soc. Rev.*, 1991, **20**, 1, and references therein.

10. Gabriel, C., Gabrield, S., Grant, E.H., Halstead, B.D.J. and Mingos, D.M.P., Dielectric parameters relevent to microwave dielectric heating, *Chem. Soc. Rev.*, 1998, **27**, 213, and references therein.

11. Hill, N.E., Vaughan, W.E., Price, A.U. and Davies, M., *Dielectric Properties and Molecules*, Van Nostrad, New York, 1969, and references therein.

12. Craig, D.Q.M., *Dielectric Aspects of Pharmacological Systems*, Taylor and Francis, London, 1995.

13. Pereux, L. and Loupy, A., Alternative rationalisation of microwave effects in organic molecules according to their reaction medium and mechanistic considerations, *Tetrahedron*, 2001, **57**, 9199.

14. Baghurst, D.R. and Mingos, D.M.P., Super heating effects associated with microwave dielectric heating, *J. Chem. Soc., Chem. Commun.*, 1992, 674.

15. Tinga, W.R., Design principles for microwave heating and sintering, *Electromagn. Energy Rev.*, 1988, **1**, 1.

16. Tinga, W.R., Microwave systems design and dielectric property measurements, *Mater. Res. Soc. Bull.*, 1994, **347**, 1.

17. Whittaker, A.G. and Mingos, D.M.P., The applications of mirowave heating to chemical syntheses, *J. Microwave Power Electromagn. Energy*, 1994, **29**, 195; Whittaker, A.G., Harrison, A., Oakley, G.S., Youngson, I.D., King, S. and Heenan, R., *Rev. Sci. Instrum.*, 2001, **72**, 172.

18. Baghurst, D.R. and Mingos, D.M.P., Applications of microwave heating techniques for the synthesis of solid state inorganic compounds, *J. Chem. Soc., Chem. Commun.*, 1988, 829.

19. Baghurst, D.R., Chippendale, A.R. and Mingos, D.M.P., Microwave syntheses for superconducting ceramics, *Nature*, 1988, **332**, 311.

20. Baghurst, D.R. and Mingos, D.M.P., The application of microwaves in the processing of inorganic materials, *British Ceramic, Transactions and Journal*, 1992, **91**, 124.

21. Cheng, J., Roy, R. and Agrawal, D., Radically different effects on materials by separated microwave electric and magnetic fields, *Mater. Res.*, 2002, **5**, 170; Roy, R., Peelamedu, P.D., Cheng, J.P., Grimes, C. and Agrawal, D., Major phase transformations and magnetic property changes caused by electromagnetic fields at microwave frequencies, *J. Mater. Res.*, 2002, **17**, 3008.

22. Whittaker, A.G. and Mingos, D.M.P., Synthetic reactions using metal powders under microwave irradiation, *J. Chem. Soc., Dalton Trans.*, 2002, 3967.

23. Whittaker, A.G. and Mingos D.M.P., Microwave assisted solid state reactions involving metal powders, *J. Chem. Soc., Dalton Trans.*, 1992, 2751.

24. Whittaker, A.G. and Mingos, D.M.P., Arcing and other microwave characteristics of metal powders in liquid systems, *J. Chem. Soc., Dalton Trans.*, 2000, 1521.

25. Zhang, X., Hayward, D.O. and Mingos, D.M.P., Effects of microwave dielectric heating on heterogeneous catalysis, *Catal. Lett.*, 2003, **88**, 33.

2 Microwave-accelerated metal catalysis: organic transformations at warp speed

KRISTOFER OLOFSSON and MATS LARHED

2.1. Introduction

Organometallic catalysis in general and palladium-catalysed reactions in particular have emerged as the success stories in the growing, heterogeneous field of microwave-assisted chemistry[1,2]. The reasons behind this are many, but few other disciplines in chemistry have adapted as readily as these to the special demands of high-temperature microwave reactions[3]. Numerous protocols have been developed that take advantage of the convenience and special characteristics of microwave chemistry. In many cases, results have been reported that surpass the results emblematic of standard modes of heating[4]. Reaction rates are generally very high and yields in many cases are improved, as competing side reactions can be minimised in the short intervals of time with the focused, *in situ* heating that are typical of this new method of superheating. Microwave-assisted chemistry has also been seen as an environmentally friendly alternative to standard heating methodologies because of the smaller volumes of solvents used in many reactions, the possibility of employing milder and less toxic reagents and the reduced energy consumption[5,6].

The special requirements of microwave flash heated organic chemistry, which in the beginning limited its use, have lately resulted in a number of innovations in the field. For example, methods have been presented for the *in situ* generation of reaction gases in sealed reaction vessels, obviating the problem of introducing gas from external sources[7]. In addition, the endeavours and resourcefulness of several research groups have led to catalytic protocols in stereoselective reactions that deliver impressive enantiomeric excesses (ee) even at extreme temperatures[8–10]. However, it is also true that many of the problems that are related to the use of transition-metal catalysis (e.g., air and moisture sensitivity, relatively expensive ligands and transition-metal catalysts) are still present, even though the chemical community is now beginning to address these issues[9,11]. Furthermore, notoriously slow metal-catalysed transformations on polymer supports have been microwave accelerated without any detected decomposition of either the resin or the linker[12–14].

The recent progress in microwave chemistry has been significantly aided by the development of heating equipment especially designed for chemical applications. Temperature and pressure can now be controlled and monitored, which has led to significantly safer and more reproducible results. One of the major challenges that remains is the limited reaction scale of many modern microwave cavities. The value of microwave heating in small-scale optimisations or high-speed medicinal chemistry reactions[15] have been widely appreciated, but large-scale applications have been few[16], and it is likely that the key to the widespread use of microwave-assisted synthesis in multi-step synthetic strategies will be the access to large-scale reactors (see Chapter 9). In combinatorial chemistry, true high-throughput methodologies that couple the short reaction times associated with rapid microwave heating with equally fast and/or automated work-up and separation techniques are also waiting to be disclosed.

Due to the widespread acceptance of the technique of microwave heating, clarification of the reasons behind the increased reaction rate is now becoming important[4,16]. Most research teams seem to favour a purely heat-derived mechanism, which leaves little room for the potential non-thermal effect of microwaves on the outcome of homogeneous reactions[4,16]. However, a recent report has suggested that it may not be possible to generalise this assumption, as transition states in bimolecular reactions may be polar and thus could be susceptible to the influence of microwave irradiation[17]. Another suggestion concerns the notion that simultaneous cooling of a reaction mixture under microwave irradiation leads to an increased reaction rate[18].

Lately, organometallic catalysis has enjoyed a renaissance with the appearance of several important papers significantly broadening the appeal of the established methods[19,20]. One of the most important of the recent breakthroughs can be considered to be the increasing ease with which aryl chlorides can now be implemented in coupling reactions catalysed by late transition metals[21]. The possibility of using these relatively cheap and readily available aryl chlorides instead of other halogens or pseudo-halides widens both the scope of commercially available starting materials and also presents new options for performing chemoselective coupling reactions. The reaction conditions associated with aryl chlorides are often quite different than the older protocols utilised for the corresponding bromides, triflates and iodides.

The introduction of microwave chemistry in itself can hardly assume the same significance as the discovery of new, valuable synthetic reactions, but it does address several other important issues. Some relatively new additions to the arsenal of microwave-assisted reaction conditions that hold great promise will be discussed in this chapter, such as ionic liquids[22–25]. These ionic solvents are promising not only for their special applications in microwave chemistry because of their high-boiling points, low-vapour pressure, thermal stability and effective interaction and heating with microwaves, but also for environmental reasons as they, to some extent, may reduce the use of standard organic solvents, as well as allow for effective catalyst recycling procedures.

This chapter concentrates on selected recent findings in the field of microwave-assisted metal-catalysed coupling reactions. (Two thorough reviews of this field have recently been published[1,2].)

2.2. Stille couplings

Lately, a number of papers have dealt with microwave-assisted reactions on palladium-doped Al_2O_3. Villemin reported on Stille, Suzuki, Heck and Trost–Tsuji reactions where potassium fluoride on alumina was used as the base[26]. The reactions were carried out without solvent or stabilising phosphine ligands in single-mode reactors. The Stille reactions were noteworthy as the toxic organotin residue remained adsorbed on the solid support, thus allowing a simplified work-up procedure for the otherwise unpleas-ant, and toxic, stannous by-products. Both the Stille and the Suzuki reactions could be performed under air. Furthermore, it was noted that with experiments where the single-mode cavity was replaced with a domestic microwave oven, non-reproducible results were obtained[26] (Scheme 2.1).

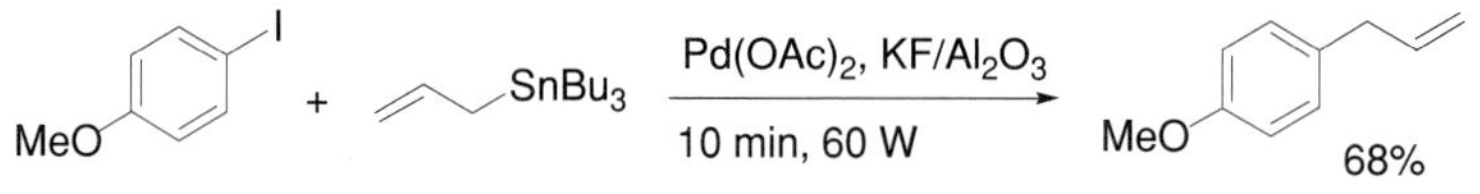

Scheme 2.1   Solvent-free Stille reaction on palladium-doped alumina.

A 'one pot' Stille coupling and Diels–Alder reaction were performed in a domestic microwave oven, producing steroidal adducts[27]. Both diethyl fumarate and diethyl malonate were tested as dienophiles, but only diethyl malonate furnished an improved ratio of product to side products, using microwave heating as compared to classical heating. The favoured formation of the thermodynamically more stable product was suggested to be the result of direct rapid heating and superheating under microwave irradiation. However, a direct analogy between the microwave and oil-bath heating methods may be difficult as different solvents were used in the two reaction conditions[27] (Scheme 2.2).

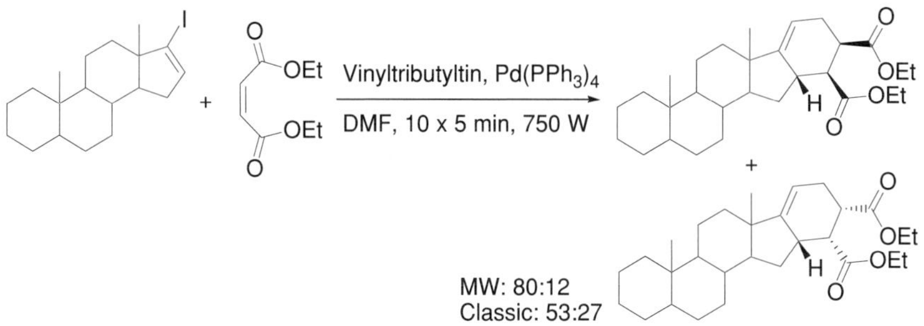

Scheme 2.2   'One pot' Stille and Diels–Alder reaction generating steroidal substrates.

Microwave-assisted Stille couplings have also recently been utilised in the preparation of radiopharmaceuticals, containing short-lived positron-emitting isotopes, for use in positron emission tomography, PET – one of the pioneering fields of microwave chemistry[28]. 1-(2′-deoxy-2′-fluoro-β-D-arabinofuranosyl)-[*Methyl*-^{11}C] thymine ([^{11}C]FMAU) was prepared in 3 min in modest radiochemical yields. The ratio of palladium to ligand, in the range of 1:2 to 1:6, did not influence the reaction considerably[29] (Scheme 2.3).

2.3. Suzuki couplings

The microwave-promoted palladium-catalysed phenylation of aroyl chlorides utilising sodium tetraphenylborate to yield unsymmetrical ketones was reported by Wang. The use of potassium fluoride as a base was found to give higher yields than potassium carbonate or triethylamine under the reaction conditions tested[30] (Scheme 2.4).

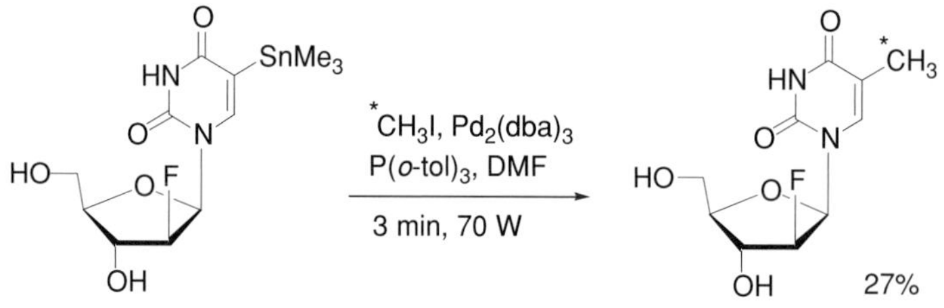

Scheme 2.3   Synthesis of [^{11}C]FMAU by Stille coupling.

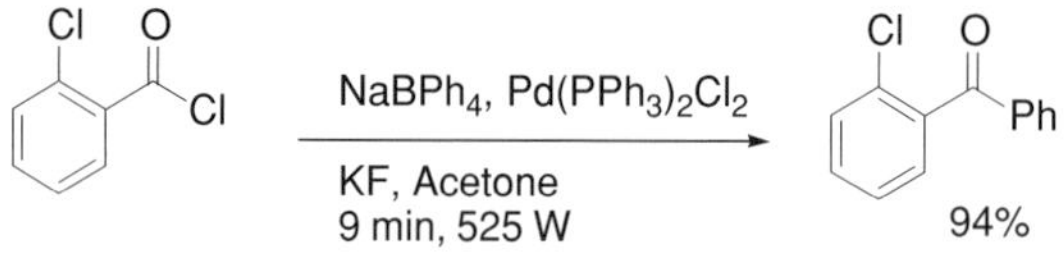

Scheme 2.4   Phenylation of aroyl chlorides yielding unsymmetrical ketones.

The same author also found the combination of potassium fluoride on alumina to be the most efficient in the phenylation reaction of aroyl chlorides. Five different bases and different power outputs were screened to find the optimal conditions[31] (Scheme 2.5).

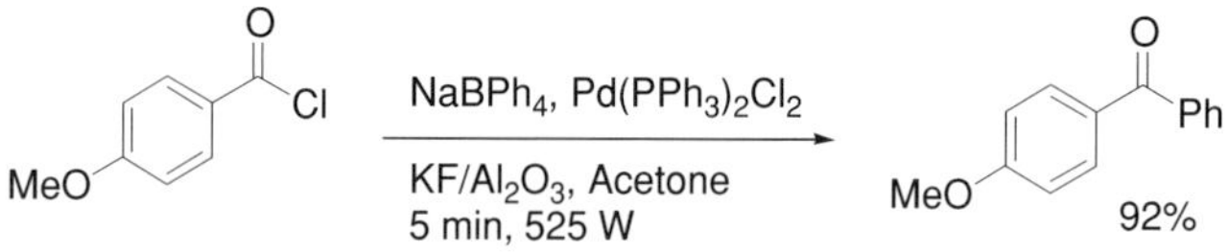

Scheme 2.5   Phenylation of aroyl chlorides using potassium fluoride on alumina.

A similar reaction was reported by Kabalka *et al.* where ligandless and solvent-free Suzuki couplings were performed with potassium fluoride on alumina. This reaction is very interesting as the catalyst used was palladium powder, the least expensive form of palladium available[32]. The authors demonstrated the simplicity of the procedure by efficient isolation of the biaryl products *via* a simple filtration. This could be done as the palladium catalyst remains adsorbed on the alumina surface. A small amount of water in the matrix was beneficial for the outcome of the reactions. Recycling of the catalyst was possible by adding fresh potassium fluoride to the palladium/alumina surface and the catalytic system remained effective at least through six reaction cycles (Scheme 2.6).

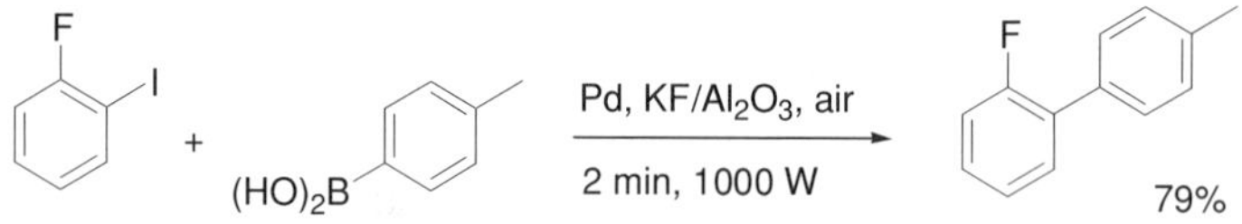

Scheme 2.6   Solvent-free and ligandless Suzuki coupling on potassium fluoride impregnated alumina.

An interesting application of the Suzuki reaction was the direct synthesis of unprotected 4-aryl phenylalanines under microwave irradiation. In addition to the excellent conversions and yields, no significant racemisation of the product was observed. Reactions with a preheated oil bath were reported to give substantially lower yields than did the corresponding single-mode microwave reactions[33] (Scheme 2.7).

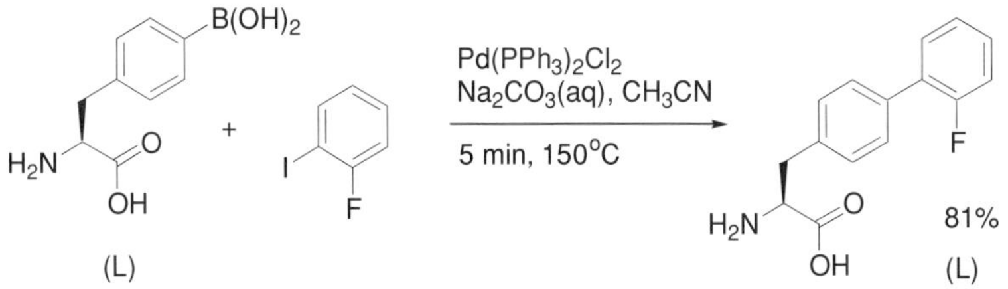

Scheme 2.7   Direct synthesis of unprotected 4-aryl phenylalanines.

The syntheses of plasmepsin inhibitors have been reported, where the final arylations of the multi-step syntheses were microwave-heated Suzuki couplings. Aryl triflates and bromides were both used as starting materials, and when triflates were used, a method without aqueous additives was chosen to suppress hydrolysis of the triflate ester[34] (Scheme 2.8). The same scaffold has also been used for the generation of three targeted hydroxyethylamine-based libraries with different P3 and P1' substituents. Microwave heating has been shown to reduce the time needed for the overall synthesis of the libraries, as the reaction times for the Suzuki couplings were reduced from hours to minutes[35] (Scheme 2.9). This combinatorial optimisation protocol afforded plasmepsin inhibitors with not only K_i values in the low-nanomolar range, but also with high selectivity versus the human protease cathepsin D.

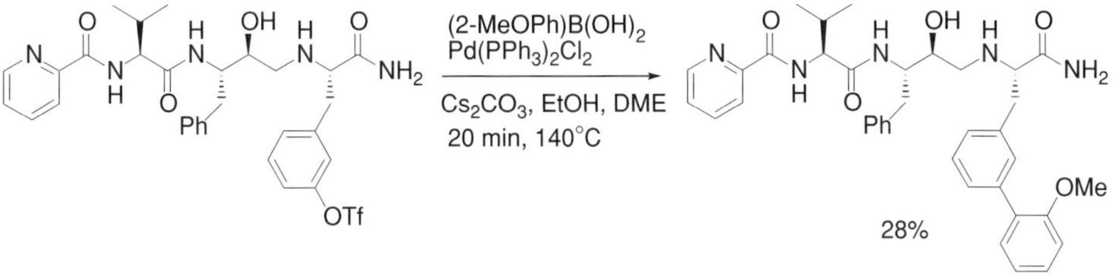

Scheme 2.8   Synthesis of plasmepsin inhibitors by microwave-assisted Suzuki couplings.

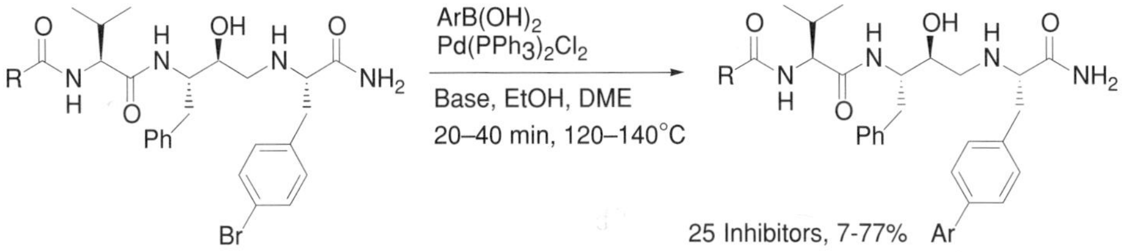

Scheme 2.9   Library synthesis of plasmepsin inhibitors with diverse P1' side chains from arylboronic acid reagents.

The quest of developing methods where organic solvents are exchanged for water is driven by the necessity of limiting the amount of waste produced by both large- and small-scale organic synthesis. Water is in this respect perhaps the ultimate substitute to organic solvents because of its non-toxic character and ready availability. Ligand-free Suzuki reactions in water[36] have been effectuated by Leadbeater with isolated yields ranging from 68 to 96% for aryl bromides and from 45 to 50% for electron rich or neutral aryl chlorides[37] (Scheme 2.10). Addition of the phase-transfer catalyst tetra-butylammonium bromide (TBAB) was found to increase the yields of the reactions. Aryl iodides that under the reported reaction conditions gave incomplete conversions by conventional heating due to solubility problems in aqueous media were successfully coupled in good yields under microwave heating.

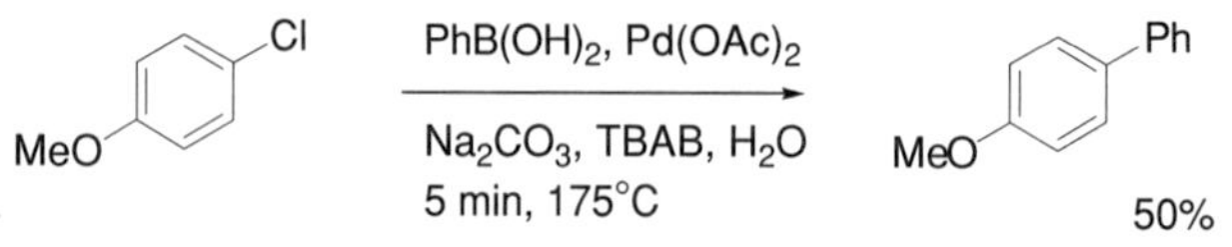

Scheme 2.10   Ligand-free Suzuki reaction in water.

Leadbeater and Marco also undertook to optimise the reaction conditions for ligand-free Suzuki reactions in water using both aryl bromides and chlorides. Comparisons between reactions performed under microwave irradiation and oil-bath heating led to the conclusion that the yields were identical or better with oil baths when aryl bromides were used as aryl precursors. However, comparisons between microwave and oil-bath heating with aryl chlorides as starting material clearly favoured the microwave technique[38].

For some applications it was found that the Suzuki couplings could be performed under microwave irradiation without transition-metal additives[11]. Using TBAB as a solvating and activating agent, Suzuki couplings were reported to proceed at 150°C with focussed microwave heating (Scheme 2.11). Oil-bath heating at 150°C for 2 h worked well with the activated 4-bromoacetophenone, but un- or deactivated aryl bromides did not couple efficiently even after extended periods of time.

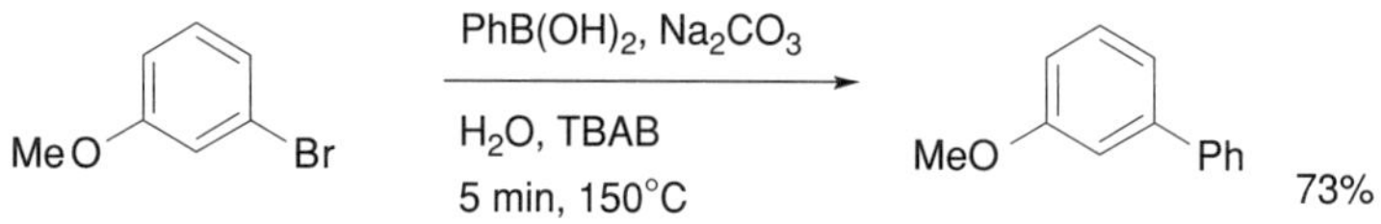

Scheme 2.11   Suzuki coupling in water without a transistion-metal catalyst.

Polyethylene glycol (PEG) is similar to water in the sense that it is a non-toxic and inexpensive solvent, which has been shown to be applicable to Suzuki couplings under microwave conditions[39]. In a domestic microwave oven, PEG-400 in combination with potassium fluoride gave high yields of biaryl products using both aryl bromides and iodides as starting materials. Oil-bath reactions conducted in 15 min were found to give yields similar to that from the microwave-heated reactions (Scheme 2.12).

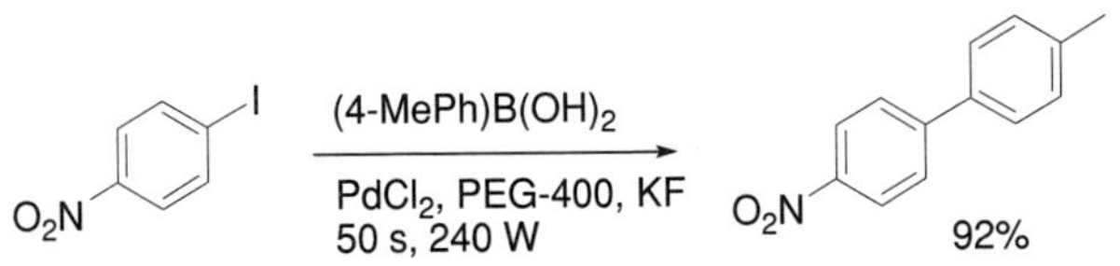

Scheme 2.12 Microwave-accelerated Suzuki couplings in polyethylene glycol.

The arylation of halopyrimidines has also been successfully performed under microwave heating. The Suzuki coupling of this class of substrates was reported to have been published only once before using classical heating and long reaction times[40] (Scheme 2.13).

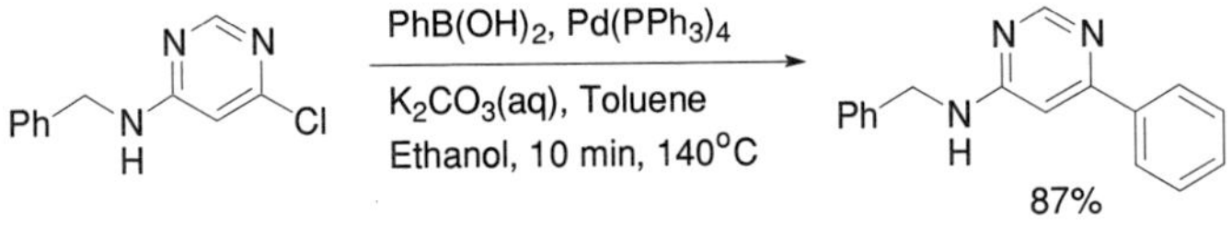

Scheme 2.13 Microwave-assisted synthesis of aminopyrimidines.

Recently, microwave-assisted Suzuki couplings under solvent-free conditions resulting in thiophene oligomers have also been reported[41].

2.4. Negishi couplings

The first microwave-assisted Negishi couplings were recently published; both aryl and alkylzinc halides were used as substrates[42] (Scheme 2.14).

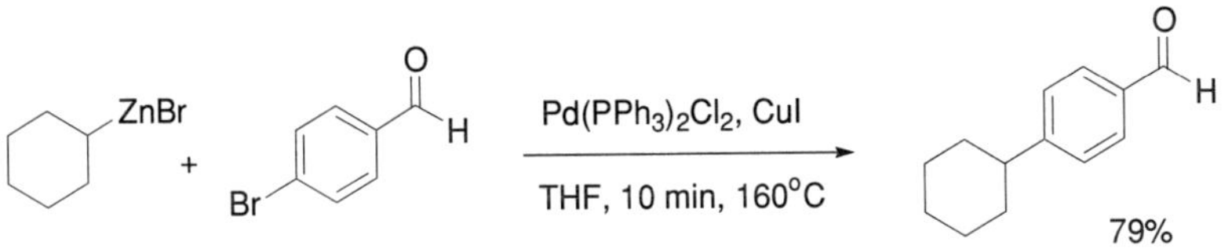

Scheme 2.14 Microwave-assisted Negishi coupling.

2.5. Heck couplings

The use of ionic liquids in microwave chemistry has great potential and a number of research groups have introduced the use of ionic liquids in synthetic approaches[43,44]. 1-Butyl-3-methylimidazolium hexafluorophosphate (bmimPF$_6$) was recently evaluated as a solvent for the microwave-promoted Heck reaction[45]. Terminal arylations of electron-poor olefins were carried out rapidly with good-to-excellent yields, using the old-fashioned catalyst palladium chloride (Scheme 2.15). As an example, the

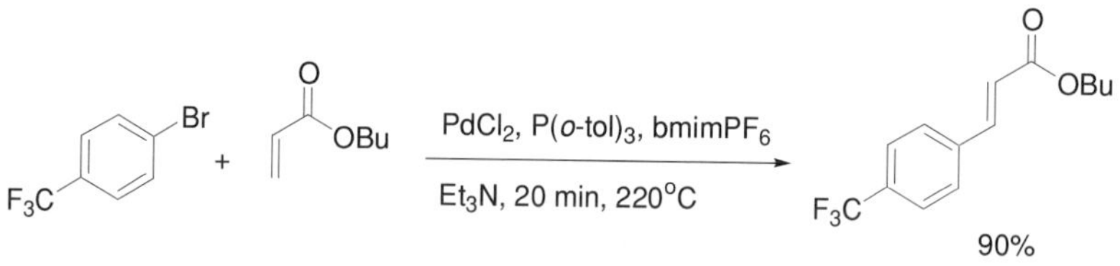

Scheme 2.15   Terminal Heck arylation with an ionic liquid as solvent.

terminal Heck arylation of butyl acrylate, conventionally a notoriously slow reaction, was finished after 20 min of single-mode microwave heating at 220°C. Another noteworthy twist was encountered in the internal arylation of butyl vinyl ether by using 1-bromo naphthalene as an aryl precursor (Scheme 2.16). Usually, when aryl or vinyl bromides are used in Heck reactions proceeding through the cationic pathway, it is normally necessary to add toxic thallium salts to scavenge the halide[46]. However, using reaction conditions including palladium acetate, ionic liquids and the bidentate ligand 1,3-diphenylphosphinopropane (DPPP), excellent regioselectivity was obtained without thallium additives[45]. The unusually high regioselectivity of this reaction was suggested to be caused by the high polarity of the reaction medium[47] (Scheme 2.16). Methods for the solvent-free preparation of ionic liquids under microwave-heating have also been reported[48,49].

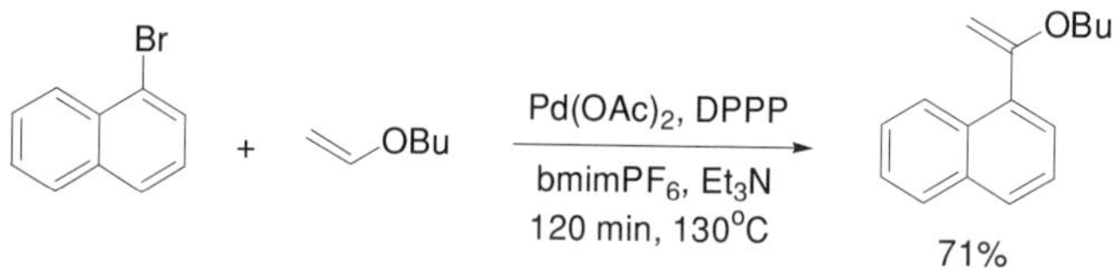

Scheme 2.16   Internal Heck arylation with an ionic liquid as solvent.

In 2002, an impressive paper was reported which dealt with Heck couplings performed with both activated and deactivated chloroarenes in TBAB with nanopalladium immobilised layered double hydroxide (LDH-Pd(0)) catalysts under microwave heating[50]. A number of heterogeneous catalytic systems were tested, however LDH-Pd(0) furnished the highest yields. In fact, some of the turnover frequencies were the highest ever recorded by any palladium catalyst for deactivated chloroarenes. The coupling between styrene and 4-chloroanisole was achieved in 1 h at 130°C when microwave heating was applied, while the reaction with standard heating at the same temperature took 40 h. The yields in the two reactions were close to identical (Scheme 2.17).

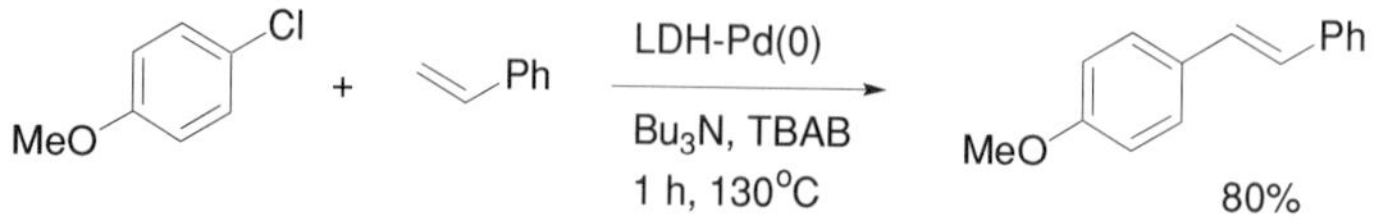

Scheme 2.17   Heck coupling of aryl chlorides with a nanopalladium catalyst.

Microwave-heated Heck reactions in water have also been performed using phase-transfer catalysis. A number of phase-transfer systems were investigated with TBAB, once more giving the best results[51] (Scheme 2.18).

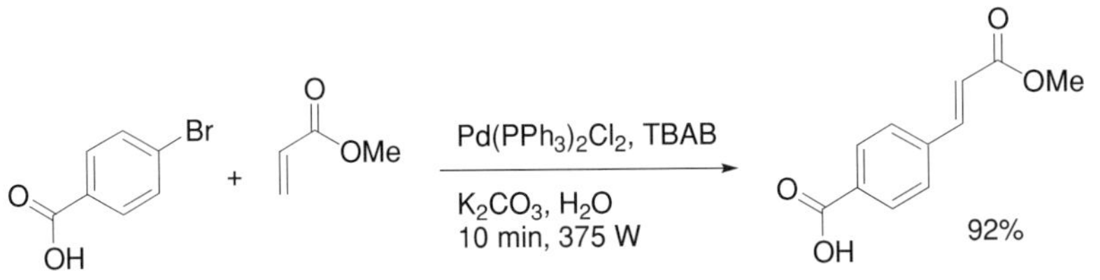

Scheme 2.18   Heck reaction in water employing phase-transfer catalysts.

Enantiomeric control in Heck couplings becomes relevant when using cyclic olefins. Forming a single new chiral compound with high selectivity by using an asymmetric Heck reaction can, however, be problematic because of the difficulties in controlling the double bond migration after arylation[52]. One excellent solution to this problem is to employ P–N ligands of the phosphinoaryl oxazoline type[53]. The strong amine base proton sponge (2,8-bis(dimethylamino)naphthalene) was used under high-temperature microwave conditions to accelerate this very slow palladium-catalysed asymmetric Heck reaction[54] (Scheme 2.19). Significant enantioselectivities of up to 92% ee were reported under air utilising the thermostable palladium-phosphineoxazoline catalytic system.

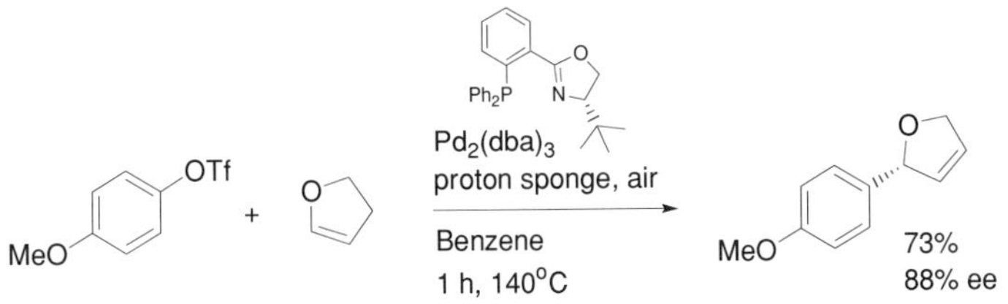

Scheme 2.19   Enantioselective Heck arylation with a phosphineoxazoline ligand.

2.6. Cyanation and Sonogashira reactions

There are several synthetic transformations available for the generation of aryl nitriles; among these, considerable research efforts have been directed towards the direct metal-catalysed coupling of aryl halides with cyanide ions. In 2000, Hallberg *et al.* reported a direct microwave-assisted preparation of aryl and vinyl nitriles from organic bromides and zinc cyanide[55]. Zhang and Neumeyer further modified this transformation to include the use of aryl triflates as arylpalladium precursors[56]. The usefulness of the method was demonstrated in the preparation of 3-cyano-3-desoxy-morphinans (Scheme 2.20).

In 2003, Leadbeater and co-workers reported a related copper iodide mediated cyanation of aryl iodides in water with TBAB as an essential additive[57]. Stoichiometric quantities of CuI were needed in this protocol, as the use of catalytic quantities resulted in significantly lower yields.

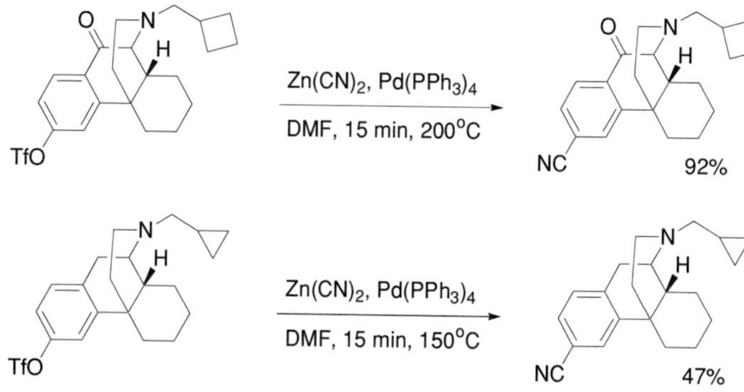

Scheme 2.20 Palladium-catalysed cyanation of morphinan triflates.

Few methods have been devised as alternatives to the palladium-catalysed Sonogashira couplings. In response to this, Wang and co-workers have developed a microwave heated and purely copper-catalysed version of this reaction using 10% CuI[58]. As shown in Scheme 2.21, high yields were achieved after just 10 min of heating in a commercial multi-mode microwave oven.

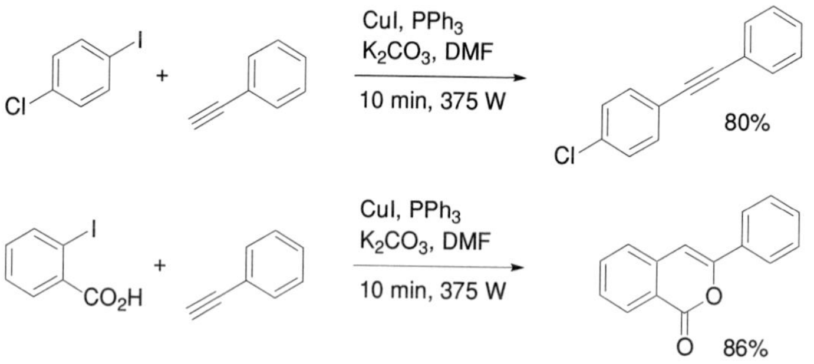

Scheme 2.21 Copper-catalysed Sonogashira coupling reactions.

2.7. Carbon–heteroatom coupling reactions

The direct palladium- or nickel-catalysed reaction of aryl halides (or pseudo-halides) with nucleophilic amines (the Buchwald–Hartwig amination[59,60]) is a highly useful tool for the preparation of aromatic amines. Due to the importance of aromatic amines, especially in medicinal chemistry, the reaction has since 1994 been extensively studied and continuously developed[59,60]. Surprisingly, synthetic organic chemists have been slow to accelerate this reaction using microwave flash heating. Thus, the first published microwave-induced Buchwald–Hartwig couplings were reported in early 2002 by Sharifi et al.[61]. Their procedure used atmospheric conditions and a domestic microwave oven (Scheme 2.22). The same year, researchers from Uppsala reported a more comprehensive study including 14 different examples[62] (Scheme 2.23). It should be noted that despite the high-reaction temperatures (130–180°C) and the presence of the strong

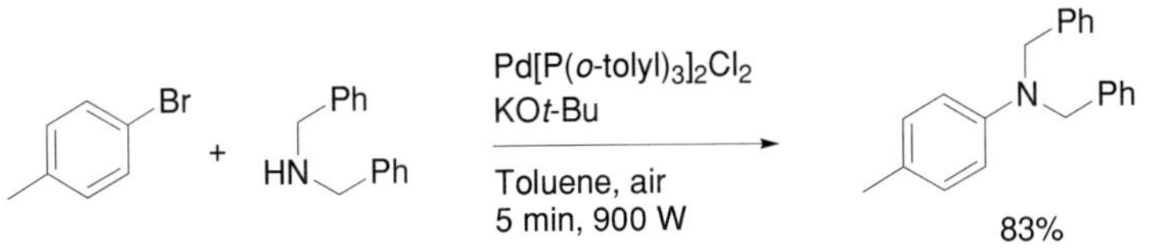

Scheme 2.22 Amination of aryl bromides employing microwave irradiation.

base KOt-Bu, decomposition of DMF (the solvent) and subsequent generation of nucle-ophilic dimethyl amine was not a preparative problem. Hence, the aminations, shown in Scheme 2.23, could be smoothly conducted in a Smith Synthesizer under air in only 4 min with racemic BINAP as the palladium ligand. The synthesis of an N-arylimidazole required a modification of the original procedure; DPPF was utilised as the ligand (Scheme 2.24).

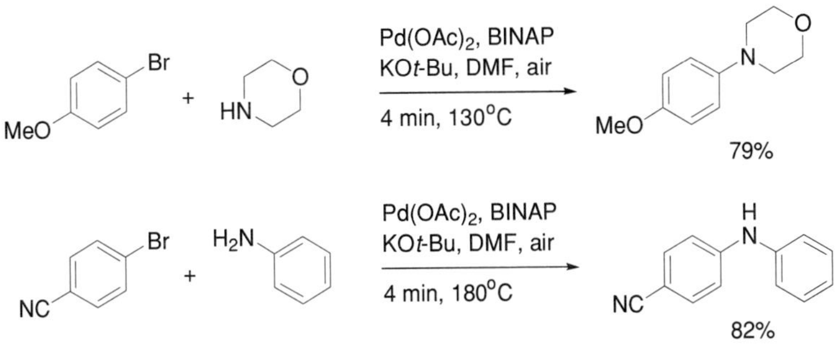

Scheme 2.23 Amination of aryl bromides employing temperature-controlled microwave heating (BINAP = 2,2'-bis (diphenylphosphino)-1,1'-binaphthyl).

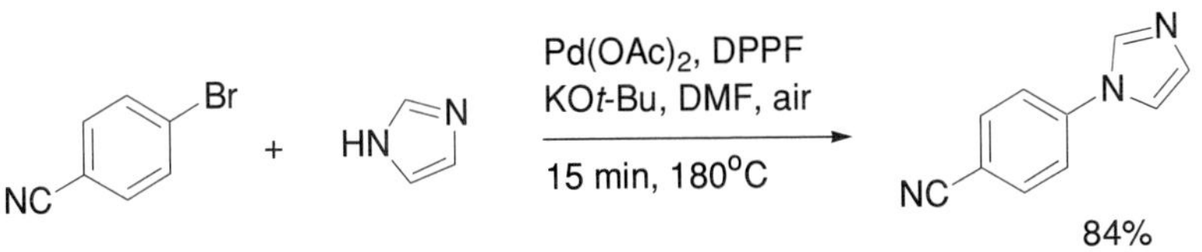

Scheme 2.24 Palladium-catalysed synthesis of a N-arylimidazole (DPPF = [1,1'-bis(diphenylphosphino) ferrocene]).

Very recently 5- and 8-aminoquinolines have been prepared from the respective aryl bromides by microwave-assisted Buchwald–Hartwig couplings in 10 min at 120°C[63]. Enhanced yields were observed under microwave conditions as compared to the corre-sponding classical oil-bath heated reactions (Scheme 2.25).

Triarylphosphine ligands are of utmost importance in transition-metal catalysis. Thus, direct palladium-catalysed high-speed generation of these ligands from secondary phosphines and aryl halide building blocks would be a suitable and convergent syn-thetic route with high flexibility. Stadler and Kappe have shown that this method for the

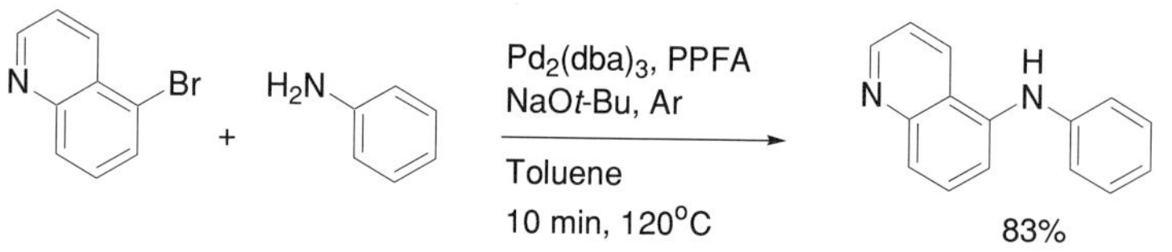

Scheme 2.25   Amination of 5-bromoquinoline employing temperature-controlled microwave heating.

generation of triarylphosphines is indeed compatible with microwave heating[64]. Short reaction times (3–30 min) and useful yields were reported from aryl iodides, bromides and triflates (Scheme 2.26).

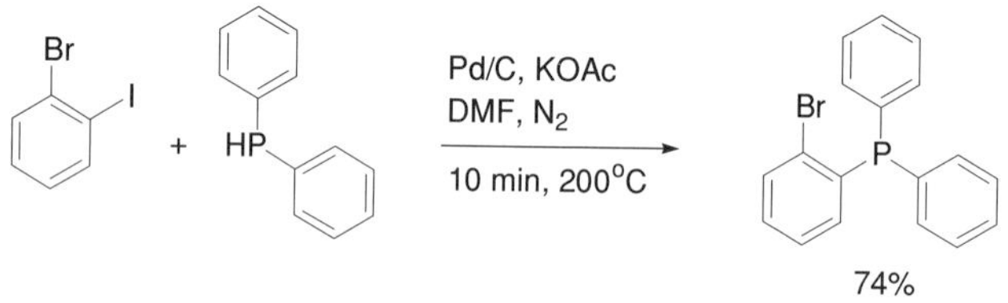

Scheme 2.26   Palladium-catalysed formation of triarylphosphines.

2.8. Asymmetric molybdenum-catalysed allylic alkylations

Asymmetric allylic substitutions have developed rapidly during the last years. Notable among the transition metals capable of mediating such reactions are palladium and molybdenum. Interestingly, molybdenum catalysis in allylic alkylations gives rise to a regioselectivity favouring the most hindered product (the internal product)[65], which is in direct contrast to allylic palladium catalysis (affording mainly the terminal product)[66]. The use of microwave heating for acceleration of asymmetric allylic substitutions has received increased attention lately[10,67,68], particularly in the case of molybdenum catalysis[9,69]. In fact, Trost has observed an enhanced efficacy of the inexpensive pre-catalyst $Mo(CO)_6$, under focused microwave conditions as compared to standard oil-bath heating[70].

The development of the robust $Mo(CO)_6$-based high-temperature protocol[9] encouraged Prof. Moberg's research group to synthesise and evaluate the reactivity of a set of new 4- and 6-substituted bis-pyridylamide ligands under microwave heating[71]. In accordance with earlier molybdenum-catalysed allylic substitutions[69], bis-pyridylamide ligands containing an electron-donating substituent in the 4-position furnished very high stereo- and regioselectivity starting from linear 3-phenylprop-2-enyl carbonate and sodium dimethyl malonate (Scheme 2.27). A set of related reactions using differently decorated bis-pyridylamide ligands and 4-substituted cinnamyl substrates were thereafter performed. One example of a rapid internal alkylation using a general and easily available chloro-substituted ligand is presented in Scheme 2.27. Finally, having identified new and interesting ligands, three racemic and branched 1-arylprop-2-en-1-ols were investigated with the chlorinated ligand (Scheme 2.28). Compared with the non-chlorinated parent ligand, the regioselectivity was improved, 69:1 versus 13:1, but the enantioselectivity was reduced, 86% ee versus 96% ee.

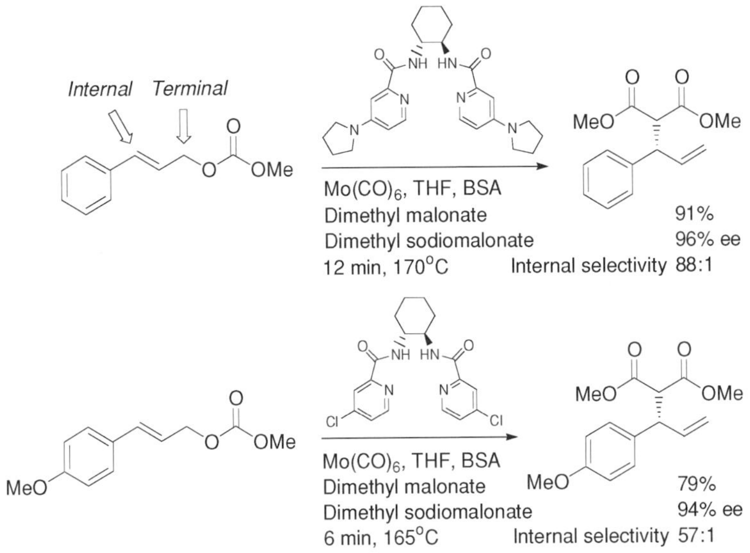

Internal Terminal

Mo(CO)$_6$, THF, BSA
Dimethyl malonate   91%
Dimethyl sodiomalonate   96% ee
12 min, 170°C   Internal selectivity 88:1

Mo(CO)$_6$, THF, BSA
Dimethyl malonate   MeO   79%
Dimethyl sodiomalonate   94% ee
6 min, 165°C   Internal selectivity 57:1

Scheme 2.27 Enantioselective molybdenum-catalysed allylic alkylation of linear arylpropenyl carbonates (BSA = N, O-bis(trimethylsilyl)acetamide).

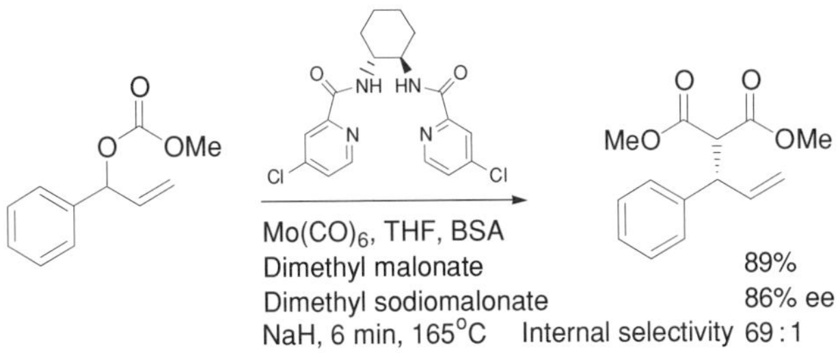

Mo(CO)$_6$, THF, BSA
Dimethyl malonate   89%
Dimethyl sodiomalonate   86% ee
NaH, 6 min, 165°C   Internal selectivity 69:1

Scheme 2.28 Enantioselective molybdenum-catalysed allylic alkylation of branched phenylpropenyl carbonate.

2.9. Carbonylative couplings

Today there is a large demand for efficient synthetic high-throughput methods in the pursuit of new drug-like molecules in lead identification and lead optimisation projects[15]. The combinatorial and medicinal chemist is therefore under constant pressure to improve the high quality compound production in terms of both time and diversity. Furthermore, the procedures presently used by the combinatorial chemists are limited to manipulation of solid and liquid reagents, unless the most advanced synthesisers are utilised. Mainstream microwave chemistry has also tended to rely heavily on non-gaseous protocols mostly because the real benefits of microwave synthesis are obtainable only in sealed vessels, where the available headspace is limited. Thus, the development of alternative gas sources; reagents that in a controlled manner release reactive gases upon heating or irradiation, is of paramount importance for high-speed microwave chemistry using gaseous reactants.

Palladium-catalysed coupling reactions under carbon monoxide have been extensively used in traditional medicinal chemistry[72]. Despite this, these crucial transformations have hardly been employed in combinatorial chemistry. These shortcomings have recently been recognised and a series of microwave-heated carbonylative transformations with solid or liquid CO-sources have been reported.

2.9.1. *Molybdenum hexacarbonyl as a solid CO-releasing reagent*

Scheme 2.29 depicts two of the first examples of microwave-assisted carbonylation reactions[7]. In these reactions, the temperature controls the rate of the CO release. Thus, during heating at 150°C in sealed vessels, carbon monoxide was smoothly emitted from the molybdenum carbonyl complex into the reaction mixture (Fig. 2.1, Profile A). As a result, aryl iodides and bromides underwent efficient aminocarbonylation with non-hindered, aliphatic, primary and secondary amines in only 15 min, using Herrmann's palladacycle as pre-catalyst[7] (Scheme 2.29). In contrast, at a reaction temperature of 210°C, carbon monoxide was liberated almost instantaneously (Fig. 2.1, Profile B).

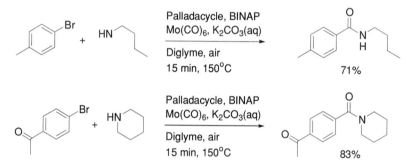

Scheme 2.29 *In situ* aminocarbonylations with $Mo(CO)_6$ as the carbon monoxide source (palladacycle = *trans*-di(acetate)bis[*o*-(di-*o*-tolylphosphino)benzyl]dipalladium(II)).

An improved *in situ* aminocarbonylation protocol was recently reported by the same Uppsala group, again using $Mo(CO)_6$ as a CO-source[73]. With this new and more general method employing DBU as a strong base, successful amidations could also be accomplished with amino acids, sluggish anilines and hindered *tert*-butyl amine (Scheme 2.30). The reactions were carried out in solution under a non-inert atmosphere with both electron-rich and electron-poor aryl iodides as well as bromides, or alternatively with resin-bound 4-iodobenzenesulphonamide. Scheme 2.30 illustrates two examples out of 24 successful reactions.

To increase the scope of the microwave $Mo(CO)_6$ procedure, oxygen nucleophiles were investigated. The omission of amine together with the addition of ethylene glycol as co-solvent, turned the aminocarbonylation method in Scheme 2.29 into a convenient hydroxycarbonylation, generating the corresponding benzoic acid derivative[7] (Scheme 2.31). With ethylene glycol present, the glycol ester is a plausible intermediate, which could subsequently undergo hydrolysis into the benzoate product.

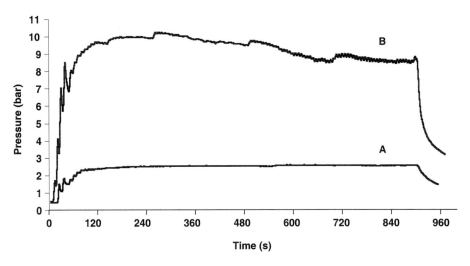

Figure 2.1 Pressure profiles for aminocarbonylations at (A) 150°C and (B) 210°C using Mo(CO)$_6$ as the carbon monoxide source (Smith Synthesizer).

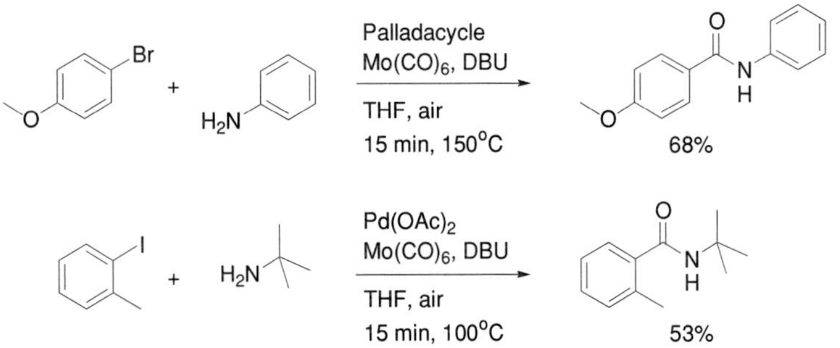

Scheme 2.30 *In situ* aminocarbonylations with sluggish amines and Mo(CO)$_6$ as the carbon monoxide source (DBU = 1,8-diazabicyclo[5.4.0]undec-7-ene).

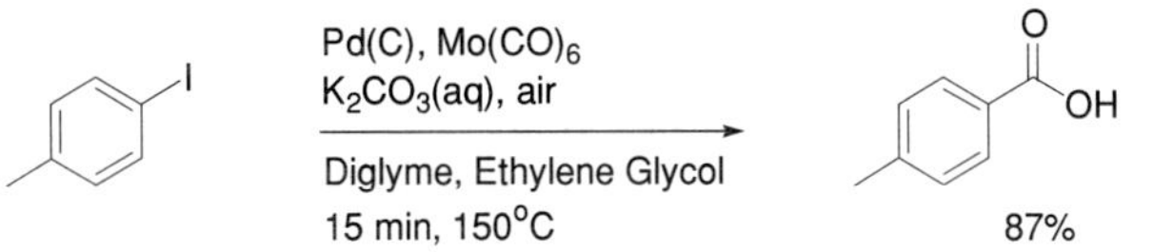

Scheme 2.31 *In situ* hydroxycarbonylation with Mo(CO)$_6$ as the carbon monoxide source.

A series of rapid microwave-mediated ester syntheses using Mo(CO)$_6$ as the carbon monoxide source were published in 2003[74]. In this paper, a range of valuable ester-protected acids (butyl-, benzyl- and trimethylsilylethyl esters) were smoothly produced both in solution (Scheme 2.32) and on solid phase (TentaGel S RAM-resin, Scheme 2.33) after 15–20 min of single-mode microwave irradiation. The use of DMAP as a nucleophilic additive increased the product yields slightly. Unfortunately, the sterically hindered *tert*-butanol furnished little or no product formation at all.

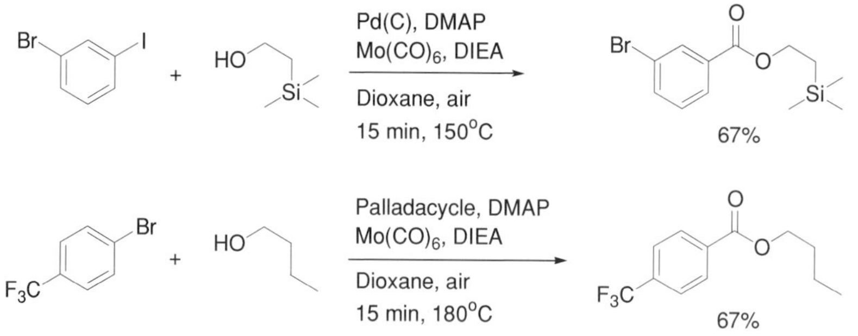

Scheme 2.32 Palladium-catalysed synthesis of esters from aryl halides with Mo(CO)$_6$ as the carbon monoxide source (DMAP = 4-dimethylaminopyridine).

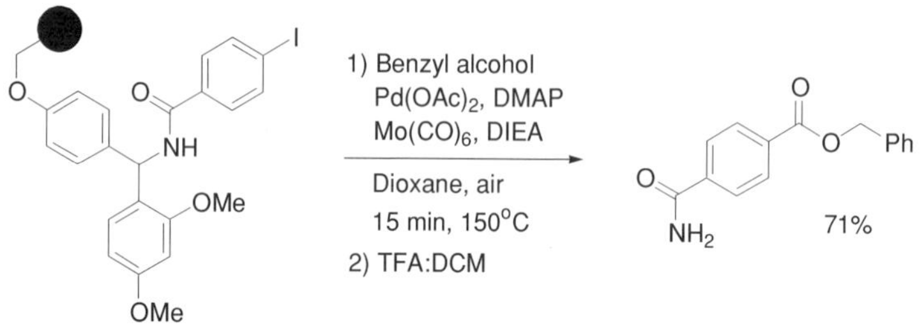

Scheme 2.33 Palladium-catalysed solid-phase ester synthesis with Mo(CO)$_6$ as the carbon monoxide source.

2.9.2. Formamides as liquid CO-releasing reagents

The concept of *in situ* liberation of carbon monoxide would be even more attractive if a metal-free material could serve as the carbon monoxide source. In the ideal carbonylation method, the organic solvent itself could be exploited for controlled generation of carbon monoxide. In 2002, Wan *et al.* addressed this issue and developed a microwave-promoted carbamoylation process based on the commonly used solvent dimethylformamide (DMF) as the carbon monoxide precursor[75]. Firstly, it was discovered that aryl dimethyl amides were accessible from the corresponding bromides in the presence of a nucleophilic catalyst, imidazole (Scheme 2.34). Secondly, tertiary benzamides other than dimethylamides were synthesised by addition of 3 equiv of an external amine (Scheme 2.34).

The continued development of microwave methods for *in situ* carbonylations is further illustrated in Alterman's recent procedure for microwave syntheses of primary benzamides from haloarenes[76]. Owing to the required use of both ammonia and carbon monoxide, the synthesis of primary amides is recognised as considerably more difficult than the corresponding synthesis of functionalised amides. In similarity to DMF, formamide is known to undergo thermal decomposition. However, formamide simultaneously generates both ammonia and carbon monoxide[77]. Therefore formamide was selected as a suitable combined solvent, ammonia synthon and CO source for the synthesis of primary benzamides. Scheme 2.35 presents two selected examples from the reported results[77] where imidazole was used as an essential activating agent. Good isolated yields were demonstrated with different types of aryl bromides after only 400 s

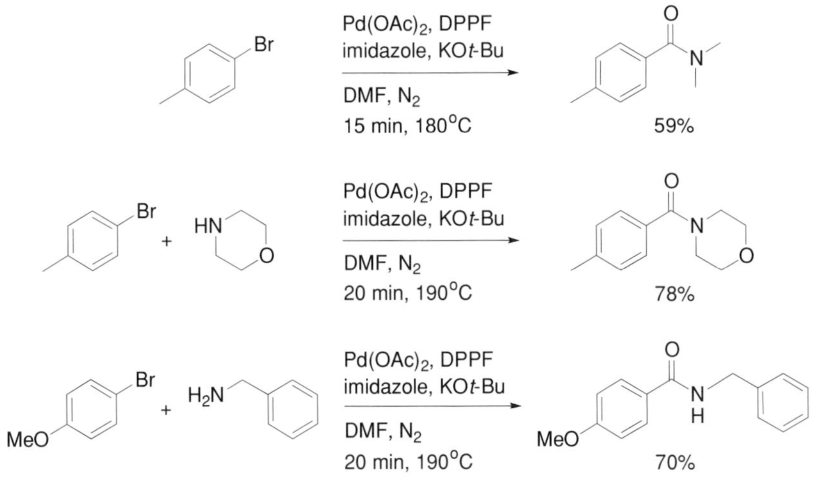

Scheme 2.34 *In situ* aminocarbonylation with DMF as the carbon monoxide source.

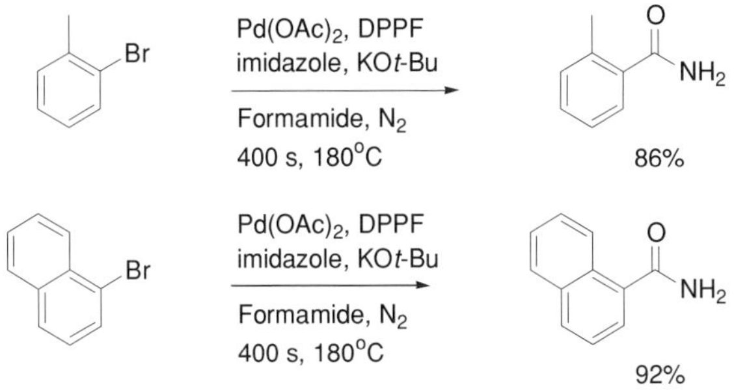

Scheme 2.35 *In situ* aminocarbonylation with formamide as a combined carbon monoxide and ammonia-releasing solvent.

of microwave heating at 180°C. Addition of the strong base KOt-Bu was necessary for efficient formamide decarbonylation and gas release as illustrated in Fig. 2.2. In the presence of only formamide and KOt-Bu (A), the reaction overpressure developed to almost 4 bar upon heating to 180°C[76]. For the preparative aminocarbonylation with 4-bromotoluene, the reaction pressure initially increased rapidly (B). The subsequent decrease in pressure with irradiation time indicated the consumption of ammonia and carbon monoxide in the carbonylation process. A control experiment (C) with pure formamide afforded an overpressure below 2 bar.

2.10. Outlook

An increasing number of microwave-assisted metal catalysis applications have been reported recently both for high-speed production of new chemical entities in early drug discovery and for other more general areas of organic synthesis. Although the number of examples on the application of solution-phase catalysis is steadily increasing, it is

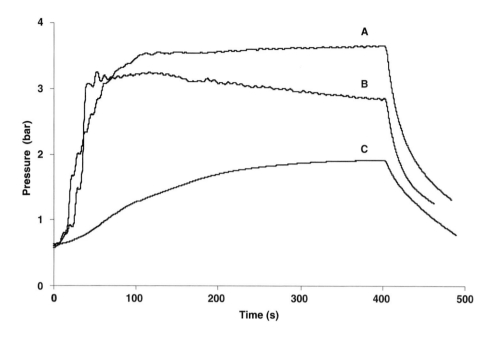

Figure 2.2 Pressure profiles recorded from microwave heating at 180°C of (A) formamide and KO*t*-Bu; (B) a preparative aminocarbonylation reaction with 4-bromotoluene and formamide and (C) pure formamide (Smith Synthesizer).

notable that many solvent-free examples are also prevalent in the literature. The reduced reaction times, the high compatibility with non-inert conditions and the experimental convenience enable high productivity. This is facilitated by modern automated single-mode cavities which feature almost parallel throughput with full sequential reaction control. The possibility to pre-programme both liquid dispensing, heating time and reaction temperature makes these microwave synthesisers very powerful for reaction optimisations in the field of transition-metal catalysis. It is likely that the future development of microwave synthesisers will result in new high-throughput systems affording parallel single-mode applicators with individual temperature control. A complementary future of microwave instrumentation may lie in the development of truly personal reactors, designed for manual operation in the hood, but connected to a central computer enabling data storage and database searches.

In the twenty-first century, pressure on the chemical community is growing to meet the demand for environmentally benign reaction processes. The evaluation of new alternative ways to perform organic synthesis will be critical in meeting the green challenge. Mild reagents and non-toxic solvents such as water can be foreseen to be employed more often in the future. We believe the combination of metal catalysis and energy-efficient microwave heating constitutes a new direction, which will be of importance not only for accelerating the chemical compound production in the fields of drug discovery and drug development, but also in the search for general and green laboratory-scale organic synthesis.

2.11. Acknowledgement

We thank Mr Gunnar Wikman and Dr Peter Nilsson for help with the manuscript. We also thank the Swedish Research Council and Knut and Alice Wallenberg Foundation. We acknowledge Personal Chemistry (now Biotage AB) for provision of the Smith Synthesizer.

2.12. References

1. Larhed, M., Moberg, C. and Hallberg, A., Microwave-accelerated homogeneous catalysis in organic chemistry, *Acc. Chem. Res.*, 2002, **35**, 717–727.
2. Olofsson, K., Hallberg, A. and Larhed, M., Transition metal catalysis and microwave flash heating in organic chemistry, In: Loupy, A. (Ed.), *Microwaves in Organic Synthesis*, Wiley: Weinheim, 2002, pp. 379–403.
3. Larhed, M. and Hallberg, A., Microwave-promoted palladium-catalysed coupling reactions, *J. Org. Chem.*, 1996, **61**, 9582–9584.
4. Lidström, P., Tierney, J., Wathey, B. and Westman, J., Microwave assisted organic synthesis: a review, *Tetrahedron*, 2001, **57**, 9225–9283.
5. Strauss, C.R., Invited review. A combinatorial approach to the development of environmentally benign organic chemical preparations, *Aust. J. Chem.*, 1999, **52**, 83–96.
6. Strauss, C.R., Application of microwaves for environmentally benign organic chemistry, In: Clark, J. and Duncan, M. (Eds.), *Handbook of Green Chemistry and Technology*, Blackwell Science: Oxford, 2002, pp. 397–415.
7. Kaiser, N.F.K., Hallberg, A. and Larhed, M., In situ generation of carbon monoxide from solid molybdenum hexacarbonyl: a convenient and fast route to palladium-catalysed carbonylation reactions, *J. Comb. Chem.*, 2002, **4**, 109–111.
8. Bremberg, U., Larhed, M., Moberg, C. and Hallberg, A., Rapid microwave-induced palladium-catalysed asymmetric allylic alkylation, *J. Org. Chem.*, 1999, **64**, 1082–1083.
9. Kaiser, N.F.K., Bremberg, U., Larhed, M., Moberg, C. and Hallberg, A., Fast, convenient, and efficient molybdenum-catalysed asymmetric allylic alkylation under noninert conditions: an example of microwave-promoted fast chemistry, *Angew. Chem., Int. Ed. Engl.*, 2000, **39**, 3596–3598.
10. Evans, P.A. and Brandt, T.A., Enantioselective palladium-catalysed allylic alkylation using *E*- and *Z*-vinylogous sulfonates, *Org. Lett.*, 1999, **1**, 1563–1565.
11. Leadbeater, N.E. and Marco, M., Transition-metal-free Suzuki-type coupling reactions, *Angew. Chem., Int. Ed. Engl.*, 2003, **115**, 1445–1447.
12. Larhed, M., Lindeberg, G. and Hallberg, A., Rapid microwave-assisted Suzuki coupling on solid-phase, *Tetrahedron Lett.*, 1996, **37**, 8219–8222.
13. Lew, A., Krutzik, P.O., Hart, M.E. and Chamberlin, A.R., Increasing rates of reaction: microwave-assisted organic synthesis for combinatorial chemistry, *J. Comb. Chem.*, 2002, **4**, 95–105.
14. Kappe, C.O., High-speed combinatorial synthesis utilizing microwave irradiation, *Curr. Opin. Chem. Biol.*, 2002, **6**, 314–320.
15. Larhed, M. and Hallberg, A., Microwave-assisted high-speed chemistry: a new technique in drug discovery, *Drug Discov. Today*, 2001, **6**, 406–416.
16. Strauss, C.R. and Trainor, R.W., Invited review. Developments in microwave-assisted organic chemistry, *Aust. J. Chem.*, 1995, **48**, 1665–1692.
17. Perreux, L. and Loupy, A., A tentative rationalization of microwave effects in organic chemistry according to the reaction medium, and mechanistic considerations, *Tetrahedron*, 2001, **57**, 9199–9223.
18. Hayes, B.L., *Microwave Synthesis. Chemistry at the Speed of Light*, CEM Publishing: Matthews, NC, 2002.
19. Cho, J.Y., Tse, M.K., Holmes, D., Maleczka, R.E. and Smith, M.R., Remarkably selective iridium catalysts for the elaboration of aromatic C—H bonds, *Science*, 2002, **295**, 305–308.
20. Ishiyama, T., Takagi, J., Hartwig, J.F. and Miyaura, N., A stoichiometric aromatic C—H borylation catalysed by iridium(I)/2,2 '-bipyridine complexes at room temperature, *Angew. Chem., Int. Ed. Engl.*, 2002, **41**, 3056–3058.
21. Littke, A.F. and Fu, G.C., Palladium-catalysed coupling reactions of aryl chlorides, *Angew. Chem., Int. Ed. Engl.*, 2002, **41**, 4176–4211.
22. Welton, T., Room-temperature ionic liquids: solvents for synthesis and catalysis, *Chem. Rev.*, 1999, **99**, 2017–2083.

23. Wasserscheid, P. and Keim, W., Ionic liquids: new "solutions" for transition metal catalysis, *Angew. Chem., Int. Ed. Engl.,* 2000, **39**, 3772–3789.
24. Seddon, K.R., Ionic liquids for clean technology, *J. Chem. Tech. & Biotech.,* 1997, **68**, 351–356.
25. Sheldon, R., Catalytic reactions in ionic liquids, *Chem. Comm.,* 2001, 2399–2407.
26. Villemin, D. and Caillot, F., Microwave mediated palladium-catalysed reactions on potassium fluoride/alumina without use of solvent, *Tetrahedron Lett.,* 2001, **42**, 639–642.
27. Skoda-Foldes, R., Pfeiffer, P., Horvath, J., Tuba, Z. and Kollar, L., Microwave-assisted stille-coupling of steroidal substrates, *Steroids,* 2002, **67**, 709–713.
28. Elander, N., Jones, J.R., Lu, S.Y. and Stone-Elander, S., Microwave-enhanced radiochemistry, *Chem. Soc. Rev.,* 2000, **29**, 239–249.
29. Samuelsson, L. and Långström, B., Synthesis of 1-(2'-deoxy-2'-fluoro-β-D-arabinofuranosyl)-[Methyl-^{11}c]Thymine ([^{11}c]Fmau) *via* a Stille cross-coupling reaction with [^{11}c]Methyl Iodide, *J. Labelled Compd: Radiopharm.,* 2002, **46**, 263–272.
30. Wang, J.X., Wei, B.G., Hu, Y.L., Liu, Z.X. and Yang, Y.H., Microwave promoted palladium-catalysed phenylation of aroyl chlorides and sodium tetraphenylborate, *Synth. Commun.,* 2001, **31**, 3885–3890.
31. Wang, J.X., Yang, Y.H., Wei, B.G., Hu, Y.L. and Fu, Y., Microwave-assisted cross-coupling reaction of sodium tetraphenylborate with aroyl chlorides on palladium-doped KF/Al$_2$O$_3$, *Bull. Chem. Soc. Jpn.,* 2002, **75**, 1381–1382.
32. Kabalka, G.W., Wang, L., Pagni, R.M., Hair, C.M. and Namboodiri, V., Solventless Suzuki coupling reactions on palladium-doped potassium fluoride alumina, *Synthesis,* 2003, 217–222.
33. Gong, Y. and He, W., Direct synthesis of unprotected 4-aryl phenylalanines via the Suzuki reaction under microwave irradiation, *Org. Lett.,* 2002, **4**, 3803–3805.
34. Nöteberg, D., Hamelink, E., Hultén, J., Wahlgren, M., Vrang, L. *et al.*, Design and synthesis of Plasmepsin I and Plasmepsin Ii inhibitors with activity in *Plasmodium Falciparum*-infected cultured human erythrocytes, *J. Med. Chem.,* 2003, **46**, 734–746.
35. Nöteberg, D., Schaal, W., Hamelink, E., Vrang, L. and Larhed, M., High-speed optimization of inhibitors of the malarial proteases Plasmepsin I and Ii., *J. Comb. Chem.,* 2003, **5**, 456–464.
36. Blettner, C.G., Konig, W.A., Stenzel, W. and Schotten, T., Microwave-assisted aqueous Suzuki cross-coupling reactions, *J. Org. Chem.,* 1999, **64**, 3885–3890.
37. Leadbeater, N.E. and Marco, M., Ligand-free palladium catalysis of the Suzuki reaction in water using microwave heating, *Org. Lett.,* 2002, **4**, 2973–2976.
38. Leadbeater, N.E. and Marco, M., Rapid and amenable Suzuki coupling reaction in water using microwave and conventional heating, *J. Org. Chem.,* 2003, **68**, 888–892.
39. Namboodiri, V.V. and Varma, R.S., Microwave-accelerated Suzuki cross-coupling reaction in polyethylene glycol (PEG), *Green Chem.,* 2001, **3**, 146–148.
40. Luo, G.L., Chen, L. and Poindexter, G.S., Microwave-assisted synthesis of aminopyrimidines, *Tetrahedron Lett.,* 2002, **43**, 5739–5742.
41. Melucci, M., Barbarella, G. and Sotgiu, G., Solvent-free, microwave-assisted ynthesis of thiophene oligomers via Suzuki coupling, *J. Org. Chem.,* 2002, **67**, 8877–8884.
42. Öhberg, L. and Westman, J., One-pot three-step solution phase syntheses of thiohydantoins using microwave heating, *Synlett,* 2001, 1893–1896.
43. Ley, S.V., Leach, A.G. and Storer, R.I., A polymer-supported thionating reagent, *J. Chem. Soc., Perkin Trans. 1,* 2001, 358–361.
44. Van der Eycken, E., Appukkuttan, P., De Borggraeve, W., Dehaen, W., Dallinger, D. *et al.*, High-speed microwave-promoted hetero-diels-alder reactions of 2(1*H*)-pyrazinones in ionic liquid doped solvents, *J. Org. Chem.,* 2002, **67**, 7904–7907.
45. Vallin, K.S.A., Emilsson, P., Larhed, M. and Hallberg, A., High-speed Heck reactions in ionic liquid with controlled microwave heating, *J. Org. Chem.,* 2002, **67**, 6243–6246.
46. Cabri, W. and Candiani, I., Recent developments and new perspectives in the Heck reaction, *Acc. Chem. Res.,* 1995, **28**, 2–7.
47. Xu, L.J., Chen, W.P., Ross, J. and Xiao, J.L., Palladium-catalysed regioselective arylation of an electron-rich olefin by aryl halides in ionic liquids, *Org. Lett.,* 2001, **3**, 295–297.
48. Varma, R.S. and Namboodiri, V.V., An expeditious solvent-free route to ionic liquids using microwaves, *Chem. Comm.,* 2001, **7**, 643–644.
49. Law, M.C., Wong, K.Y. and Chan, T.H., Solvent-free route to ionic liquid precursors using a water-moderated microwave process, *Green Chem.,* 2002, **4**, 328–330.
50. Choudary, B.M., Madhi, S., Chowdari, N.S., Kantam, M.L. and Sreedhar, B., Layered double hydroxide supported nanopalladium catalyst for Heck-, Suzuki-, Sonogashira-, and Stille-type coupling reactions of chloroarenes, *J. Am. Chem. Soc.,* 2002, **124**, 14127–14136.
51. Wang, J.X., Liu, Z.X., Hu, Y.L., Wei, B.G. and Bai, L., Palladium and phase transfer catalysed heck cross coupling reaction in water under microwave irradiation, *Synth. Commun.,* 2002, **32**, 1607–1614.

52. Larhed, M. and Hallberg, A. Scope, mechanism, and other fundamental aspects of the intermolecular Heck reaction. In Negishi, E.-i., (Ed.) *Handbook of Organopalladium Chemistry for Organic Synthesis*, Wiley & sons Inc., New York, 2002, Vol 1, pp. 1133–1178.

53. Loiseleur, O., Meier, P. and Pfaltz, A., Chiral phosphanyldihydrooxazoles in asymmetric catalysis: enantioselective Heck reactions, *Angew. Chem., Int. Ed. Engl.*, 1996, **35**, 200–202.

54. Nilsson, P., Gold, H., Larhed, M. and Hallberg, A., Microwave-assisted enantioselective Heck reactions: expediting high reaction speed and preparative convenience, *Synthesis*, 2002, 1611–1614.

55. Alterman, M. and Hallberg, A., Fast microwave-assisted preparation of aryl and vinyl nitriles and the corresponding tetrazoles from organo-halides, *J. Org. Chem.*, 2000, **65**, 7984–7989.

56. Zhang, A. and Neumeyer, J.L., Microwave promoted Pd-catalysed cyanation of aryl triflates: a fast and versatile access to 3-cyano-3-desoxy-10-ketomorphinans, *Org. Lett.*, 2003, **5**, 201–203.

57. Arvela, R.K., Leadbeater, N.E., Torenius, H.M. and Tye, H., Rapid cyanation of aryl iodides in water using microwave promotion, *Org. Biomol. Chem.*, 2003, **1**, 1119–1121.

58. Wang, J.X., Liu, Z.X., Hu, Y.L., Wei, B.G. and Kang, L.Q., Microwave-assisted copper catalysed coupling reaction of aryl halides with terminal alkynes, *Synth. Commun.*, 2002, **32**, 1937–1945.

59. Yang, B.H. and Buchwald, S.L., Palladium-catalysed amination of aryl halides and sulfonates, *J. Organomet. Chem.*, 1999, **576**, 125–146.

60. Hartwig, J.F., Transition metal catalysed synthesis of arylamines and aryl ethers from aryl halides and triflates: scope and mechanism, *Angew. Chem., Int. Ed. Engl.*, 1998, **37**, 2047–2067.

61. Sharifi, A., Hosseinzadeh, R. and Mirzaei, M., Rapid microwave induced palladium catalysed amination of aryl bromides, *Mon. Chem.*, 2002, **133**, 329–332.

62. Wan, Y.Q., Alterman, M. and Hallberg, A., Palladium-catalysed amination of aryl bromides using temperature-controlled microwave heating, *Synthesis*, 2002, 1597–1600.

63. Wang, T., Magnin, D.R. and Hamann, L.G., Palladium-catalysed microwave-assisted amination of 1-bromonaphthalenes and 5- and 8-bromoquinolines, *Org. Lett.*, 2003, **5**, 897–900.

64. Stadler, A. and Kappe, C.O., Rapid formation of triarylphosphines by microwave-assisted transition metal-catalysed C-P cross-coupling reactions, *Org. Lett.*, 2002, **4**, 3541–3543.

65. Trost, B.M. and Hachiya, I., Asymmetric molybdenum-catalysed alkylations, *J. Am. Chem. Soc.*, 1998, **120**, 1104–1105.

66. Trost, B.M. and VanVranken, D.L., Asymmetric transition metal-catalysed allylic alkylations, *Chem. Rev.*, 1996, **96**, 395–422.

67. Bremberg, U., Lutsenko, S., Kaiser, N.F., Larhed, M., Hallberg, A. *et al.*, Rapid and stereoselective C—C, C—O, C—N and C—S couplings via microwave accelerated palladium-catalysed allylic substitutions, *Synthesis*, 2000, 1004–1008.

68. Sebahar, H.L., Yoshida, K. and Hegedus, L.S., Effect of adjacent chiral tertiary and quaternary centers on the metal-catalysed allylic substitution reaction, *J. Org. Chem.*, 2002, **67**, 3788–3795.

69. Belda, O., Kaiser, N.F., Bremberg, U., Larhed, M., Hallberg, A. *et al.*, Highly stereo- and regioselective allylations catalysed by Mo-pyridylamide complexes: electronic and steric effects of the ligand, *J. Org. Chem.*, 2000, **65**, 5868–5870.

70. Trost, B.M. and Andersen, N.G., Utilization of molybdenum- and palladium-catalysed dynamic kinetic asymmetric transformations for the preparation of tertiary and quaternary stereogenic centers: a concise synthesis of tipranavir, *J. Am. Chem. Soc.*, 2002, **124**, 14320–14321.

71. Belda, O. and Moberg, C., Substituted pyridylamide ligands in microwave-accelerated Mo(0)-catalysed allylic alkylations, *Synthesis*, 2002, 1601–1606.

72. Larsen, R.D., Palladium catalysis in the synthesis of medicinal agents, *Curr. Opin. Drug Disc. Dev.*, 1999, **2**, 651–667.

73. Wannberg, J. and Larhed, M., Increasing rates and scope of reactions: sluggish amines in microwave-heated aminocarbonylation reactions under air, *J. Org. Chem.*, 2003, **68**, 5750–5753.

74. Georgsson, J., Hallberg, A. and Larhed, M., Rapid palladium-catalysed synthesis of esters from aryl halides utilizing Mo(CO)$_6$ as a solid carbon monoxide source, *J. Comb. Chem.*, 2003, **5**, 350–352.

75. Wan, Y.Q., Alterman, M., Larhed, M. and Hallberg, A., Dimethylformamide as a carbon monoxide source in fast palladium-catalysed aminocarbonylations of aryl bromides, *J. Org. Chem.*, 2002, **67**, 6232–6235.

76. Wan, Y.Q., Alterman, M., Larhed, M. and Hallberg, A., Formamide as a combined ammonia synthon and carbon monoxide source in fast palladium-catalysed aminocarbonylations of aryl halides, *J. Comb. Chem.*, 2003, **5**, 82–84.

77. Serp, P., Hernandez, M., Richard, B. and Kalck, P., A facile route to carbonylhalogenometal complexes (M = Rh, Ir, Ru, Pt) by dimethylformamide decarbonylation, *Eur. J. Inorg. Chem.*, 2001, 2327–2336.

3 Heterocyclic chemistry using microwave-assisted approaches

THIERRY BESSON and CHRISTOPHER T. BRAIN

3.1. Introduction

Despite the area of microwave-assisted chemistry being 20 years old, the technique has only recently received widespread global acceptance. This is a consequence of the recent availability of commercial microwave systems specific for synthesis, which offer improved opportunities for reproducibility, rapid synthesis, rapid reaction optimisation and the potential discovery of new chemistries. The beneficial effects of microwave irradiation are finding an increased role in process chemistry, especially in cases when usual methods require forcing conditions or prolonged reaction times.

Since a large number of natural products and target drug compounds contain an aromatic heterocyclic core, synthetic routes towards these molecules are usually quite challenging. The various opportunities offered by microwave technology are particularly attractive for the synthesis of aromatic heterocycles implied in drug discovery strategies, where fast high-yielding protocols and the avoidance or facilitation of purification are highly desirable.

This chapter aims to review recent developments in the synthesis of heteroaromatic compounds under conditions that include the application of microwave heating in the ring-forming step. However, functionalisation of pre-formed heteroaromatic core structures (e.g., N-alkylation of nitrogen heterocycles, heteroaromatic substitution) and the generation of heteroaromatic species from existing non-aromatic ring systems (e.g., by dehydrogenation) are not included. A comprehensive tabular appraisal of the microwave literature up to June 2000 has been published[1] and this chapter highlights developments subsequent to this. This chapter is not intended to be exhaustive in its content, but rather highlights significant examples where microwave heating has been either synthetically enabling or has provided a key advantage over conventional thermal methods. In common with other areas of microwave-promoted synthesis, early examples relied on the expedient use of microwave equipment designed for non-laboratory applications, in particular domestic microwave ovens. Many interesting reports of studies employing domestic microwave instruments have appeared in the literature; however, there is some debate over the safety and reproducibility aspects of the use of non-dedicated microwave instruments in the laboratory environment. Custom-designed microwave instruments for laboratory use are now commercially available and these offer superior control of the reaction conditions and enhanced reproducibility whilst also providing a key safety advantage. Wherever possible, this chapter focuses on chemistry carried out using the latter type of instrumentation. The use of microwaves in conjunction with other new synthetic technologies is emphasised.

The chapter is grouped according to the main heterocycle types in order of increasing complexity, commencing with five-membered ring systems containing one heteroatom and their fused ring analogues, followed by five-membered systems with more than one heteroatom and then the analogous higher-membered ring systems. Fused aromatic heterocycles are also described with a distinction between fused ring compounds and polycyclic molecules, which share at least one heteroatom. Syntheses of heterocyclic systems of particular biological or commercial interest are emphasised.

3.2. Five-membered systems with one heteroatom

3.2.1. Furans and benzofurans

2-Carboxy-substituted benzofurans have been prepared under solventless phase-transfer conditions (solid potassium carbonate/tetrabutylammonium bromide) by condensation of a substituted salicylaldehyde with chloroacetic acid esters (Scheme 3.1)[2]. Similarly, 2-carboxyaryl-substituted benzofurans were prepared by condensation of a set of salicylaldehydes with α-tosyloxyketones in the presence of solid potassium fluoride doped alumina (Scheme 3.1)[3]. In each case, a domestic microwave instrument was employed.

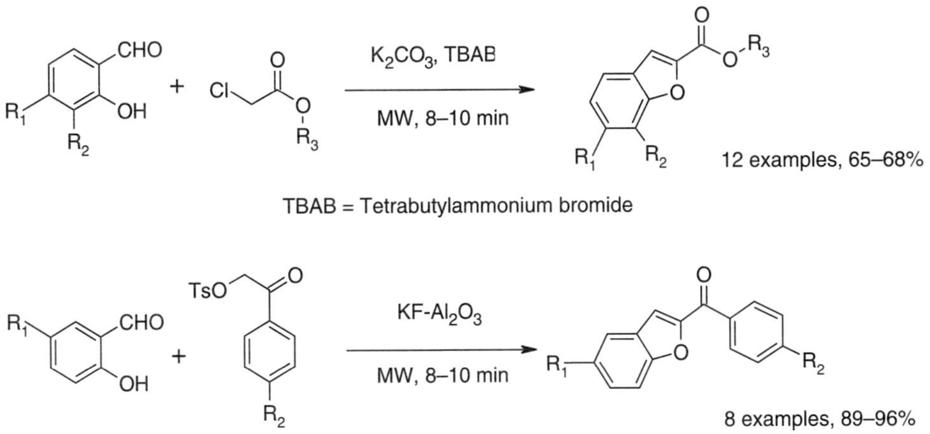

TBAB = Tetrabutylammonium bromide

Scheme 3.1

Novel 3-aryl-2-imino-4-methyl-2,5-dihydrofurans have also been efficiently synthesised using focused microwave heating by a 'one-pot' condensation from α-ketols and substituted phenylacetonitrile in the presence of sodium ethoxide in ethanol (Scheme 3.2)[4].

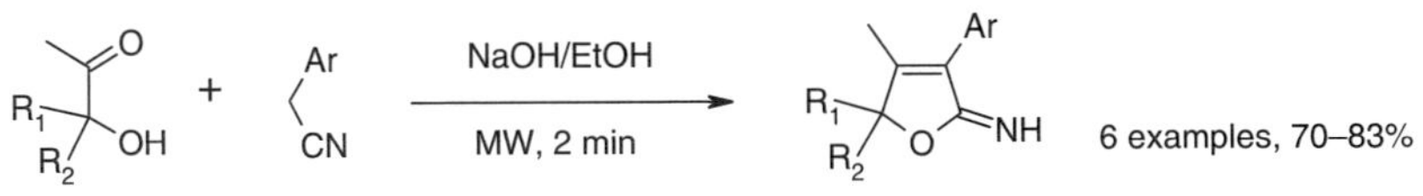

Scheme 3.2

3.2.2. Pyrroles, indoles and indolizines

Danks reported a microwave-assisted variant of the classical Paal–Knoor pyrrole synthesis (Scheme 3.3)[5]. This solvent-free method provided a considerable rate advantage (reactions complete within 2 min compared to over 12 h in conventional thermal heating) over classical procedures and even non-nucleophilic amines were condensed smoothly in the absence of Lewis acid promoters. Purification consisted of a simple silica gel filtration. The use of an early-dedicated laboratory instrument in this work is also noteworthy; however, information of how the reaction temperature was controlled is not provided.

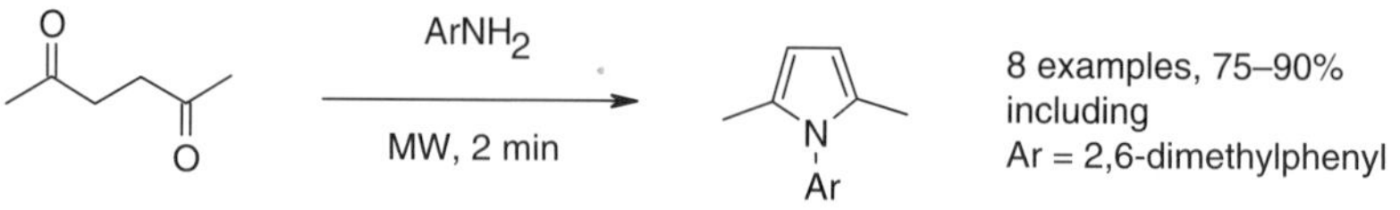

Scheme 3.3

Multi-component couplings are an attractive means of rapidly assembling densely functionalised structures from simple building blocks and have found particular application in the synthesis of heterocyclic libraries (see Chapter 5). The use of solventless reaction conditions employing inorganic supports has become a popular reaction format under microwave heating. These examples are particularly prevalent in cases where domestic instruments have been used since the hazard of utilising volatile solvents in these instruments is avoided. This approach has been used in an efficient microwave-promoted synthesis of highly substituted alkylpyrroles and fused pyrroles, which was achieved by three-component coupling of an α, β-unsaturated aldehyde/ketone, an amine and a nitroalkane on silica gel without solvent (Scheme 3.4)[6]. Alternatively, a three-component combination of an α, β-unsaturated nitroalkene, an aldehyde/ketone and an amine on alumina without solvent can be employed (Scheme 3.4)[6].

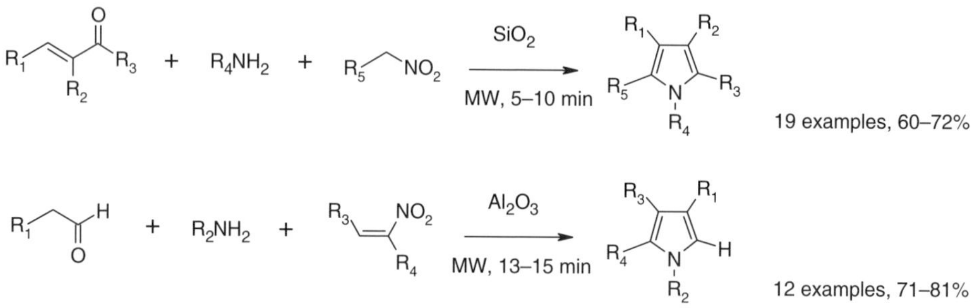

Scheme 3.4

A novel microwave-mediated three-component coupling of α-acyl bromides, pyridine and internal alkynes was carried out in the absence of a solvent on activated basic alumina to provide a collection of indolizines (Scheme 3.5)[7]. It was proposed that the reaction proceeded via *in situ* generation of a dipole from an N-acyl pyridinium salt, followed by a [3+2] cycloaddition reaction. A dedicated laboratory microwave system was used and solvent-free microwave heating conditions were demonstrated to be superior in terms of rate and yield over conventional thermal heating protocols.

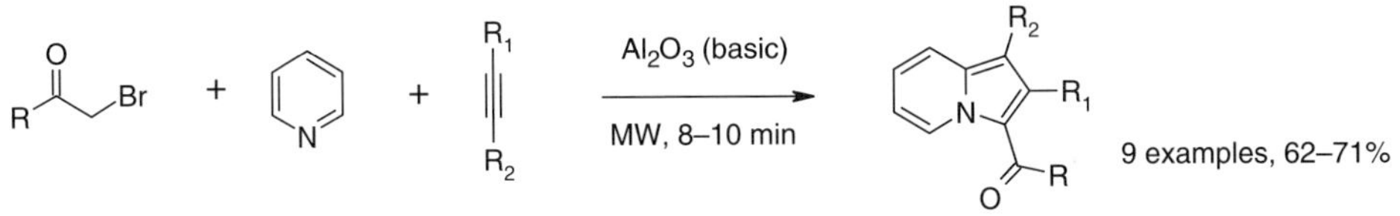

Scheme 3.5

A microwave-assisted Fischer indole synthesis under solvent-free conditions with Montmorillonite K10 clay modified with zinc chloride was employed in a key step of the synthesis of analogues of the cytostatic natural product, Sempervirine (Scheme 3.6)[8]. A dedicated laboratory microwave synthesizer was utilised.

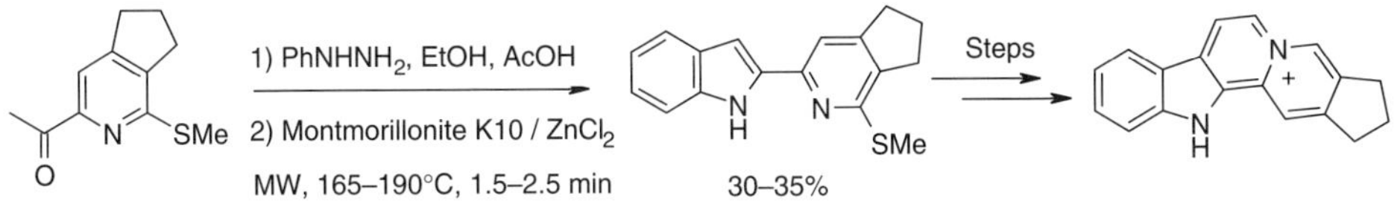

Scheme 3.6

Microwave-promoted palladium-catalysed processes have found wide general application (see Chapter 2). A Larock-type heteroannulation of an iodoaniline and an internal alkyne has been employed in the synthesis of substituted indoles[9] (Scheme 3.7). The microwave conditions were carefully optimised using a focused microwave reactor. Application of microwave heating provided clear advantages in reaction rate and yield over conventional thermal conditions. It is interesting to note that fixed microwave power input provided improved yields over constant temperature conditions (variable microwave power input). This chemistry was successfully extended to a solid-phase format (Rink amide resin)[10].

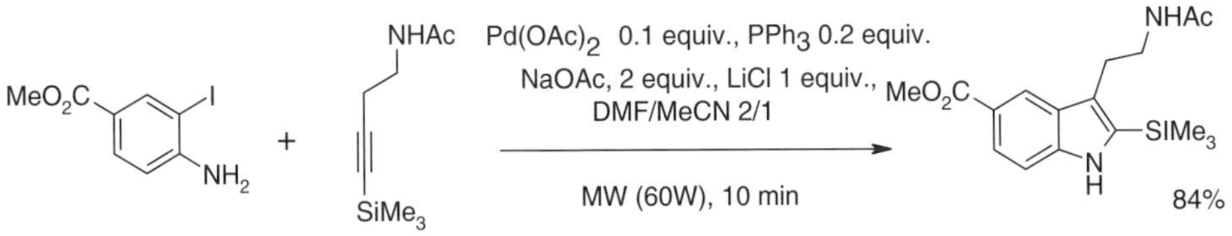

Scheme 3.7

3.2.3. Thiophenes

Continuing the trend for solventless reactions, a microwave variant of the classical Paal–Knorr thiophene synthesis has been reported[11]. Thionation–cyclisation of a range of 1,4-dicarbonyl compounds with solid Lawesson's reagent provided a series of thio-phenes rapidly in high yield and with minimal purification compared to the equivalent solution-phase reactions (Scheme 3.8).

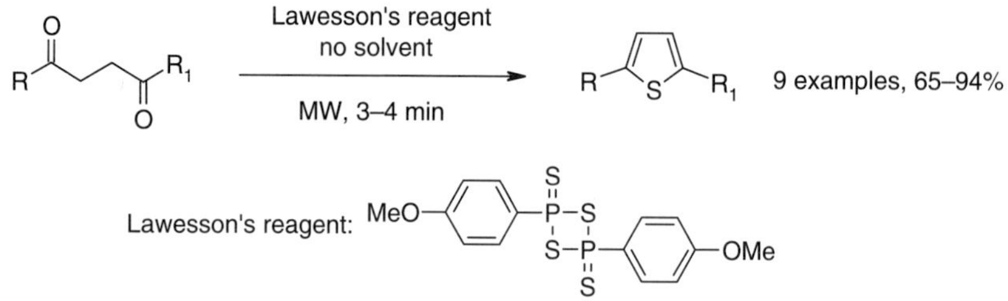

Scheme 3.8

An interesting modification of the Gewald thiophene synthesis combined a one-pot, multi-component coupling on solid-phase, with microwave-assisted synthesis[12]. Cyanoacetic acid bound Wang resin was treated with elemental sulphur and a range of aldehydes or ketones under microwave heating (see Chapter 5, Section 5.2.5). Ad-dition of an acid chloride and a second microwave cycle provided polymer-bound 2-acylaminothiophenes, which were cleaved from the resin in high yields and purities (Scheme 3.9). The chemistry was carried out in sealed vials using a laboratory-dedicated focused microwave instrument.

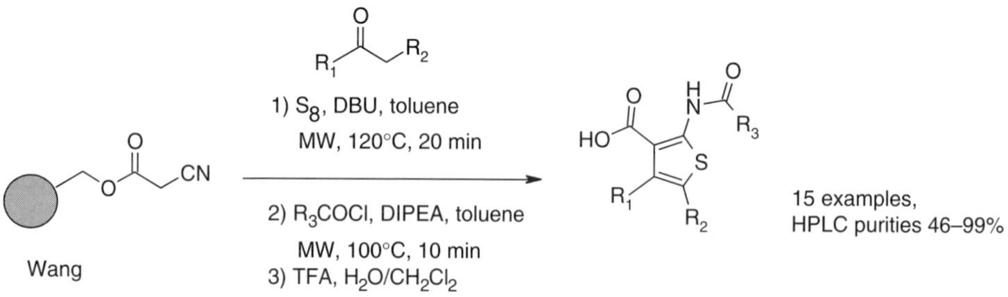

Scheme 3.9

3.3. Five-membered systems with two heteroatoms

3.3.1. *Imidazoles, pyrazoles and benzimidazoles*

Imidazoles have been synthesised by a four-component condensation of benzoin, an aromatic aldehyde, a primary amine and ammonium acetate on silica or Zeolite HY[13]. The reactions were complete in 6 min using a domestic instrument, in contrast to the many hours required for the conventionally thermal heated condensation reaction. In addition, it was found that it is possible to replace the aromatic aldehyde/ammonium acetate combination with a benzonitrile (Scheme 3.10)[14].

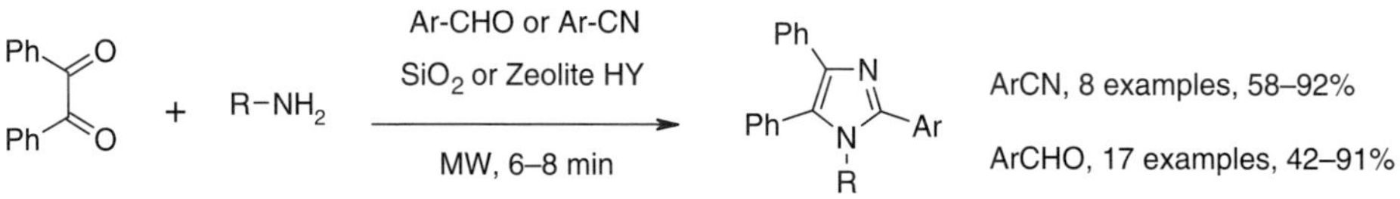
Ph
O
Ph
O
+
R−NH2
Ar-CHO or Ar-CN
SiO2 or Zeolite HY
MW, 6–8 min
Ph
N
Ph
N
Ar
R
ArCN, 8 examples, 58–92%
ArCHO, 17 examples, 42–91%

Scheme 3.10

The antifungal agent nortopsentin D was synthesised via cyclodehydration of a ketoamide by the action of ammonium acetate under focused microwave heating conditions. The reaction rate and yield were clearly improved relative to the thermal heating conditions (Scheme 3.11)[15]. In a formally identical disconnection, a 24-membered library of 4(5)-sulphanyl-1*H*-imidazoles was prepared in a three-component coupling process[16]. A noteworthy aspect of this work was the use of parallel microwave processing in a custom-designed reactor format (Scheme 3.11).

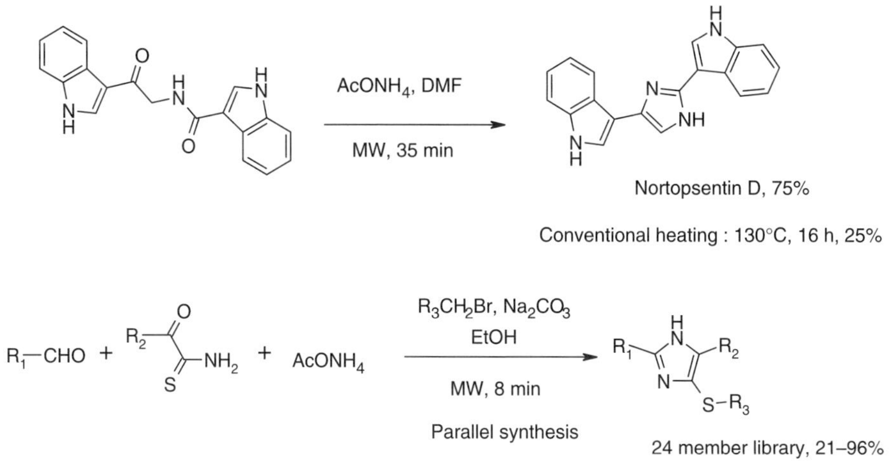
AcONH4, DMF
MW, 35 min
Nortopsentin D, 75%
Conventional heating : 130°C, 16 h, 25%
R1−CHO +
+
AcONH4
R3CH2Br, Na2CO3
EtOH
MW, 8 min
Parallel synthesis
24 member library, 21–96%

Scheme 3.11

Two closely related reports of pyrazole generation by condensation of substituted hydrazines with enamino carbonyl compounds have appeared. *In situ* formation of an enaminoketone, by treatment of a diketone with dimethylformamide dimethyl acetal, was followed by tandem Michael addition–elimination/cyclodehydration under aqueous conditions in sealed microwave vessels (Scheme 3.12)[17]. Isoxazoles and pyrimidines were also prepared by replacing the substituted hydrazine with hydroxylamine or amidines, respectively (see Chapter 5, Section 5.3.2). The overall process may be regarded as another example of a multi-component coupling. In a similar fashion, enamino propenoates were condensed with substituted hydrazines to afford substituted pyrazoles (see Chapter 5, Section 5.3.2) (Scheme 3.12)[18].

The classical condensation of substituted phenylene diamines and carboxylic acids[19–21](or aldehydes under oxidising conditions[21]) to provide benzimidazoles

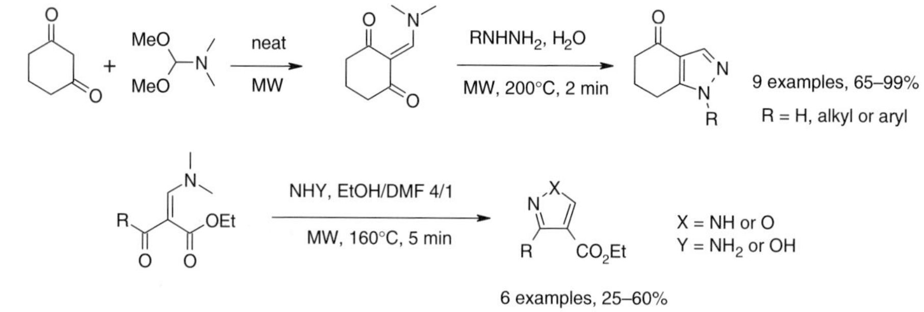

Scheme 3.12

typically requires rather forcing conditions and reagents. Consequently, microwave heating continues to find application for this transformation. A rapid solution-phase combinatorial synthesis of 2-(arylamino)benzimidazoles has been carried out, in which the key step involves a microwave-mediated cyclocondensation of a polyethylene glycol (PEG) supported substituted phenylene diamine with an isothiocyanate (Scheme 3.13)[22].

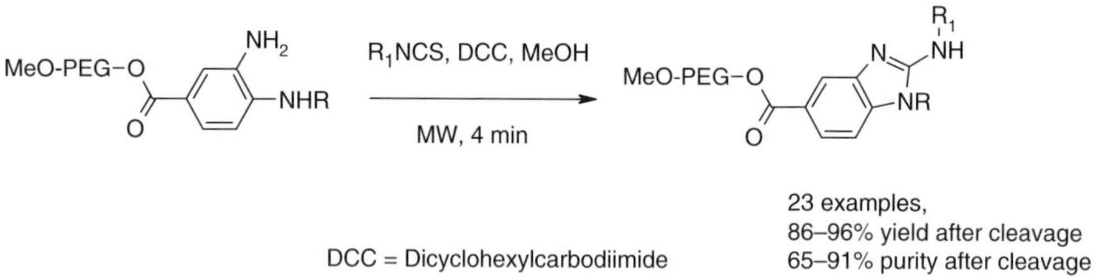

Scheme 3.13

An alternative strategy towards benzimidazole synthesis relies on the palladium-catalysed cyclisation of (2-bromophenyl)amidines. This chemistry has been reported to take place under aqueous reaction conditions, in the presence of sodium hydroxide in sealed microwave vials. The products were isolated by a 'catch and release' method using a strongly acidic ion exchange resin, thereby avoiding conventional chromatographic purification (Scheme 3.14)[23]. Selectively, N-functionalised benzimidazoles were conveniently prepared by this method.

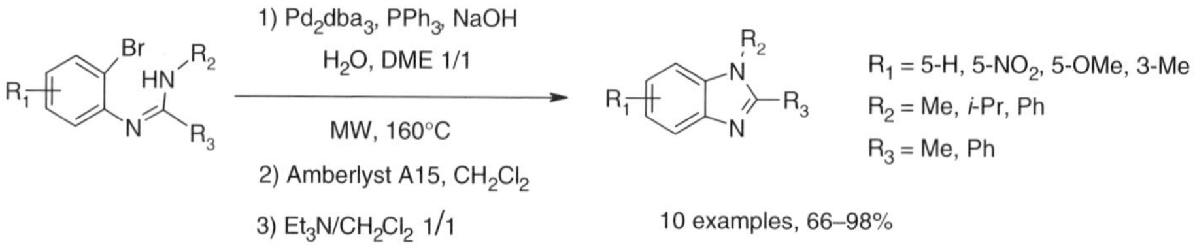

Scheme 3.14

3.3.2. *Oxazoles, isoxazoles, thiazoles, benzoxazoles and benzothiazoles*

Multi-substituted oxazoles have been prepared from simple ketone and primary amide building blocks in a microwave-mediated one-pot procedure[24]. Ketones were treated with hypervalent iodine(III) sulphonate [hydroxyl-(2,4-dinitrobenzene)-sulphonyloxy)iodo]benzene (HDNIB) under solvent-free microwave heating conditions to generate the intermediate α-[(2,4-dinitrobenzene)sulphonyl]oxy carbonyl compound. This latter species was condensed with acetamide or benzamide under microwave heating conditions (Scheme 3.15). An alternative approach towards oxazole synthesis involved a modification of the classical Robinson–Gabriel cyclisation of 2-acylamino carbonyl compounds. This cyclisation employed the Burgess reagent (including its PEG-supported analogue) as a mild dehydrating reagent, under single-mode microwave heating conditions (Scheme 3.15)[25]. The often problematic cyclisation of unstable 2-acylamino aldehydes also proceeded smoothly under these conditions.

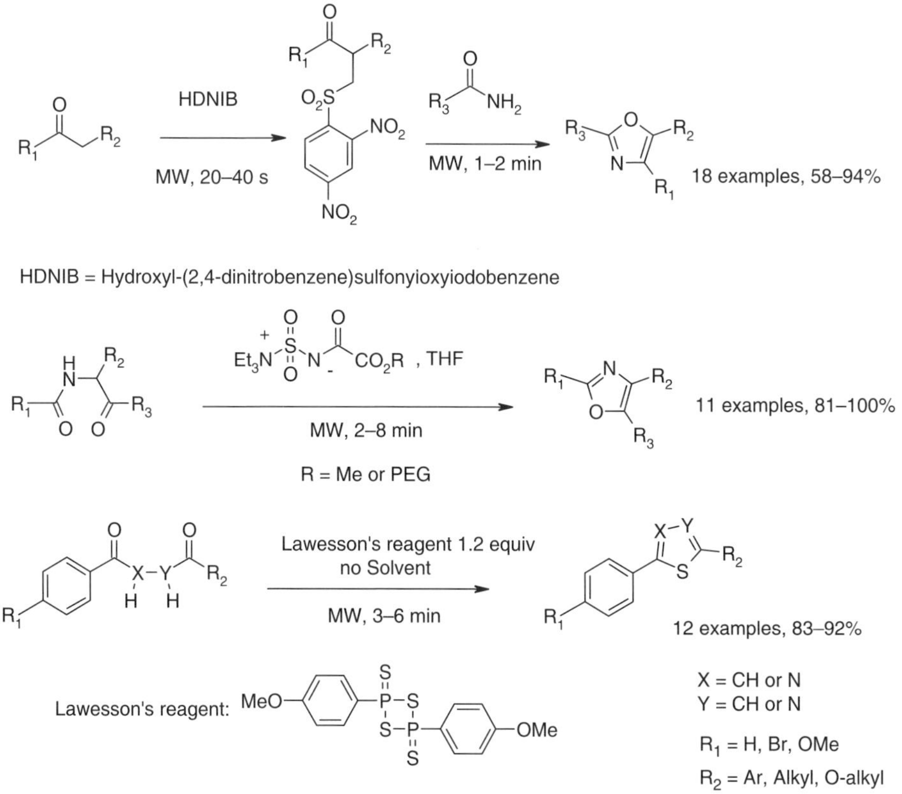

HDNIB = Hydroxyl-(2,4-dinitrobenzene)sulfonyloxyiodobenzene

R = Me or PEG

Lawesson's reagent 1.2 equiv
no Solvent
MW, 3–6 min

12 examples, 83–92%

X = CH or N
Y = CH or N
R_1 = H, Br, OMe
R_2 = Ar, Alkyl, O-alkyl

Lawesson's reagent: MeO—⟨⟩—P(=S)–S / S–P(=S)—⟨⟩—OMe

18 examples, 58–94%

11 examples, 81–100%

Scheme 3.15

The microwave-mediated condensation of hydroxylamine with enaminoketones to generate isoxazoles has been mentioned earlier (refer to Scheme 3.12)[17]. In a procedure analogous to the Paal–Knorr thiophene synthesis, 2-aminoacyl carbonyl compounds were cyclised to thiazoles by utilising Lawesson's reagent under microwave heating (Scheme 3.15)[11].

A 48-membered library of 2-arylbenzoxazoles has been prepared by the condensation of substituted 2-aminophenols with a series of acid chlorides. The reactions proceeded in the absence of a base in sealed tubes in an automated microwave instrument, which used sequential rather than parallel reaction processing. Comparisons to the conventional thermal conditions demonstrated the importance of the high temperatures and pressures achieved under microwave heating, which ensured that the reactions proceeded efficiently (Scheme 3.16)[26]. An analogous synthesis of benzoxazoles by the cyclocondensation reaction of 2-aminophenols with S-methylisothioamide hydroiodides on silica gel, under microwave irradiation, has also been reported (Scheme 3.16)[27].

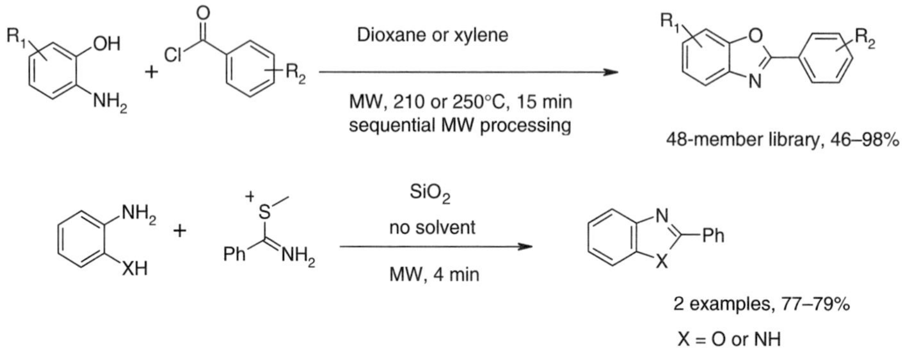

Scheme 3.16

While studying the chemistry of 4,5-dichloro-1,2,3-dithiazolium chloride (Appel's salt) and its derivatives, Besson reported the synthesis of various benzothiazoles from N-arylimino-1,2,3-dithiazoles, which could be synthesised from commercially available aromatic amines[28]. In this work, the authors explored a variety of strategies to construct the benzothiazole ring and demonstrated that in all cases the focused microwave methodologies were more productive and under well-defined conditions provided convenient methods for scale-up (Scheme 3.17)[28]. Comparisons were also made between reactions performed under solvent-free conditions and in the presence of solvent[29]. It is noteworthy that there is no general rule and some reactions performed in the presence of solvent may sometimes be more convenient than the same dry media reaction.

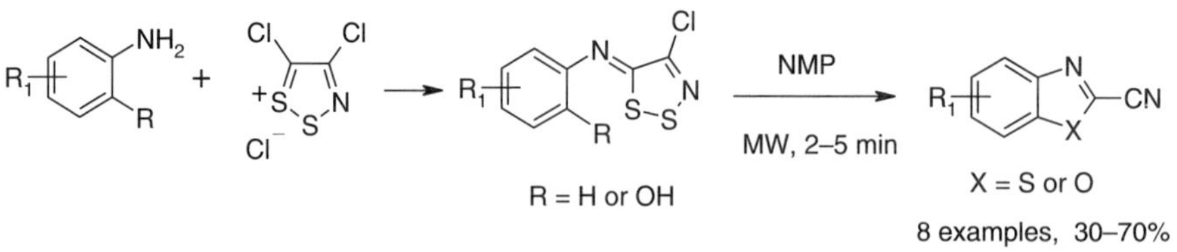

Scheme 3.17

The microwave-assisted synthesis of a benzoxazole derivative by the same approach was also described starting from 2-aminophenol (Scheme 3.17)[29].

3.4. Five-membered ring systems with more than two heteroatoms

3.4.1. Triazoles

1,2,3-Triazoles are generally prepared by the 1,3-dipolar cycloaddition of organic azides with acetylenes or acetylene equivalents. One reported example involves the reaction of α-acylphosphonium ylides with aryl azides to afford 1-aryl-5-substituted 1,2,3-triazoles (Scheme 3.18). This reaction was initially reported to take place in refluxing benzene and required several days of reaction time[30]. The same reaction was performed in a domestic microwave instrument using silica gel as the support. Adopting this microwave-mediated solvent-free protocol, the reaction time could be reduced to just 4–10 min[31].

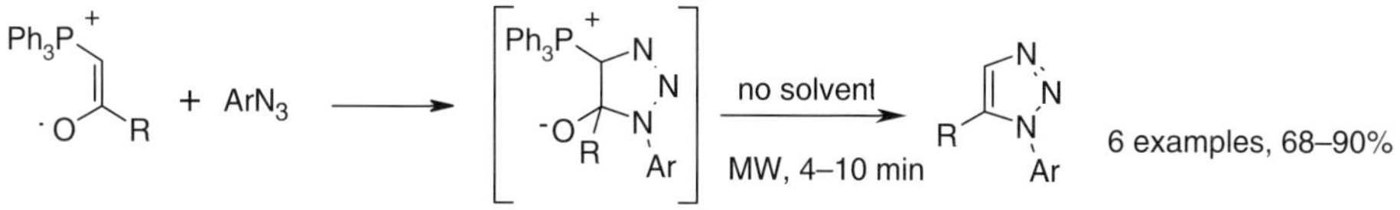

Scheme 3.18

The 1,3-dipolar cycloaddition of azides to acetylenic amides is particularly difficult under conventional thermal conditions and extended reaction times of 14 h to 1 week have been reported[32,33]. Katritzky reported a microwave-mediated solvent-free variant of this procedure to give N-substituted C-carbamoyl-1,2,3-triazoles in good to excellent yields in only 30 min (Scheme 3.19)[34].

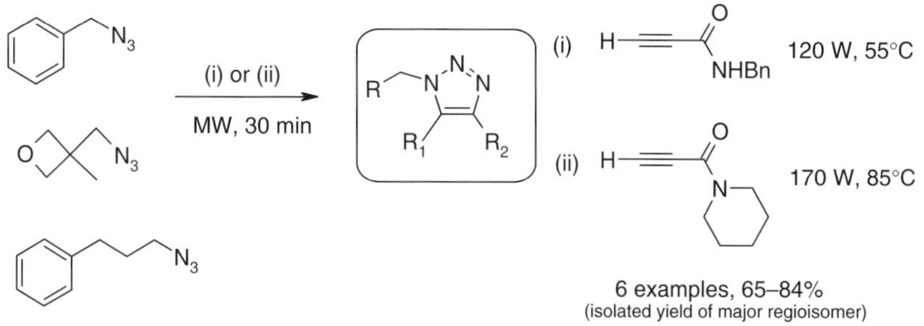

Scheme 3.19

An important classical synthesis of 1,2,4-triazoles is the Pellizzari reaction in which an acylhydrazide is condensed with an amide (or thioamide) at high temperature. Variations exist in which the amide component is first activated towards nucleophilic attack as a S-alkyl thioamide salt. Employing this type of approach, S-methyl isothioamide hydroiodides were condensed with acylhydrazides in the presence of silica gel and solid ammonium acetate in a domestic microwave instrument (Scheme 3.20)[35]. An interesting one-pot three-component synthesis of spiro-fused oxindole triazoles has been reported, which involves the condensation of substituted isatins with anilines under microwave heating, followed by cyclocondensation with thiosemicarbazide (Scheme 3.20)[36]. In this case, a dry reaction support such as Montmorillonite clay was preferable to an organic solvent. This is because the release of potentially harmful flammable solvent vapours/gaseous by-products (e.g., hydrogen sulphide) is prevented within the domestic microwave instrument.

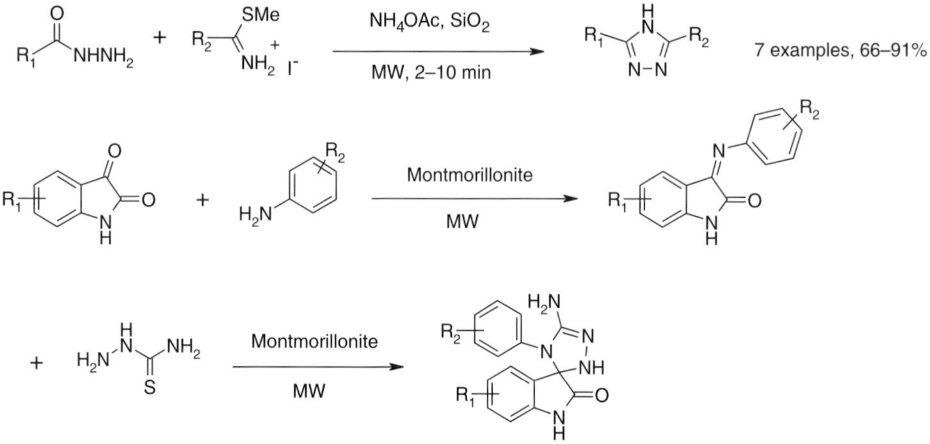

Scheme 3.20

A number of symmetrically 3,5-disubstituted 4-amino-1,2,4-triazoles were able to be rapidly synthesised via focussed microwave heating, by the reaction of aromatic nitriles with excess hydrazine hydrate in ethylene glycol, in the presence of hydrazine dihydrochloride (Scheme 3.21)[37].

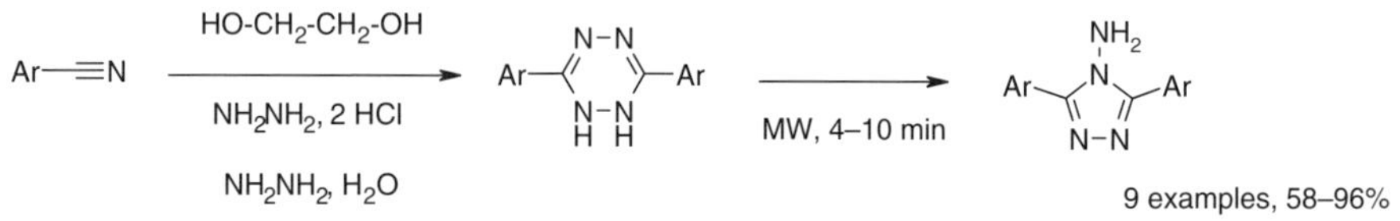

Scheme 3.21

3.4.2. Oxadiazoles

Oxadiazoles have often attracted the attention of medicinal chemists as stable bioisosteres of metabolically labile esters. 1,3,4-Oxadiazoles are generally prepared by cyclodehydration of 1,2-diacylhydrazines or their equivalents. Symmetrical 2,5-disubstituted examples were able to be rapidly prepared in a one-pot condensation–cyclodehydration of benzoic acids (2 equiv) with hydrazine dihydrochloride, in the presence of phosphorus pentoxide as the dehydrating agent[38]. In the search for milder conditions to effect this transformation, a novel polymer-supported variant of the Burgess reagent was prepared for use under microwave heating conditions (see Conditions A, Scheme 3.22)[39]. Alternatively, the cyclodehydration process could be achieved by using a combination of the commercially available polymer-supported phosphazene base PS-BEMP and p-toluenesulphonyl chloride (see Conditions B, Scheme 3.22 and Chapter 6). The combination of polymer-supported reagents with microwave heating facilitates rapid, clean transformations, whilst work-up and purification are achieved by simple filtration of the polymer (see Chapter 6 for further examples).

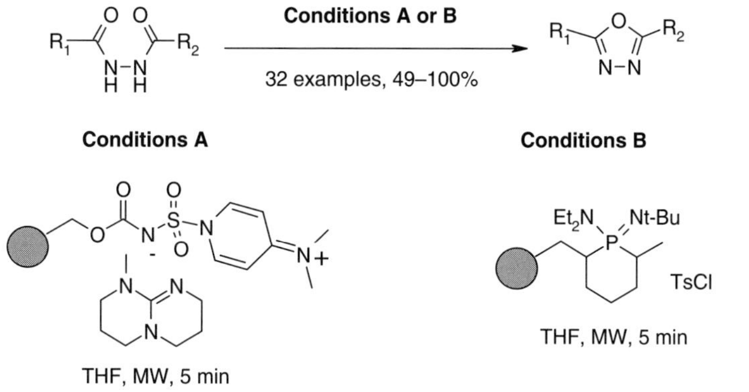

Scheme 3.22

Alternative microwave-mediated methods for the preparation of oxadiazoles based on the dethionation–cyclisation of sulphur analogues have also been reported. These include the dethionation–cyclisation of acylthiosemicarbazides (RC=ONHNHC=SNHR)[40,41] using mercury(II) acetate (Scheme 3.23) and also from acyldithiocarbazinate salts (CC=ONHNHC=SS⁻K⁺) (Scheme 3.23)[42].

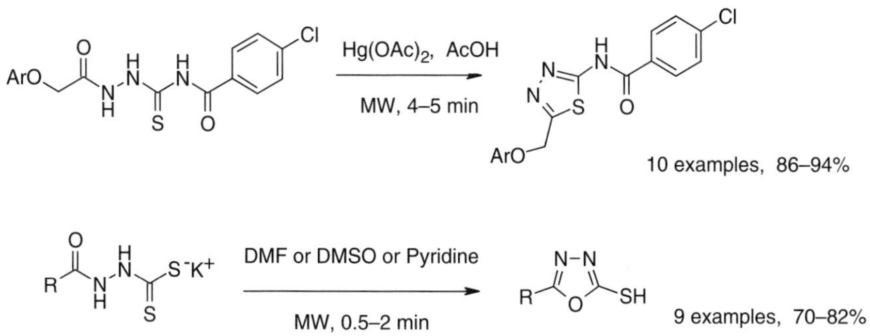

Scheme 3.23

Under acidic conditions, in the absence of a mercury(II) salt, acylthiosemicarbazides may be cyclised to 1,3,4-thiadiazoles by microwave heating (Scheme 3.24)[43].

In addition, thionation–cyclisation of 1,2-diacylhydrazines to 1,3,4-thiadiazoles has been achieved by the action of Lawesson's reagent under solvent-free microwave heating (Scheme 3.24)[11].

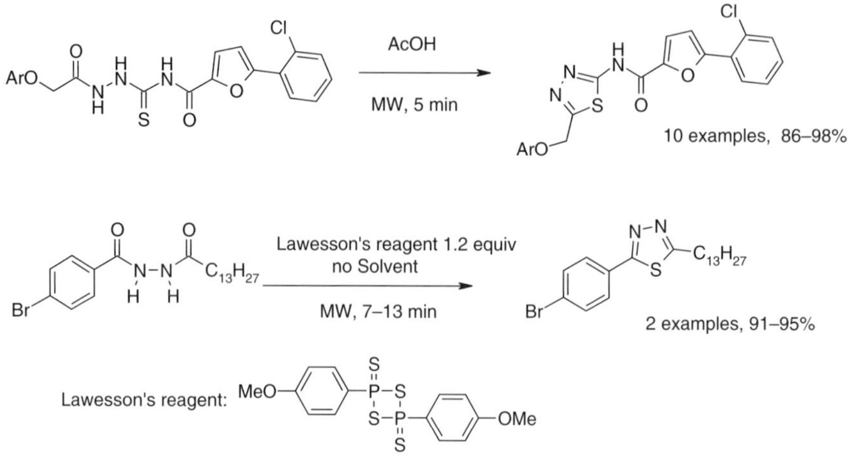

Scheme 3.24

3.4.3. *Tetrazoles*

Tetrazoles have often found application in medicinal chemistry as carboxylic acid bioisosteres. In addition, the biaryltetrazole motif defines the Sartan class of anti-hypertensive Angiotensin II antagonists (Scheme 3.25). These heterocycles are generally synthesised by the reaction of an azide with a nitrile. Tri-*n*-butyltin azide or trimethylsilylazide are commonly used as sources of azide in this procedure. Hallberg and co-workers developed an elegant procedure in which aryl or vinyl bromides are converted into nitriles under microwave-promoted palladium-catalysed conditions[44]. Microwave heating was then used to convert the nitriles into tetrazoles by treatment with sodium azide, in the presence of ammonium chloride in *N,N*-dimethylformamide (DMF) (Scheme 3.26). The reactions could also be run as a convenient one-pot process, thereby enabling direct conversion of an aryl bromide to a tetrazole.

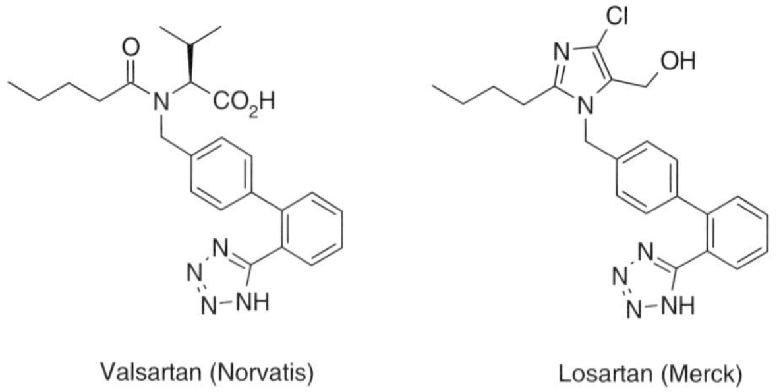

Valsartan (Norvatis)   Losartan (Merck)

Scheme 3.25

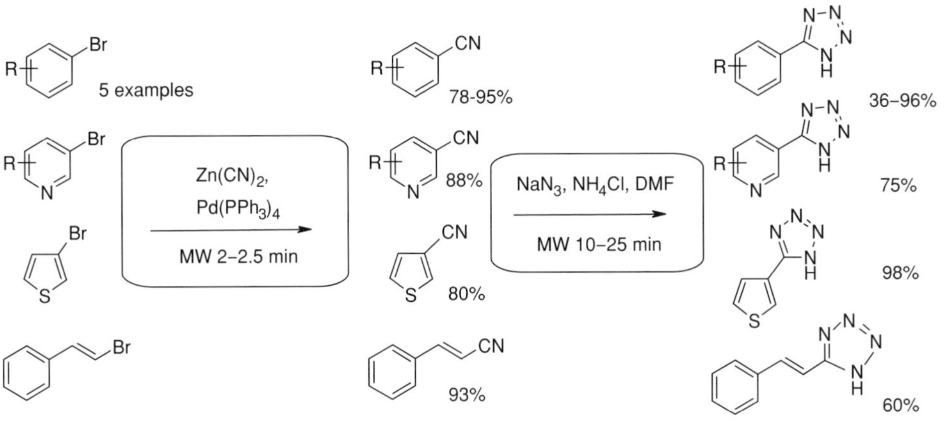

Scheme 3.26

3.5. Six-membered heterocycles containing one heteroatom

3.5.1. Pyridines, quinolines, isoquinolines and fused ring analogues

The classical Hantzch dihydropyridine synthesis has been reported under microwave flash heating conditions[45,46]. More unusually, Hantzch products have also been obtained from Biginelli reaction under microwave flash heating conditions (Scheme 3.27)[47].

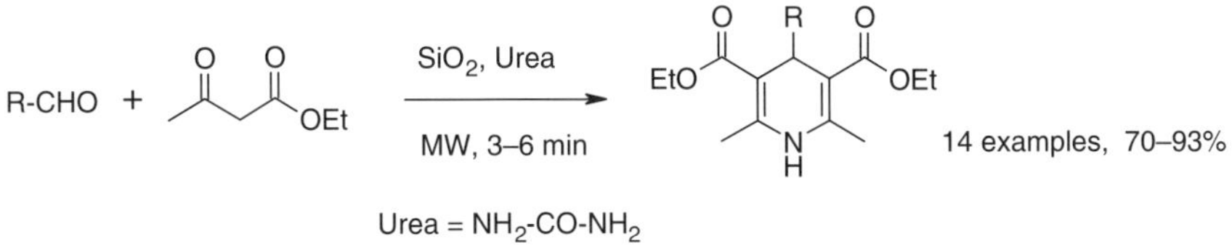

Scheme 3.27

Tri- or tetra-substituted pyridines were prepared in a one-pot Bohlmann–Rahtz heteroannulation of ethyl β-aminocrotonate and an alkynone, which involved a Michael addition–cyclodehydration sequence. The reaction proceeded within just 20 min under single-mode microwave heating conditions (Scheme 3.28)[48].

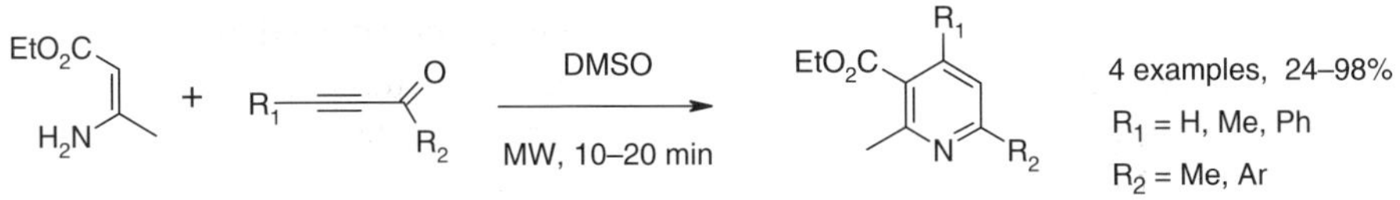

Scheme 3.28

While numerous synthetic approaches have been developed for quinoline synthesis, most are unsatisfactory for combinatorial library production since they do not tolerate a wide range of substitutions and functionalities. Since microwave heating conditions can dramatically reduce reactions times and often improve the purities and yields of products, a new efficient approach for generating 2-aminoquinolines involving a three-component condensation has been investigated[49]. The reaction involves rapid focused microwave heating of a secondary amine with an aldehyde to afford an enamine. Subsequent reaction of the resulting *in situ* formed enamine with 2-azidobenzophenone under microwave heating produces the 2-aminoquinoline derivatives (Scheme 3.29). Direct reaction comparison between conventional thermal and microwave heating conditions, using identical reagent stoichiometry and sealed reaction vessels, indicated that greatly improved yields were achievable in the case of the microwave examples. The authors also noted that the purities of the quinoline products resulting from the conventional thermal conditions were poorer because of the presence of decomposed 2-azidobenzophenone as a contaminant.

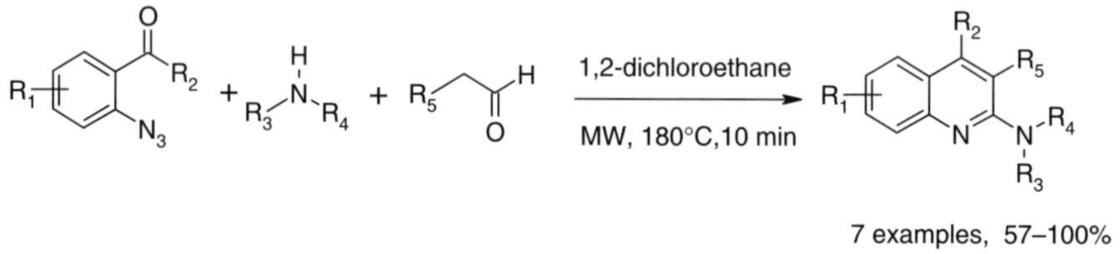

7 examples, 57–100%

Scheme 3.29

Adopting a similar approach, a small array of 12 quinoline derivatives was synthesised using microwave heating (4 min) under solvent-free conditions in the presence of 0.1–0.5 equiv of diphenylphosphate (Scheme 3.30)[50].

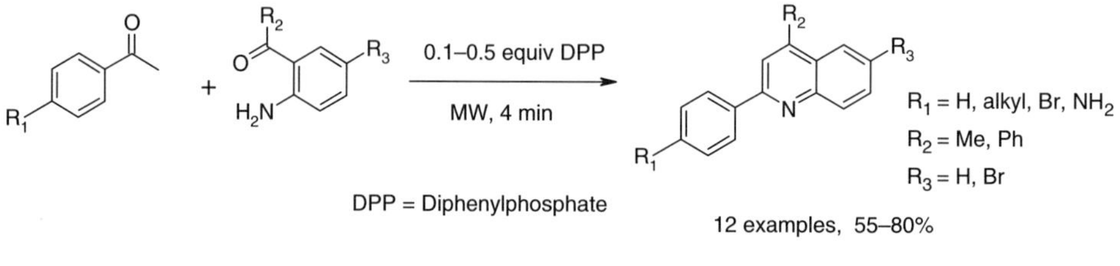

DPP = Diphenylphosphate

R_1 = H, alkyl, Br, NH_2
R_2 = Me, Ph
R_3 = H, Br

12 examples, 55–80%

Scheme 3.30

In an effort to develop an economical, rapid and safe method devoid of solvent usage, Kidwai *et al.* investigated the dry media synthesis of antibacterial quinolines utilising alumina as the support (Scheme 3.31)[51]. The products were obtained in improved yield compared to that from the conventional heating method. Furthermore, the reaction times were reduced once again from hours to seconds. In a complementary solvent-free approach under microwave heating conditions, 4-alkylquinolines were successfully prepared by a one-pot Skraup's reaction of anilines with alkyl vinyl ketones on silica gel support impregnated with indium(III) chloride (Scheme 3.31)[52].

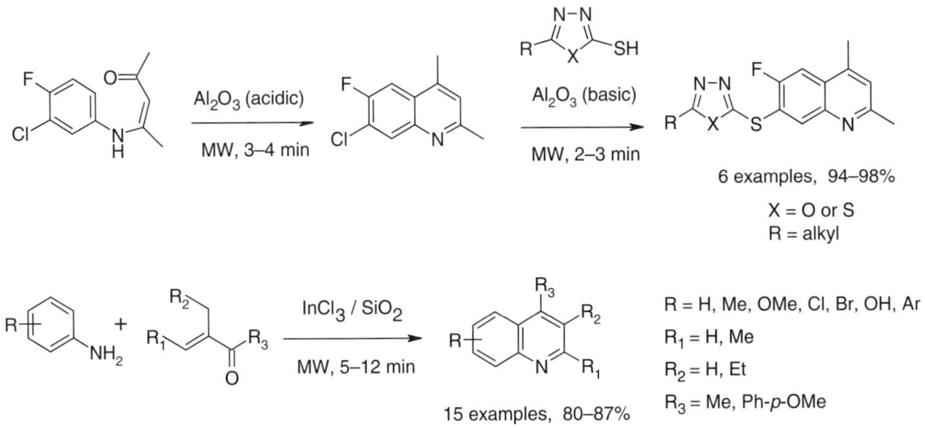

Scheme 3.31

Hetero-Diels–Alder reactions provide an attractive means of rapidly constructing complex heterocyclic ring systems. Cycloaddition of pyrazolyl imines with a variety of electron-deficient dienophiles has been used to assemble pyrazolo[3,4,*b*]pyridines in a focused microwave reactor under solvent-free conditions[53]. The reactions proceeded in modest to excellent yield, depending upon the choice of diene and dienophile (Scheme 3.32).

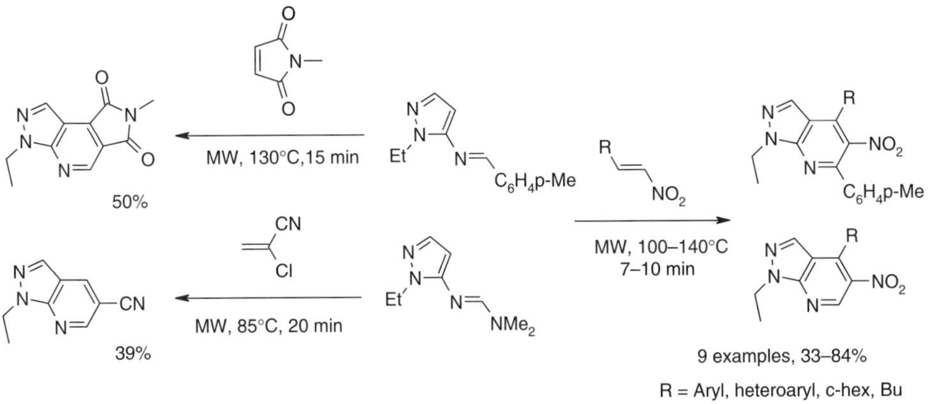

Scheme 3.32

3.5.2. *Benzopyrans*

Coumarins (also called 2*H*-1-benzopyran-2-ones or 2-oxo-2*H*-chromenes) are common in nature and find their main applications as fragrances, pharmaceuticals and agrochemicals. These compounds can be synthesised by methods such as the Claisen rearrangement, the Perkin reaction, the Knoevenagel condensation as well as the Pechmann reaction. The latter reaction involves condensation of phenols with β-ketonic esters in the presence of various reagents. An efficient microwave-assisted solvent-free synthesis of 7-aminocoumarins has been developed by Besson *et al.* on solid support (graphite/Montmorillonite K10 clay) utilising the Pechmann reaction[54]. In this convenient methodology, the strong interaction between the microwave energy and the graphite/clay support is successfully exploited for the generation of heat. In addition, the Montmorillonite clay plays the role of an acidic catalyst in the reaction (Scheme 3.33).

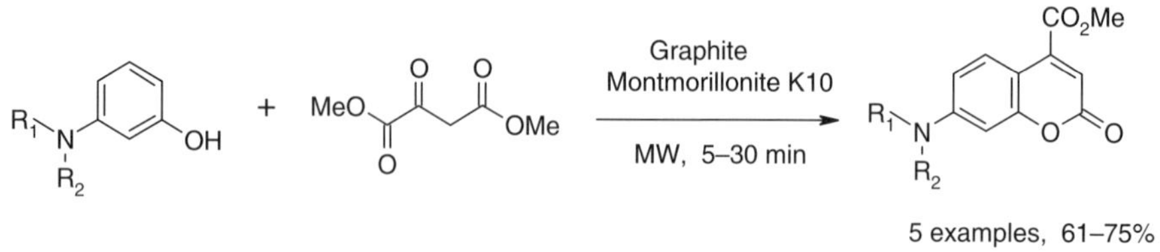

Scheme 3.33

De la Hoz *et al.* have studied the microwave-assisted synthesis of substituted coumarins by the condensation of phenol, 1,3-dihydroxybenzene/1,3,5-trihydroxybenzene and propynoic/propenoic acids[55]. The examples illustrate that the application of heterogeneous supported catalysts can eliminate the production of acidic waste streams, associated with conventional Lewis acid catalysts. This synthetic approach constitutes an interesting alternative to the Pechman reaction (Scheme 3.34).

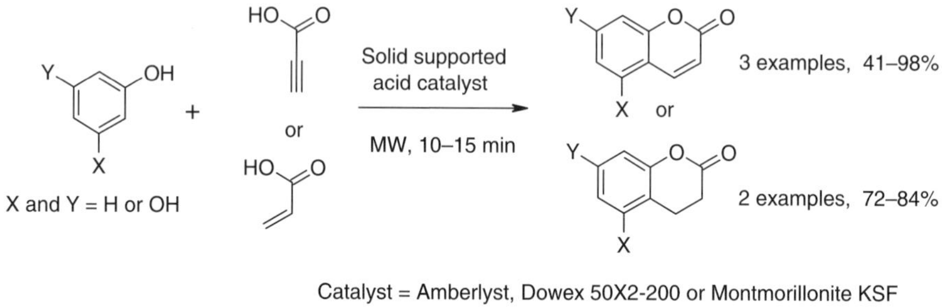

Catalyst = Amberlyst, Dowex 50X2-200 or Montmorillonite KSF

Scheme 3.34

Synthesis of the benzopyran ring has also been performed by microwave-assisted copper-catalysed cross coupling of an aryl iodide with terminal alkynes, in the presence of copper(I) iodide/triphenylphosphine (Scheme 3.35)[56]. An alternative approach involving microwave heating of mixtures of salicylaldehyde and various derivatives of ethyl acetate in the presence of piperidine has enabled rapid Knoevenagel synthesis of coumarin derivatives (Scheme 3.35)[57].

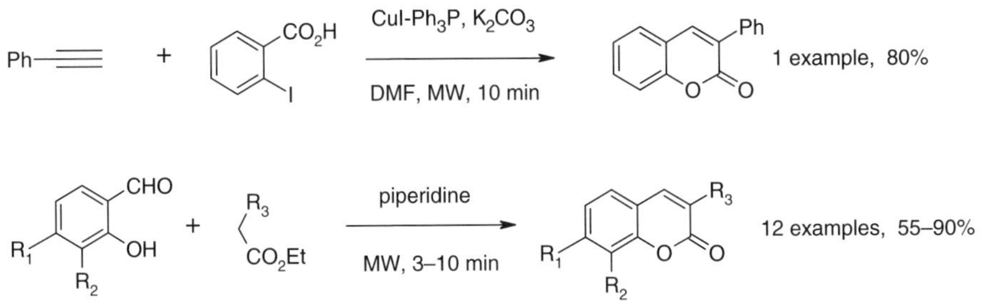

Ph———   +   [structure with CO$_2$H and I] $\xrightarrow{\text{CuI-Ph}_3\text{P, K}_2\text{CO}_3}$ [coumarin with Ph]   1 example, 80%

DMF, MW, 10 min

[aldehyde structure with CHO, OH, R$_1$, R$_2$]   +   [structure with R$_3$ and CO$_2$Et] $\xrightarrow{\text{piperidine}}$ [coumarin product with R$_1$, R$_2$, R$_3$]   12 examples, 55–90%

MW, 3–10 min

Scheme 3.35

3.6. Six-membered heterocycles containing at least two heteroatoms

3.6.1. *Pyrimidines and quinazolines*

Pyrimidines are an important class of compounds that are becoming increasingly significant due to their biological therapeutic properties. Dihydropyrimidines were synthesised in high yields by a rapid microwave-assisted one-pot cyclocondensation of an aldehyde, a β-ketoester and urea, employing various acid catalysts under solvent-free conditions[58]. This adaptation of the ancestral Biginelli multi-component reaction is simple, rapid and the resulting dihydropyrimidines can be readily isolated by a simple filtration procedure (Scheme 3.36).

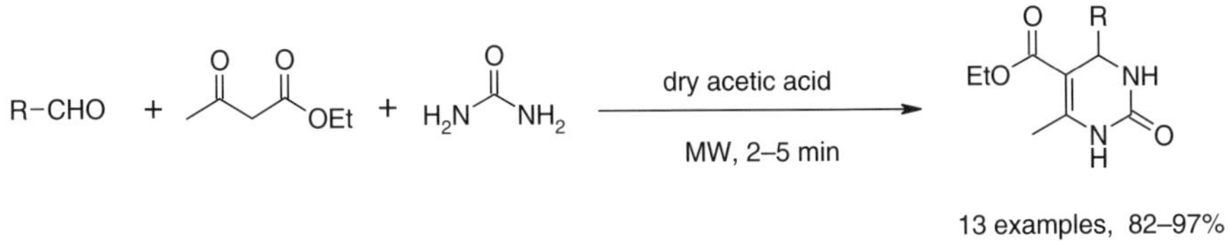

R–CHO   +   [β-ketoester structure, OEt]   +   H$_2$N–C(O)–NH$_2$ $\xrightarrow{\text{dry acetic acid}}$ [dihydropyrimidine product with EtO, O, R, NH, N–H, O]

MW, 2–5 min

13 examples, 82–97%

Scheme 3.36

Kappe *et al.* optimised the reaction conditions for the Biginelli three-component cyclocondensation to enable the automated microwave-assisted synthesis of a dihydropyrimidine compound library[59]. This approach represents a much more efficient method of synthesis compared to the traditional procedures highlighted previously. A small library of 48 pyrimidines analogues was prepared in an automated fashion, using a variety of combinatorial building blocks and subsequent sequential microwave heating of each reaction vial (see also Section 5.2.2, Chapter 5). The quinazoline skeleton, when selectively functionalised, can be utilised as a valuable intermediate for the preparation of a number of compounds with pronounced biological activities. The most common synthetic method for formation of the 3*H*-quinazolin-4-one ring is based on the Niementowski reaction. This ring formation usually involves the reaction of anthranilic acids (or a derivative, e.g., 2-aminobenzonitrile) with formamide for prolonged reaction times at high temperatures. Microwave-assisted investigation of this ancestral procedure led to very good yields of the products compared to conventional conditions with significant rate enhancements (Scheme 3.37)[60]. It is often the case that previously difficult and traditional chemical transformations can be rapidly and safely completed by performing these reactions under well controlled microwave heating conditions (reactions run in a single-mode system).

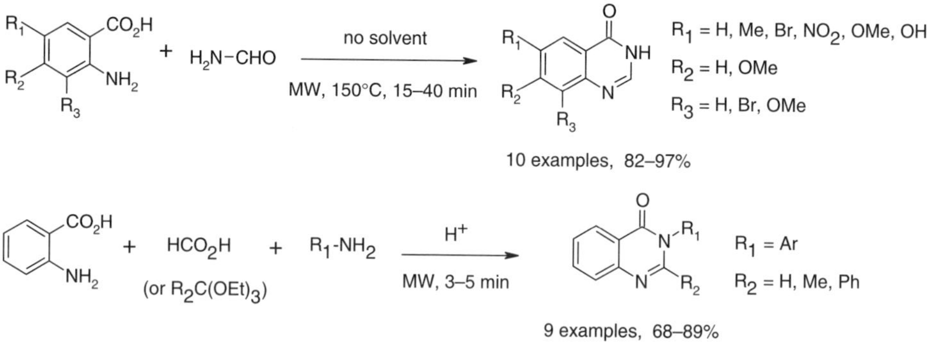

Scheme 3.37

A similar rapid microwave one-pot synthesis of substituted quinazolin-4-ones was also reported, which involved cyclocondensation af anthranilic acid, formic acid (or an orthoester) and an amine under solvent-free conditions (Scheme 3.37)[61]. A complimentary approach was adopted to synthesise 4-aminoquinazolines in very good yields, involving the reaction of aromatic nitrile compounds with 2-aminobenzonitrile in the presence of a catalytic amount of base (Scheme 3.38)[62]. The reactions were performed in a domestic microwave oven and required only a very short heating time. A microwave-assisted synthesis of a variety of new 3-substituted-2-alkyl-4-(3H)-quinazolinones using isatoic anhydride, 2-aminobenzimidazole and orthoesters has also been described (Scheme 3.38)[63].

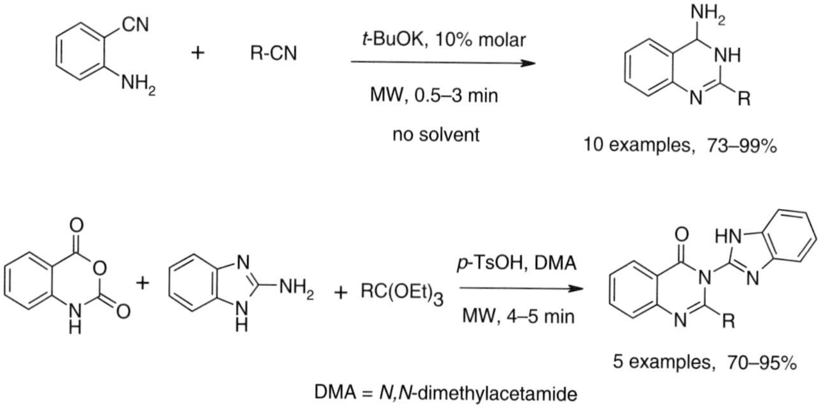

DMA = N,N-dimethylacetamide

Scheme 3.38

3.6.2. *Triazines and tetrazines*

Graphite couples strongly (used as a solid support) with microwaves via the conduction mechanism (see Chapter 6) to enable rapid heating of the surrounding reaction medium. The resulting high-temperature gradient often leads to increased reaction rates compared to conventional thermal heating procedures. This property of graphite has been exploited in the microwave-assisted one-pot synthesis of cyanuric acid ([1,3,5] triazine-2,4,6-triol), prepared via pyrolysis of urea in the absence of water and organic solvents (Scheme 3.39)[64]. The authors suggested that the higher yields and short reaction rates observed in this study were the direct consequence of localised superheating (often called 'hot spots'). These effects are often observed in other examples of microwave-heated heterogeneous reactions.

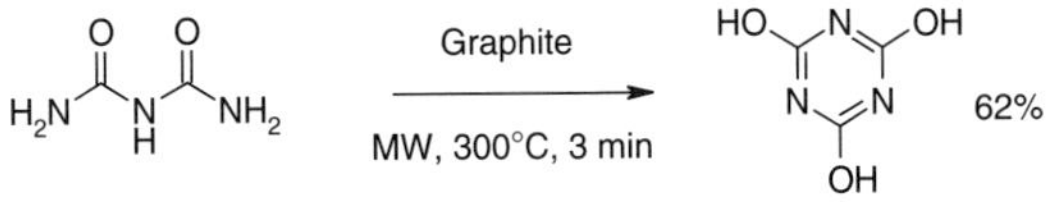

Scheme 3.39

Avalos *et al.* has reported the microwave-assisted synthesis of tetrazine derivatives by a hetero Diels–Alder reaction of homochiral 1,2-diaza-1,3-butadienes with diethyl azodicarboxylate (Scheme 3.40)[65]. Under conventional conditions, reactions could be performed in benzene solution at room temperature. However, under microwave heating conditions, the reaction was significantly accelerated (by a factor of 1000) when carried out solvent free. The observed stereoselectivity was identical for the hetero Diels–Alder reaction under both microwave-heated and conventional conditions.

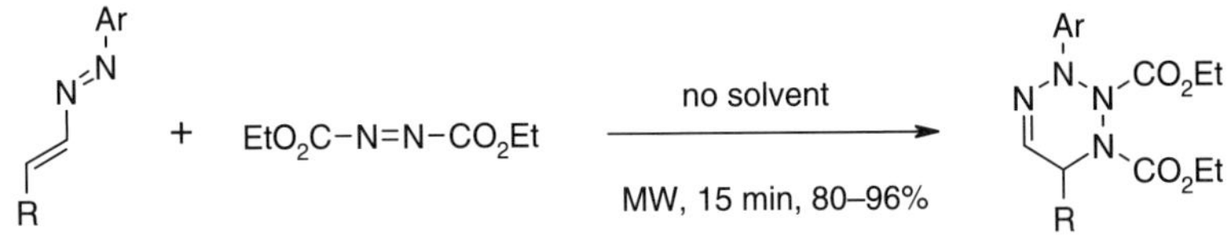

Scheme 3.40

3.7. Seven-membered heterocycles containing at least two heteroatoms: 1,4 and 1,5-benzodiazepines

The benzodiazepine nucleus is a traditional pharmacophoric scaffold, which has been extensively studied, in medicinal chemistry. 1,4-benzodiazepinones can be readily synthesised by an intramolecular cycloaddition in *N,N*-dimethylformamide (DMF) under microwave heating conditions[66]. The synthetic approach summarised in Scheme 3.41 was performed using a microwave synthesiser especially designed for organic reactions. Both conventional heating and microwave heating conditions for the reactions were compared in this particular study; the experimental conditions for the microwave-heated examples were similar to those used for the conventionally heated examples (i.e., same concentration of starting material and volume of solvent). It is noteworthy that the two microwave-assisted reactions of the synthetic sequence (first and last reaction steps of the sequence, Scheme 3.41) enabled the 2-methyl-1,4-benzodiazepin-5-ones to be obtained in significantly better overall yields than those obtained by conventional methods.

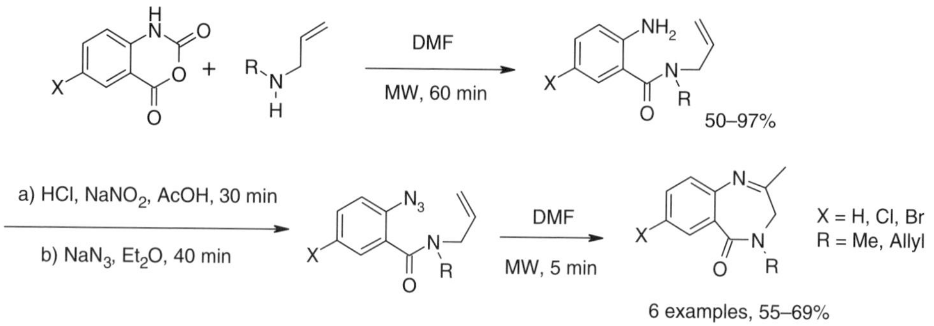

Scheme 3.41

In a simple microwave-assisted and solvent-free approach, substituted isatoic anhydrides were reacted with 4-substituted prolines to afford fused 1,4-benzodiazepine derivatives[67]. The reactions proceeded in less than 3 min and the fused 1,4-benzo-diazepine products were obtained in very good yields (Scheme 3.42). This condensation reaction represents a practical alternative approach to the typical traditional methods.

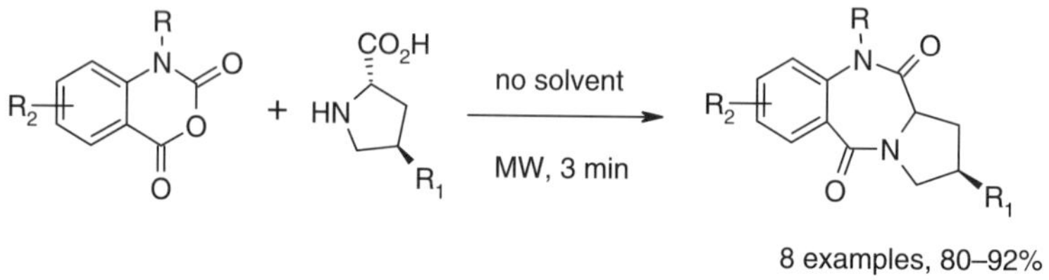

Scheme 3.42

A microwave-assisted synthesis of various 1,5-benzodiazepines was also investigated by condensation of ketones with o-phenylenediamines under dry media conditions (i.e., no solvent/solid support)[68]. The reactions were carried out by first simply mixing the o-phenylenediamine with the ketone (2.1 equiv) in the presence of a catalytic amount of acetic acid. The reactions mixtures were then irradiated in a domestic microwave oven (Scheme 3.43). The resulting benzodiazepine derivatives were obtained in almost quantitative yields.

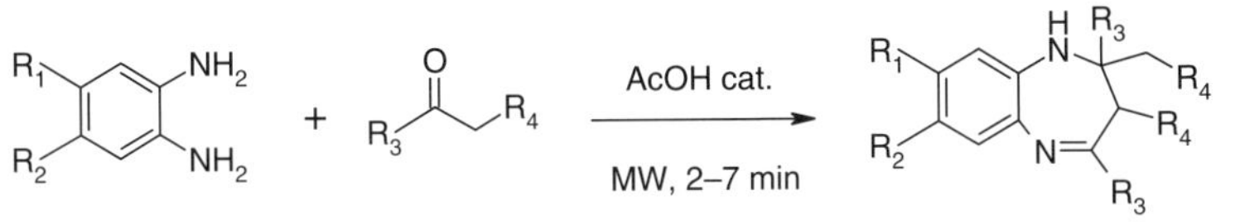

12 examples, 93–97%

Scheme 3.43

3.8. Polycyclic heterocycles

The preceding examples described in this chapter have illustrated the interest and application of microwave heating in the preparation of a wide range of heterocycles. The various possibilities offered by this technology are particularly attractive for multi-step synthesis and drug discovery, where high-yielding protocols and avoidance or facilitation of purification are highly desirable (see also Chapters 6 and 7). For all these reasons, the microwave-assisted synthesis of various polyheterocyclic structures has also been investigated. The examples of molecules described below have been mainly inspired by natural marine or terrestrial alkaloids, which were synthesised to investigate their pharmaceutical potential. In this section, the fused ring heterocycles have been separated from the fused heterocycles, which share at least one heteroatom. For the following examples, comparison of the microwave-assisted conditions with the conventional thermal conditions will illustrate how generally higher yields and faster reactions times can be achieved by using microwave heating.

3.8.1. *Fused ring heterocycles*

Besson and co-workers have investigated the microwave-assisted multi-step (seven steps) synthesis of thiazoloquinazolinone derivatives, utilising commercially available nitroanthranilic acids as the initial precursors[69]. Comparison of the conventional thermal heating and microwave heating approaches demonstrated that the overall time for the multi-step synthesis could be considerably reduced (by a factor of 8) by adopting the microwave-heated reaction methods (Scheme 3.44). In addition, the reactions were cleaner and the products could be purified rapidly. For the microwave-heated multi-step synthesis, the overall yield of the final product was increased by a factor of 2, which enabled the scale of the overall synthesis to be increased from 0.2 to 1 g.

The pharmaceutical value of the unsubstituted thiazoloquinazolinones may be limited due to a lack of substituents such as basic amino groups. However, the microwave-assisted multi-step approach (Scheme 3.44) would enable investigation of the effect of introducing various substituents on biological activity. Previously, the same group had reported another microwave multi-step synthesis (six steps) of thiazoloquinazolinone derivatives[70]. Unfortunately, this pathway was not well adapted for easy introduction of various substituents on the core and only one isomer could be obtained.

The following examples constitute a non-exhaustive list of fused heterocyclic templates, which have been successfully prepared employing similar microwave approaches and strategies.

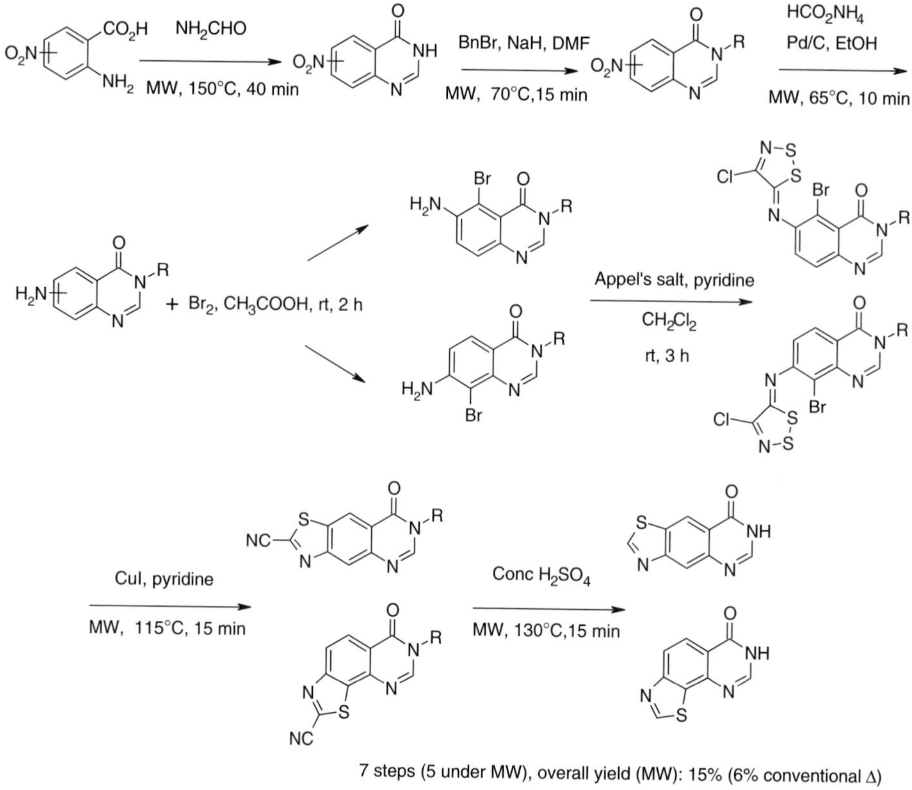

7 steps (5 under MW), overall yield (MW): 15% (6% conventional Δ)

Scheme 3.44

Kappe *et al.* reported the microwave-assisted synthesis of pyrido[2,3-*d*]pyrimidines via a one-pot three component cyclocondensation of α,β-unsaturated esters, amidines and malonitrile (or ethyl cyanoacetate) (Scheme 3.45)[71]. Quiroga *et al.* reported a similar three component cyclocondensation to synthesise regiospecifically 5,8-dihydropyrido[2,3-*d*]pyrimidines under solvent-free conditions, starting from a combination of aminopyrimidin-4-ones, benzoylacetonitrile and benzaldehyde (Scheme 3.45)[72].

Dave and Shah have reported a Gould–Jacod type reaction for the microwave-assisted synthesis of thieno[3,2-*e*]pyrimido[1,2-*c*]pyrimidines via intermediate thieno[2,3-*d*]pyrimidines (Scheme 3.46)[73]. A one-pot synthesis of pyrano[2,3-*d*]pyrimidines was also described by Kidwai and co-workers starting from thiobarbituric acids. The thiobarbituric acid intermediates were also prepared by microwave heating, using basic alumina as the solid support (Scheme 3.46)[74].

A number of methods are available for the synthesis of fused pyrazoles, the most commonly used method involves the reaction of β-chlorovinylaldehydes with hydrazine hydrate or phenyl hydrazine. This reaction has been employed by Loupy and co-workers in the microwave-assisted synthesis of pyrazolo[3,4-*b*]quinolines and pyrazolo[3,4-*c*]pyrazoles (Scheme 3.47)[75].

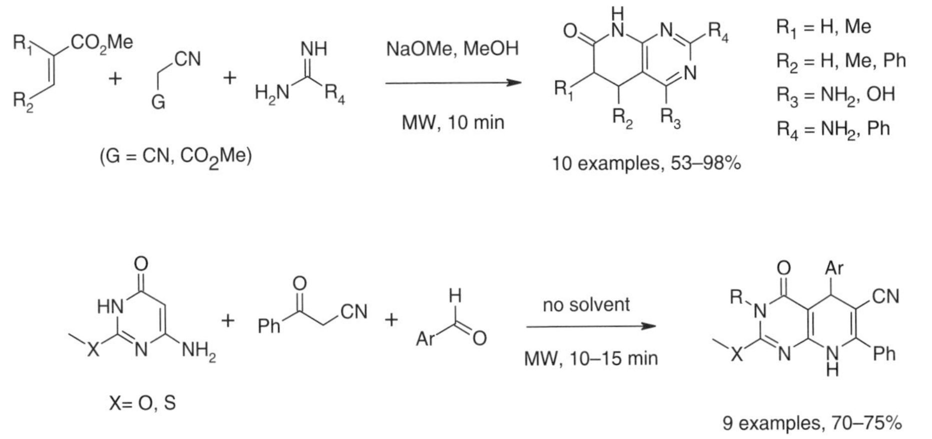

NaOMe, MeOH
MW, 10 min
(G = CN, CO2Me)
10 examples, 53–98%
R1 = H, Me
R2 = H, Me, Ph
R3 = NH2, OH
R4 = NH2, Ph
no solvent
MW, 10–15 min
X= O, S
9 examples, 70–75%

Scheme 3.45

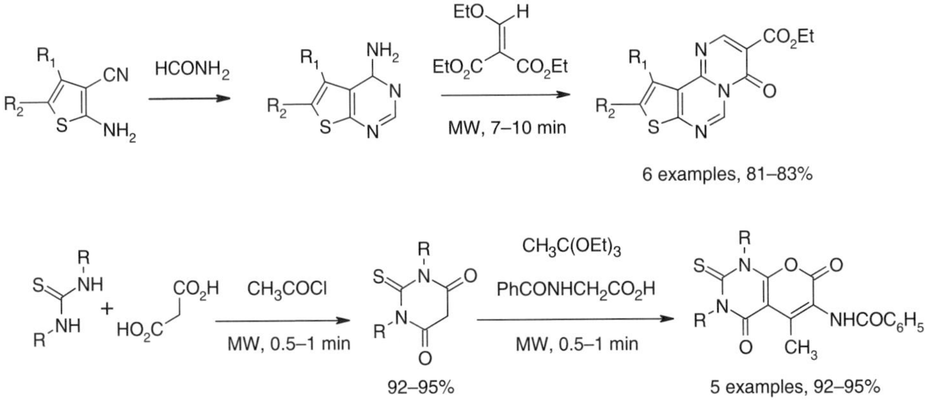

HCONH2
MW, 7–10 min
6 examples, 81–83%
CH3COCl
MW, 0.5–1 min
92–95%
CH3C(OEt)3
PhCONHCH2CO2H
MW, 0.5–1 min
5 examples, 92–95%

Scheme 3.46

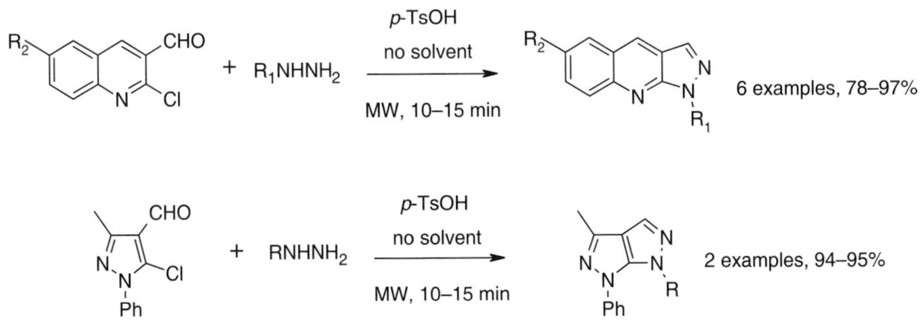

p-TsOH
no solvent
MW, 10–15 min
6 examples, 78–97%
p-TsOH
no solvent
MW, 10–15 min
2 examples, 94–95%

Scheme 3.47

[1,4]Dioxino[2,3-*f*]benzothiazoles and [1,4]dioxino[2,3-*g*]benzothiazoles have been successfully synthesised from imino-1,2-3-dithiazoles under microwave heating conditions. Introduction of the thiazole moiety into these polyheterocyclic systems was carried out employing an open focused microwave oven. This work represented the first report of a multi-step synthesis performed utilising microwave heating (Scheme 3.48)[76].

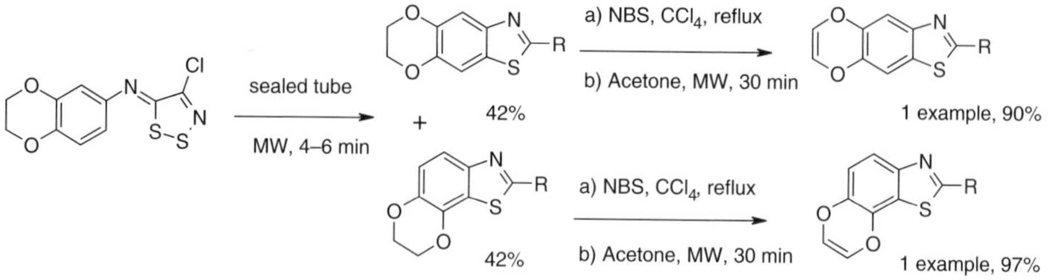

Scheme 3.48

3.8.2. *Fused heterocycles sharing at least one heteroatom*

The microwave-assisted chemistry of a variety of aromatic heterocycles has been extended to the synthesis of fused molecules which share, at least, one heteroatom. In this area, the synthesis of nitrogen containing compounds has been actively investigated. All the compounds described below have been prepared in an effort to find compounds with interesting biological activity.

An efficient microwave-assisted multi-step synthesis of 8*H*-quinazolino[4,3-*b*]quinazolin-8-one has been investigated by Besson and co-workers[77]. The synthesis involved two Niementowski condensations starting from substituted anthranilic acids (Scheme 3.49). Both homogeneous and heterogeneous conditions were studied in an effort to develop a convenient synthesis of the desired compounds. The solventless procedure allowed easier access to the quinazolino[4,3-*b*]quinazolin-8-ones and gave better yields than the method performed in the presence of solvents. However, the procedure with solvents would offer the possibility of investigating the microwave-assisted solid-phase synthesis of these quinazolinones, which would facilitate purification of the final products.

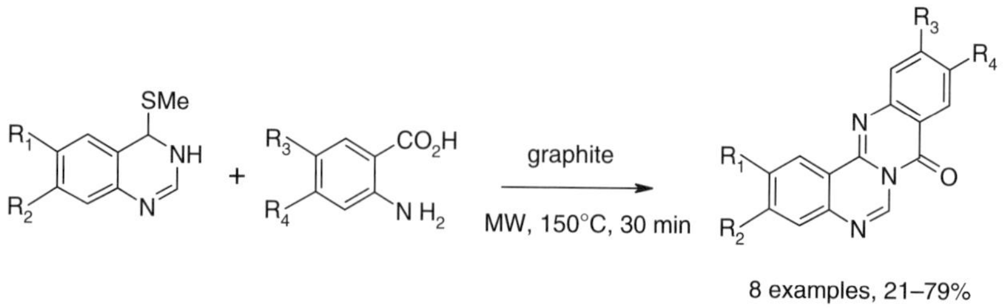

Scheme 3.49

Developing their work on the use of microwave-assisted Niementowski reactions, the same group published the synthesis of novel triaza- and tetraaza-benzo[a]-indeno[1,2-c]anthracen-5ones by the condensation of anthranilic acid with 2-(2-aminophenyl)indole or benzimidazole (Scheme 3.49)[78,79].

In the search for new bioactive heterocyclic molecules inspired by marine natural alkaloids, the rapid synthesis of various derivatives of indolo[1,2-c]quinazolines and benzimidazo[1,2-c]quinazolines has been described[29]. In these studies, the use of graphite as a 'thermal accelerator' (only 10% by weight) enabled ready access to the desired compounds, where conventional methods had failed (Scheme 3.50). Graphite is a strong absorber of microwave energy and is therefore a very good generator of heat by microwave dielectric heating.

Following a similar strategy, Besson's group has also described the microwave-assisted thermocyclisation of imino-1,2,3-dithiazoles into imidazo[4,5,1-ij]quinolines (Scheme 3.50)[80].

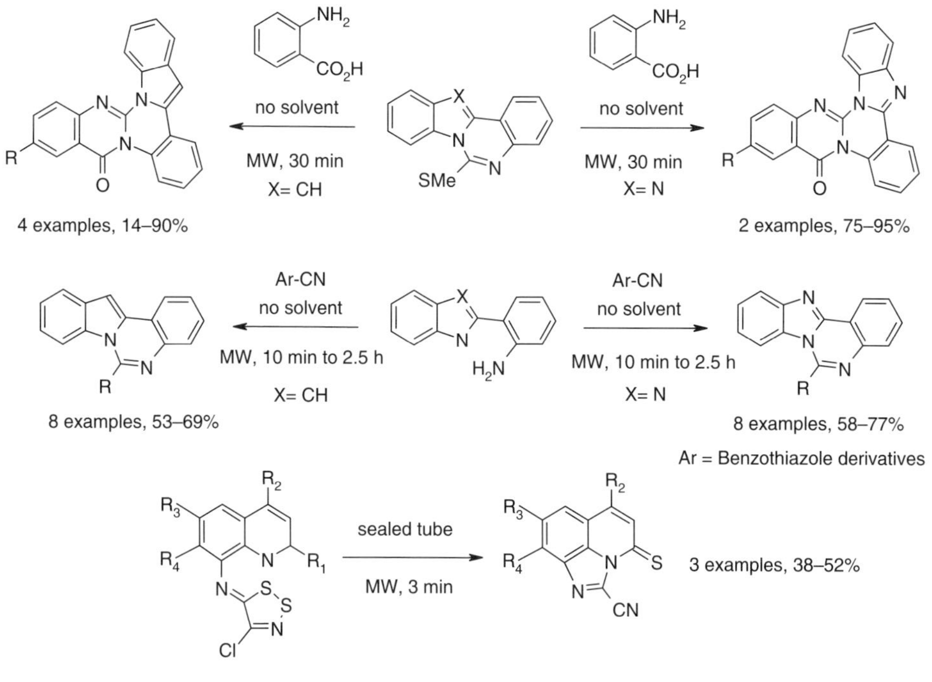

Scheme 3.50

In an effort to extend the scope of various synthetic studies towards 2-amino derivatives of 1,4-dihydropyridine, Kappe and co-workers performed a microwave-mediated regioselective construction of novel pyrimido[1,2-a]pyrimidines (Scheme 3.51)[81]. This ring system can be found in marine-derived natural products such as Crambescidin and Batzelladine alkaloids.

An improved modification of the Ganesan synthesis of Luotonin A has been developed under microwave heating for the synthesis of various annulated [1,2-b]quinazolinones.

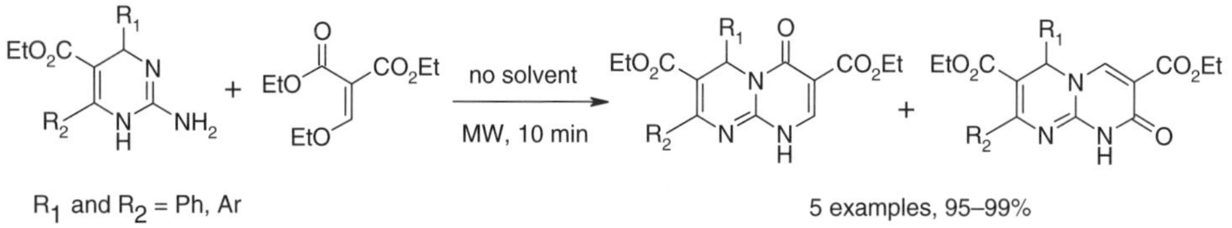

R_1 and R_2 = Ph, Ar

5 examples, 95–99%

Scheme 3.51

This modification involves the two component solvent-free condensation of substituted lactams and isatoic anhydride (Scheme 3.52)[82].

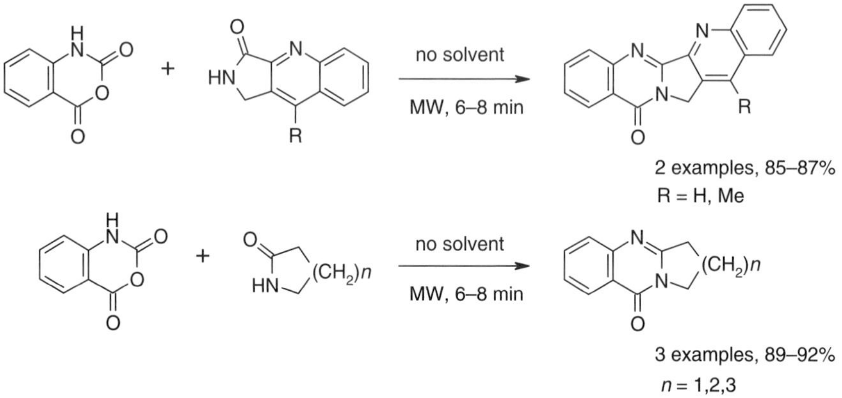

no solvent

MW, 6–8 min

2 examples, 85–87%
R = H, Me

no solvent

MW, 6–8 min

3 examples, 89–92%
n = 1,2,3

Scheme 3.52

The last example cited involves the microwave-assisted route towards novel 5-substituted-2,3-dihydroimidazo[1,2-c]thieno[3,2-e]pyrimidines described by Rao and co-workers. Employing microwave heating and an easy work-up procedure, the products were obtained in improved yields over the conventional methods (Scheme 3.53)[83].

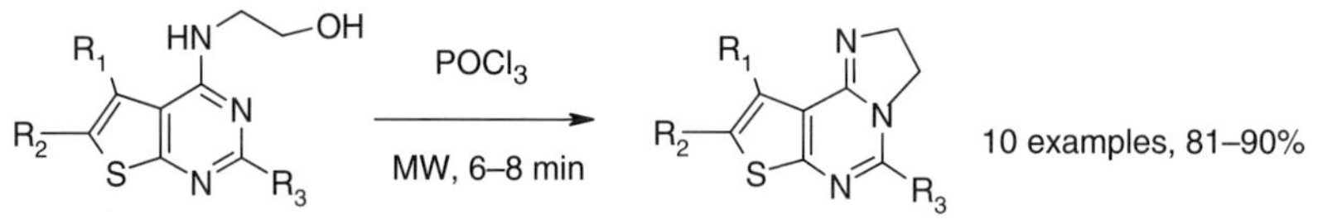

$POCl_3$

MW, 6–8 min

10 examples, 81–90%

Scheme 3.53

3.9. Conclusion

The examples described in this chapter demonstrate that microwave-assisted synthesis can allow easy and rapid access to various aromatic heterocyclic compounds, which may have interesting pharmaceutical potential. These aromatic heterocyclic compounds have been synthesised in a variety of microwave systems, including the domestic microwave oven. Despite the often associated reproducibility issues with chemistries developed in the domestic microwave oven, these examples highlight opportunities for these chemical reactions to be investigated more reproducibily in the single-mode focussed microwave synthesisers now commercially available.

We have also shown that performing microwave-heated reactions should be considered with special attention. A few of these considerations can be applied generally in conducting microwave-heated reactions and include the following: (a) the ratio between the quantity of the material and the support (e.g. graphite) or the solvent is very important; (b) for solid starting materials, the use of solid supports can offer operational, economical and environmental benefits over conventional methods. However, association of liquid/solid reactants on solid supports may lead to uncontrolled reactions, which may result in worse results than those obtained by the comparative conventional thermal reactions. In these cases, simple fusion of the products or addition of an appropriate solvent may lead to more convenient mixtures or solutions for microwave-assisted reactions.

The strategies explored and defined in the various examples presented here open the door to wider application of microwave chemistry in industry. The most important problem for chemists today (in particular, drug discovery chemists) is to scale-up microwave chemistry reactions for a large variety of synthetic reactions with minimal optimisation of the procedures for scale-up. At the moment, there is a growing demand from industry to scale-up microwave-assisted chemical reactions, which is pushing the major suppliers of microwave reactors to develop new systems. In the next few years, these new systems will evolve to enable reproducible and routine kilogram scale microwave-assisted synthesis.

3.10. References

1. Lidström, P., Tierney, J., Wathey, B. and Westman, J., Microwave-assisted organic synthesis: a review, *Tetrahedron*, 2001, **57**, 9225–9283.
2. Bogdal, D. and Warzala, M., Microwave-assisted preparation of benzo[*b*]furans under solventless phase-transfer catalytic conditions, *Tetrahedron*, 2000, **56**, 8769–8773.
3. Varma, R.S., Kumar, D. and Liesen, P.J., Solid state synthesis of 2-aroylbenzo[*b*]furans, thiazoles and 3-aryl-5,6-dihydroimidazo[2,1-*b*][1,3]thiazoles from α-tosyloxy ketones using microwave irradiation, *J. Chem. Soc., Perkin Trans. 1*, 1998, 4093–4096.
4. Villemin, D. and Liao, L., Application of microwaves in organic synthesis: a rapid and efficient synthesis of new 3-aryl-2-imino-4-methyl-2,5-dihydrofurans and 3-aryl-3-2-(5*H*)-furanones, *Synth. Commun.*, 2003, **33**, 1575–1585.
5. Danks, T.N., Microwave-assisted synthesis of pyrroles, *Tetrahedron Lett.*, 1999, **40**, 3957–3960.
6. Ranu, B.C. and Hajra, A., Synthesis of alkyl-substituted pyrroles by three-component coupling of carbonyl compound, amine and nitro-alkane/alkene on a solid surface of silica gel/alumina under microwave irradiation, *Tetrahedron*, 2001, **57**, 4767–4773.
7. Bora U., Saikia A. and Boruah R.C., A novel microwave-mediated one-pot synthesis of indolizines *via* a three-component reaction, *Org. Lett.*, 2003, **5**, 435–438.
8. Lipinska, T., General route to the total synthesis of sempervirine analogues containing modified *E* rings, potential cytostatics, *Tetrahedron Lett.*, 2002, **43**, 9565–9567.
9. Finaru, A., Berthault, A., Besson, T., Guillaumet, G. and Berteina-Raboin, S., Microwave-assisted synthesis of 5-carboxymethoxy-*N*-acetyltryptamine derivatives, *Tetrahedron Lett.*, 2002, **43**, 787–790.
10. Finaru, A., Berthault, A., Besson, T., Guillaumet, G. and Berteina-Raboin, S., Microwave-assisted solid-phase synthesis of 5-carboxamido-*N*-acetyltryptamine derivatives, *Org. Lett.*, 2002, **4**, 2613–2615.

11. Kiryanov, A.A., Sampson, P. and Seed, A.J., Synthesis of 2-alkoxy-substituted thiophenes, 1,3-thiazoles, and related *S*-heterocycles *via* Lawesson's reagent-mediated cyclization under microwave irradiation: applications for liquid crystal synthesis, *J. Org. Chem.,* 2001, **66**, 7925–7929.
12. Hoener, A.P.F., Henkel, B. and Gauvin, J.C., Novel one-pot microwave-assisted Gewald synthesis of 2-acyl amino thiophenes on solid support, *Synlett,* 2003, 63–66.
13. Balalaie, S. and Arabanian, A., One-pot synthesis of tetrasubstituted imidazoles catalyzed by zeolite HY and silica gel under microwave irradiation, *Green Chem.,* 2000, **2**, 274–276.
14. Balalaie, S., Hashemi, M.M. and Akhbari, M., A novel one-pot synthesis of tetrasubstituted imidazoles under solvent-free conditions and microwave irradiation, *Tetrahedron Lett.,* 2003, **44**, 1709–1711.
15. Fresneda, P.M., Molina, P. and Sanz, M.A., Microwave-assisted regioselective synthesis of 2,4-disubstituted imidazoles: nortopsentin D synthesized by minimal effort, *Synlett,* 2001, 218–221.
16. Coleman, C.M., MacElroy, J.M.D., Gallagher, J.F. and O'Shea, D.F., Microwave parallel library generation: comparison of a conventional- and microwave-generated substituted 4(5)-sulfanyl-1*H*-imidazole library, *J. Comb. Chem.,* 2002, **4**, 87–93.
17. Molteni, V., Hamilton, M.M., Mao, L., Crane, C.M., Termin, A.P. and Wilson, D.M., Aqueous one-pot synthesis of pyrazoles, pyrimidines and isoxazoles promoted by microwave irradiation, *Synthesis,* 2002, 1669–1674.
18. Westman, J., Lundin, R., Stalberg, J., Ostbye, M., Franzen, A. and Hurynowicz, A., Alkylaminopropenones and alkylamino-propenoates as efficient and versatile synthons in microwave-assisted combinatorial synthesis, *Comb. Chem. High Throughput Screen.,* 2002, **5**, 565–570.
19. Lu, J., Yang, B. and Bai, Y., Microwave irradiation synthesis of 2-substituted benzimidazoles using PPA as a catalyst under solvent-free conditions, *Synth. Commun.,* 2002, **32**, 3703–3709.
20. Reddy, G.V., Rama Rao, V.V.V.N.S., Narsaiah, B. and Rao, P.S., A simple and efficient method for the synthesis of novel trifluoromethyl benzimidazoles under microwave irradiation conditions, *Synth. Commun.,* 2002, **32**, 2467–2476.
21. Bougrin, K., Loupy, A., Petit, A., Daou, B. and Soufiaoui, M., Novel route for synthesis of 2-trifluoromethyl-arylimidazoles on K10 montmorillonite in a microwave-irradiated "dry medium", *Tetrahedron,* 2001, **57**, 163–168.
22. Bendale, P.M. and Sun, C.-M., Rapid microwave-assisted liquid-phase combinatorial synthesis of 2-(arylamino)benzimidazoles, *J. Comb. Chem.,* 2002, **4**, 359–361.
23. Brain, C.T. and Steer, J.T., An improved procedure for the synthesis of benzimidazoles, using palladium-catalysed aryl-amination reaction chemistry, *J. Org. Chem.,* 2003, **68**, 6814–6816.
24. Lee, J.C., Choi, H.J. and Lee, Y.C., Efficient synthesis of multi-substituted oxazoles under solvent-free microwave irradiation, *Tetrahedron Lett.,* 2003, **44**, 123–125.
25. Brain, C.T. and Paul, J.M., Rapid synthesis of oxazoles under microwave conditions, *Synlett,* 1999, 1642–1644.
26. Pottorf, R.S., Chadha, N.K., Katkevics, M., Ozola, V., Suna, E., Ghane, H., Regberg, T. and Player, M.R., Parallel synthesis of benzoxazoles *via* microwave-assisted dielectric heating, *Tetrahedron Lett.,* 2003, **44**, 175–178.
27. Rostamizadeh, S. and Derafshian, E., A simple route to the preparation of benzimidazoles and benzoxazoles, *J. Chem. Res. (S),* 2001, 227–228.
28. Frère, S., Thiéry, V. and Besson T., Eco-friendly microwave-assisted scaleable synthesis of 2-cyanobenzo-thiazoles *via* *N*-arylimino-1,2,3-dithiazoles, *Synth. Commun.,* 2003, **33**, 3789–3798.
29. Frère, S., Thiéry, V., Bailly, C. and Besson T., Novel 6-substituted benzothiazol-2-yl indolo[1,2-*c*]quinazolines and benzimidazo[1,2-*c*]quinazolines, *Tetrahedron,* 2003, **59**, 773–779.
30. Ykman, P., L'Abbe, G. and Smets, G., Reactions of aryl azides with α-keto phosphorus ylides, *Tetrahedron,* 1971, **27**, 845–849.
31. Tao, L., Zhang, L.L., Shen, S.J. and Han, X.P., Microwave-promoted rapid synthesis of 1-aryl-1,2,3-triazoles, *Chin. Chem. Lett.,* 2001, **12**, 763–764.
32. Habich, D., Barth, W. and Rosner, M., Synthesis of 3′-(1,2,3-triazol-1-yl)-3′-deoxythymidines, *Heterocycles,* 1989, **29**, 2083–2088.
33. Mearman, R.C., Newall, C.E. and Tonge, A.P., Structure–activity relationships in 1,2,3-triazol-1-yl derivatives of clavulanic acid, *J. Antibiot.,* 1984, **37**, 885–891.
34. Katritzky, A.R. and Singh, S.K., Synthesis of C-carbamoyl-1,2,3-triazoles by microwave-induced 1,3-dipolar cycloaddition of organic azides to acetylenic amides, *J. Org. Chem.,* 2002, **67**, 9077–9079.
35. Rostamizadeh, S., Tajik, H. and Yazdanfarahi, S., Solid phase synthesis of 1,2,4-triazoles under microwave irradiation, *Synth. Commun.,* 2003, **33**, 113–117.
36. Dandia, A., Singh, R., Sachdeva, H. and Arya, K., Microwave-assisted one pot synthesis of a series of trifluoromethyl substituted spiro [indole-triazoles], *J. Fluorine Chem.,* 2001, **111**, 61–67.
37. Bentiss, F., Lagrenee, M. and Barbry, D., Accelerated synthesis of 3,5-disustitued 4-amino-1,2,4-triazoles under microwave irradiation, *Tetrahedron Lett.,* 2000, **41**, 1539–1541.
38. Bentiss, F., Lagrenee, M. and Barbry, D., Rapid synthesis of 2,5-disubstituted 1,3,4-oxadiazoles under microwave irradiation, *Synth. Commun.,* 2001, **31**, 935–938.

39. Brain, C.T. and Brunton, S.A., Synthesis of 1,3,4-oxadiazoles using polymer-supported reagents, *Synlett*, 2001, 382–384.
40. Wang, X., Li, Z.G., Wei, B. and Yang, J., Synthesis of 2-(4-methoxyphenyloxyacetyl amido)-5-aryloxymethyl-1,3,4-oxadiazoles under microwave irradiation, *Synth. Commun.*, 2002, **32**, 1097–1103.
41. Wang, X., Li, Z., Zhang, Z. and Da, Y., Synthesis of 2-(4-chloro benzoylamido)-5-aryloxymethyl-1,3,4-oxadiazoles under microwave irradiation. *Synth. Commun.*, 2001, **31**, 1907–1911.
42. Joshi, S. and Karnik, A.V., Facile conversion of acyldiithiocarbazinate salts to 1,3,4-oxadiazole derivatives under microwave irradiation, *Synth. Commun.*, 2002, **32**, 111–114.
43. Li, Z., Wang, X. and Da, Y., Synthesis of 2-(5-(2-chlorophenyl)-2-furoylamido)-5-aryloxymethyl-1,3,4-thiadiazoles under microwave irradiation, *Synth. Commun.*, 2001, **31**, 1829–1836.
44. Alterman, M. and Hallberg, A., Fast microwave-assisted preparation of aryl and vinyl nitriles and the corresponding tetrazoles from organo-halides, *J. Org. Chem.*, 2000, **65**, 7984–7989.
45. Kidwai, M., Saxena, S., Mohan, R. and Venkataramanan, R., A novel one pot synthesis of nitrogen containing heterocycles: an alternate methodology to the Biginelli and Hantzsch reactions, *J. Chem. Soc., Perkin Trans. 1*, 2002, 1845–1846.
46. Ohberg, L. and Westman, J., An efficient and fast procedure for the Hantzsch dihydropyridine synthesis under microwave conditions, *Synlett*, 2001, **8**, 1296–1298.
47. Yadav, J.S., Reddy, B.V.S. and Reddy, P.T., Unprecedented synthesis of Hantzsch 1,4-dihydropyridines under Biginelli reaction conditions, *Synth. Commun.*, 2001, **31**, 425–430.
48. Bagley, M.C., Lunn, R. and Xiong, X., A new one-step synthesis of pyridines under microwave-assisted conditions, *Tetrahedron Lett.*, 2002, **43**, 8331–8334.
49. Wilson, N.S., Sarko, C.R. and Roth, G.P., Microwave-assisted synthesis of 2-aminoquinolines, *Tetrahedron Lett.*, 2002, **43**, 581–583.
50. Song, S.J., Cho, S.J., Park, D.K., Kwon, T.W. and Jenekhe, S.A., Microwave enhanced solvent-free synthesis of a library of quinoline derivatives, *Tetrahedron Lett.*, 2003, **44**, 255–257.
51. Kidwai, M., Bhushan, K.R., Sapra, P., Saxena, R.K. and Gupta, R., Alumina-supported synthesis of antibacterial quinolines using microwaves, *Bioorg. Med. Chem.*, 2000, **8**, 69–72.
52. Ranu, B.C., Hajra, A. and Jana, U., Microwave-assisted simple synthesis of quinolines from anilines and alkyl vinyl ketones on the surface of silica gel in the presence of indium(III) chloride, *Tetrahedron Lett.*, 2000, **41**, 531–533.
53. Diaz-Ortiz, A., De la Hoz, A. and Langa, F., Microwave irradiation in solvent-free conditions: an eco-friendly methodology to prepare indazoles, pyrazolopyridines and bipyrazoles by cycloaddition reactions, *Green Chem.*, 2000, **2**, 165–172.
54. Frère, S., Thiéry, V. and Besson, T., Microwave accelaration of the Pechmann reaction on graphite/-montmorillonite K10: application to the preparation of 4-substituted 7-aminocoumarins, *Tetrahedron Lett.*, 2001, **42**, 2791–2794.
55. De la Hoz, A., Moreno, A. and Vazquez, E., Use of microwave irradiation and solid acid catalyst in an enhanced and environmentaly friendly synthesis of coumarin derivatives, *Synlett*, 1999, 608–610.
56. Wang, J-.X., Hu, Z.L.Y. and Wei, B., Copper-catalysed cross coupling reaction under microwave irradiation conditions, *J. Chem. Res. (S)*, 2000, 536–537.
57. Bogdal, D., Coumarins: fast synthesis by knoevenagel condensation under microwave irradiation, *J. Chem. Res. (S)*, 1998, 468–469.
58. Yaddav, J.S., Reddy, B.V.S., Reddy, E.J. and Ramalingam, T., Microwave-assisted efficient synthesis of dihydropyrimidines: improved high yielding protocol for the Biginelli reaction, *J. Chem. Res. (S)*, 2000, 354–355.
59. Stadler, A. and Kappe, C.O., Automated library generation using sequential microwave-assisted chemistry. Application toward the Biginelli multicomponent condensation, *J. Comb. Chem.*, 2001, 3, 624–630.
60. Alexandre, F.R., Berecibar, A. and Besson, T., Microwave-assisted Niementowski reaction. Back to the roots, *Tetrahedron Lett.*, 2002, **43**, 3911–3913.
61. Rad-Moghadam, K. and Khajavi, M.S., One-pot synthesis of substituted quinazolin-4(3H)-ones under microwave irradiation, *J. Chem. Res. (S)*, 1998, 702–703.
62. Seijas, J. A., Vazquez-Tato, M.P. and Martinez, M.M., Microwave enhanced synthesis of 4-aminoquinazolines, *Tetrahedron Lett.*, 2000, **41**, 2215–2217.
63. Hazarkhani, H. and Kamiri, B., A facile synthesis of new 3-(2-benzimidazolyl)-2-alkyl-4-(3H)-quinazolinones under microwave irradiation, *Tetrahedron*, 2003, **59**, 4757–4760.
64. Chemat, F. and Poux, M., Microwave-assisted pyrolysis of urea supported on graphite under solvent-free conditions, *Tetrahedron Lett.*, 2001, **42**, 3693–3695.
65. Avalos, M., Babiano, R., Cintas, P., Clemente, F.R., Jimenz, J.L., Palacios, J.C. and Sanchez, J.B., Hetero-Diels-Alder reactions of homochiral 1,2-diaza-1,3-butadienes with diethyl azodicarboxylate under microwave irradiation. Theoretical rationale of the stereochemical outcome, *J. Org. Chem.*, 1999, **64**, 6297–6305.
66. Santagada, V., Perissutti, E., Fiorino, F., Vivenzio, B. and Caliendo, G., Microwave enhanced solution synthesis of 1,4-benzodiazepin-5-ones, *Tetrahedron Lett.*, 2001, **42**, 2397–2400.

67. Kamal, A., Reddy, B.S.N. and Reddy, G.S.K., Microwave-assisted synthesis of Pyrrolo[2,1-c] [1,4]benzo-diazepine-5-11-diones, *Synlett*, 1999, **8**, 1251–1252.
68. Pozarentzi, M., Stephanidou-Stephanatou, J. and Tsoleridis, C.A., An efficient method for the synthesis of 1,5-benzodiazepine derivatives under microwave irradiation without solvent, *Tetrahedron Lett.*, 2002, **43**, 1755–1758.
69. Alexandre, F.R., Berecibar, A., Wrigglesworth, R. and Besson, T., Efficient synthesis of thiazoloquinazolinone derivatives, *Tetrahedron Lett.*, 2003, **44**, 4455–4458.
70. Besson, T., Guillard, J. and Rees, C.W., Multistep synthesis of thiazoloquinazolines under microwave irradiation in solution, *Tetrahedron Lett.*, 2000, **41**, 1027–1030.
71. Mont, N., Teixido, J., Borell, J.I. and Kappe, C.O., A three-component synthesis of pyrido[2,3-d]pyrimidines, *Tetrahedron Lett.*, 2003, **44**, 5385–5387.
72. Quiroga, J., Cisnero, C., Insuasty, B., Abonia, R., Nogueras, M. and Sanchez, A., A regiospecific three-component one-step cyclocondensation to 6-cyano-5, 8-dihydropyrido[2,3-d]pyrimidin-4-(3H)ones using microwaves and solvent-free conditions, *Tetrahedron Lett.*, 2001, **42**, 5625–5627.
73. Dave, C.G. and Shah, R.D., Gould-jacob type of reaction in the synthesis of thieno[3,2-e]pyrimido [1,2-c]pyrimidines: a comparison of classical heating vs solvent-free microwave irradiation, *Heterocycles*, 1999, **51**, 1819–1826.
74. Kidwai, M., Venkataramanan, R., Garg, R.K. and Bhushan, K.R., Novel one pot synthesis of new pyranopy-rimidines using microwaves, *J. Chem. Res. (S)*, 2000, 586–587.
75. Paul, S., Gupta, M., Gupta, R. and Loupy, A., Microwave-assisted solvent-free synthesis of pyrazolo [3,4-b]quinolines and pyrazolo[3,4-c]pyrazoles using p-TsOH, *Tetrahedron Lett.*, 2001, **42**, 3827–3829.
76. Guillard, J. and Besson, T., Synthesis of novel dioxinobenzothiazole derivatives, *Tetrahedron*, 1999, **55**, 5139–5144.
77. Alexandre, F.R., Berecibar, A., Wrigglesworth, R. and Besson, T., Novel series of of 8H-quinazolino [4,3-b]quinazolin-8-ones *via* two Niementowski condensation, *Tetrahedron*, 2003, **59**, 1413–1419.
78. Domon, L., Le Coeur, C., Grelard, A., Thiéry, V. and Besson, T., Efficient modified von Niementowski synthesis of novel derivatives of 5a, 14b, 15-triazabenzo[a]indeno[1,2-c]anthracen-5-one from indolo [1,2-c]quinazoline, *Tetrahedron Lett.*, 2001, **42**, 6671–6674.
79. Soukri, M., Guillaumet, G., Besson, T., Aziane, D., Aadil, M., Essassi, El M. and Akssira, M., Synthesis of novel 5a,10,14b,15-tetraaza-benzo[a]indeno[1,2-c] anthracen-5-one and benzimidazo[1,2-c]quinazoline derivatives under microwave irradiation, *Tetrahedron Lett.*, 2000, **41**, 5857–5860.
80. Besson, T., Rees, C.W., Roe, D.G. and Thiéry, V., Imidazoquinolinethiones from 8-aminoquinolines by a novel *peri*-participation, *J. Chem. Soc., Perkin Trans. 1*, 2000, 555–561.
81. Vanden Eynde, J.J., Hecq, N., Kataeva, O. and Kappe, C.O., Microwave-mediated regioselective synthesis of novel pyrimido[1,2-a]pyrimidines under solvent-free conditions, *Tetrahedron*, 2001, **57**, 1785–1791.
82. Yadav, J.S. and Reddy, B.S.V., Microwave-assisted rapid synthesis of the cytotoxic alkaloid luotonin A, *Tetrahedron Lett.*, 2002, **43**, 1005–1907.
83. Prasad, M.R., Rao, A.R.R., Rao, P.S. and Rajan, K.S., Microwave-assisted synthesis of novel 5-substituted-2,3-dihydroimidazo[1,2-c]thieno[3,2-e] pyrimidines, *Synthesis*, 2001, 2119–2123.

4 Microwave-assisted reductions

TIMOTHY N. DANKS and GABRIELE WAGNER

4.1. Introduction

Reductions of organic functional groups play a central role in synthesis. However, their application is often limited and is poorly compatible with the synthesis of sensitive products because of the harsh conditions used. Similar to many other types of transformations in organic chemistry, high temperature and long reaction times are common, and many reactions suffer from the formation of by-products. The improvement of currently existing reduction schemes towards faster, more selective, economically and ecologically acceptable procedures is therefore an important issue. In this context, the application of microwave heating was shown to have a strongly beneficial impact on a large variety of synthetic procedures. Often reaction times can be drastically reduced, thus leading to overall milder reaction conditions, higher yields and purer products. In many examples the application of microwave irradiation has also allowed for the use of milder and more environmentally acceptable reagents[1−7]. The use of microwave irradiation is thus expected to improve reductions of functional groups in a similar sense.

A number of reductions in organic chemistry require the use of gases under high pressure, for example, catalytic hydrogenation reactions with hydrogen. Although the use of gases in the laboratory and on a small scale does not necessarily create a severe safety hazard, the reactions require dedicated equipment, such as autoclaves or gas cylinders, and an oxygen-free atmosphere. The introduction of solid hydrogen donors allows for hydrogenation reactions being performed in normal reaction vessels, and this has facilitated the use of microwave heating in such type of reactions and allowed to benefit from the general advantages of microwave-assisted reactions.

Solvent-free reactions play an important role in microwave-enhanced chemistry[8−11], and especially the development of 'dry' reactions on mineral supports in conjunction with microwave irradiation have had a strong impact on the development of safe and efficient reduction schemes. In dry reaction conditions, the microwave energy is directly absorbed by the solid support or the reagent itself, thus allowing for fast and homogenous heating of the sample, which would not be feasible with thermal heating because of poor thermal conduction in powders.

The aim of this chapter is to present an overview of microwave-assisted reductions used in organic chemistry. The chemistry highlighted in this section is subdivided for convenience and describes the reduction of carbon–carbon multiple bonds, carbonyl groups, nitrogen-bearing functionalities and hydrodehalogenation reaction. A description of debenzylation and other reactions acting as deprotection protocols have not been included since in these cases, although reducing conditions are used, the isolated products are not a result of a true reduction process.

4.2. Reduction of carbon–carbon multiple bonds

Carbon–carbon multiple bonds are most commonly reduced to saturated hydrocarbons by hydrogenation or by other methods, such as the use of hydrides. While a large number of alkenes have been reduced under microwave irradiation, surprisingly little work has been published on microwave-assisted reduction of alkynes.

4.2.1. C—C multiple bond reduction using transfer hydrogenation

The use of hydrogen gas in hydrogenation reactions is often considered inconvenient, because of the necessity to carry out the reactions in closed vessels in an oxygen-free atmosphere. Additionally, reactions with hydrogen gas often require high pressure and the use of autoclaves.

Instead, catalytic transfer hydrogenation, in which the hydrogen gas is replaced by a solid hydrogen donor, has become increasingly popular for safety issues and the much simpler operations involved with the reaction. The substitution of hydrogen gas by a solid hydrogen donor reagent also renders the reactions more suitable to microwave applications. Formate salts are most frequently applied as hydrogen donors in both thermal and microwave reactions, although dihydropyridines, hydrazine, hypophosphites, cyclohexene or cyclohexadiene may also be used with success. Furthermore, the introduction of solid-supported variants of these hydrogen sources help to facilitate the work-up procedures, thus making the overall synthetic process more efficient.

Bose was the first to report the use of microwaves to promote transfer hydrogenation reactions in organic synthesis. A series of β-lactam derivatives were hydrogenated using formates and Pd/C or Raney Ni as a catalyst (Scheme 4.1). C=C bonds in the side chains can be saturated, thus giving easy access to a number of β-lactam derivatives. Under certain conditions and depending on the choice of the catalyst, cleavage of the β-lactam ring by hydrogenolysis of the N—C$_4$ bond can occur[12–14].

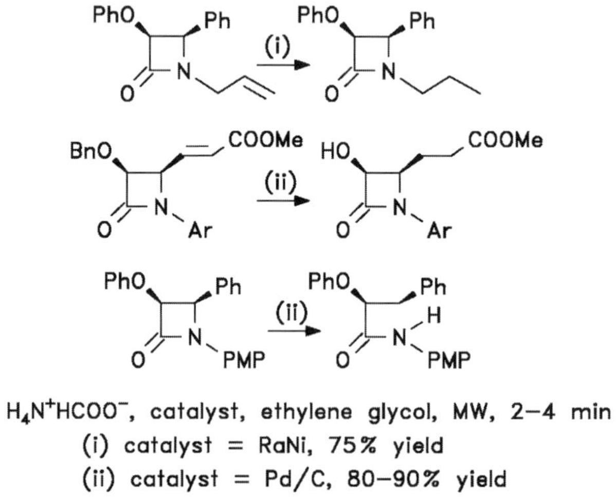

$H_4N^+HCOO^-$, catalyst, ethylene glycol, MW, 2–4 min
(i) catalyst = RaNi, 75% yield
(ii) catalyst = Pd/C, 80–90% yield

Scheme 4.1

Similar reaction conditions as those by Bose were used for a range of other applications, for example, the synthesis of heterocycles. A combination of a microwave-assisted Paal–Knorr reaction[15] with a transfer hydrogenation takes place in the preparation of 2,5-di- and 1,2,5-trisubstituted pyrroles from *E*-1,4-diaryl-2-butene-1,4-diones in a 'one-pot' operation. Hydrogenation was achieved with ammonium formates and 10% Pd/C as catalyst in PEG-200. Yields of up to 92% were obtained within 0.5–2 min (Scheme 4.2)[16].

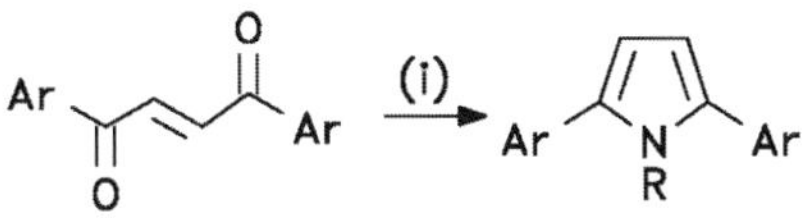

(i) $RNH_3^+HCOO^-$, Pd/C, PEG–200, MW, 0.5–2 min
9 examples, 56–92%
R = H, ^nBu, Ph, CH_2Ph
Ar = Ph, $4\text{-}ClC_6H_4$, $4\text{-}BrC_6H_4$, $4\text{-}MeC_6H_4$, $4\text{-}MeOC_6H_4$

Scheme 4.2

Unsaturated sterols such as cholesterol, campesterol, sitosterol and bile alcohols with unsaturated side chains can be transfer hydrogenated efficiently and with high yields under microwave irradiation, using ammonium formate and a Pd/C catalyst in methylene chloride/propylene glycol solvents (Scheme 4.3)[17].

Catalytic transfer hydrogenation with sodium formate and 10% Pd/C was also applied to the hydrogenation of soybean oil. The reaction rates were up to eight times greater using microwave heating than with conventional heating at the same temperature. The effect was attributed to enhanced transport kinetics at the catalyst and oil–water interface under microwave conditions[18].

With solid deuterium or tritium donors, the hydrogen-transfer methodology can be used for selective isotopic labelling of pharmacologically relevant compounds. The use of solid reagents in a small reaction volume is much preferable to handling HD_2 or T_2 gas in a conventional hydrogenation[19], especially when radioactive labelling with tritium is required. Also the combination of transfer hydrogenation with aromatic dehalogenation under microwave irradiation provides a rapid route to deuterium-labelled compounds with enhanced isotopic incorporation[20].

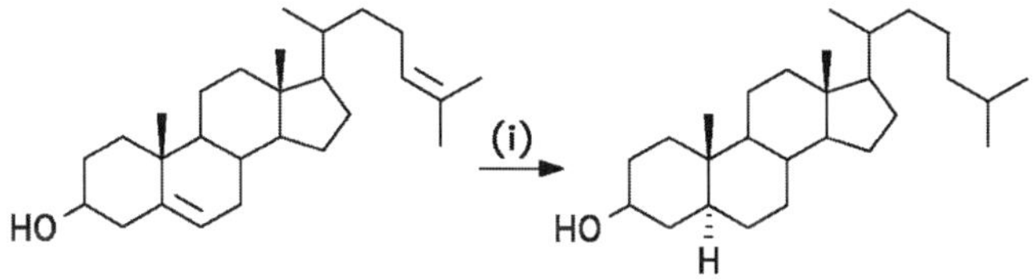

(i) $NH_4^+HCOO^-$, Pd/C, MW, 4–7 min
8 examples, 80–95%

Scheme 4.3

Ammonium formate sublimates easily and thus tends to leave the reaction mixture at elevated temperature and condense in colder parts of the apparatus. This may produce uncontrollable pressure due to the formation of ammonia gas or block narrow parts of the reaction vessel. Thus the development of non-volatile formate salts is an issue of interest. A combination of formate bound to an ion-exchange resin and Wilkinson's catalyst was used successfully in the transfer hydrogenation of electron-deficient alkenes. The solid-supported reagent is cheap and easily recyclable and allows for easy work-up of the reaction by filtration. Reactions were complete upon microwave irradiation for 30 s, and the products were obtained in almost quantitative yields[21]. Analogously, ammonium formate supported on alumina was used in conjunction with Wilkinson's catalyst for alkene hydrogenation. This source of hydrogen offers the additional advantage of scavenging the transition-metal catalyst after use, thus further facilitating the work-up of the reaction[22]. Compared to thermal conditions, shorter reaction times were required and purer products obtained (Scheme 4.4)[21,22].

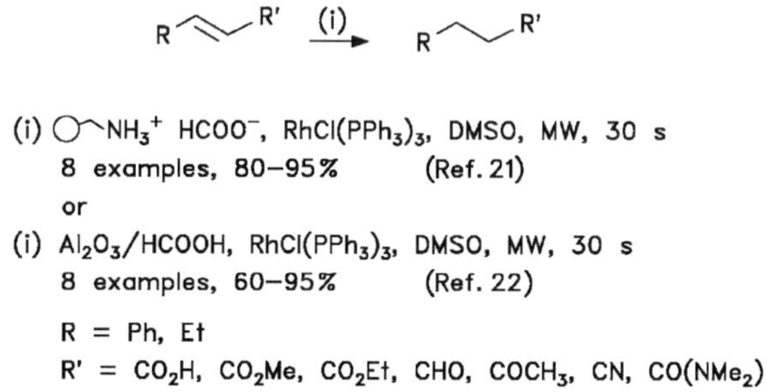

(i) ⬡︎⌢NH$_3^+$ HCOO$^-$, RhCl(PPh$_3$)$_3$, DMSO, MW, 30 s
    8 examples, 80–95%   (Ref. 21)

or

(i) Al$_2$O$_3$/HCOOH, RhCl(PPh$_3$)$_3$, DMSO, MW, 30 s
    8 examples, 60–95%   (Ref. 22)

    R = Ph, Et
    R' = CO$_2$H, CO$_2$Me, CO$_2$Et, CHO, COCH$_3$, CN, CO(NMe$_2$)

Scheme 4.4

A further step towards optimised conditions in the catalytic transfer hydrogenation of alkenes was achieved with the introduction of the ionic liquid N-butyl-N'-methylimidazolium hexafluorophosphate (BMIMPF$_6$) as a solvent. The reduction of alkenes with formates and Pd/C in BMIMPF$_6$ leads to saturated hydrocarbons in high yields. With an alkyne, a mixture of *cis/trans* alkenes and the saturated alkane was obtained (Scheme 4.5). Sufficiently pure products were isolated by a simple liquid–liquid

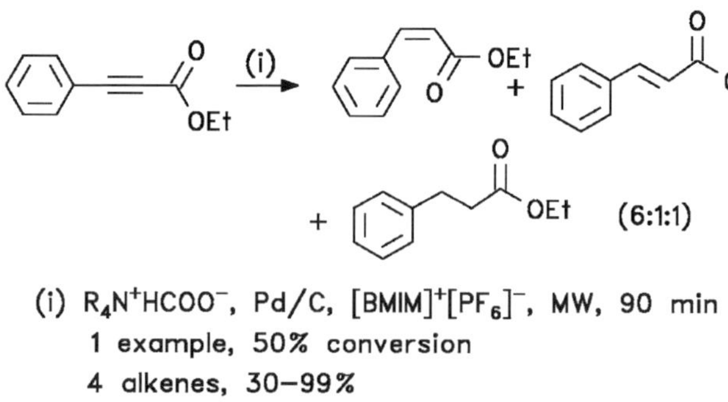

(i) R$_4$N$^+$HCOO$^-$, Pd/C, [BMIM]$^+$[PF$_6$]$^-$, MW, 90 min
    1 example, 50% conversion
    4 alkenes, 30–99%

Scheme 4.5

extraction with methyl *tert*-butyl ether. Furthermore, the solvent system and catalyst could be recycled[23].

Until now, hydrogen sources other than formates have been rarely reported in microwave-assisted transfer hydrogenations of carbon–carbon multiple bonds. An exception is a transfer hydrogenation of electron-deficient alkenes where a series of 1,4-dihydropyridines supported on silica gel were used as the hydrogen source (Scheme 4.6). The influences of electronic effects of the alkene, steric effects of the dihydropyridine and type and power of the microwave irradiation were studied[24].

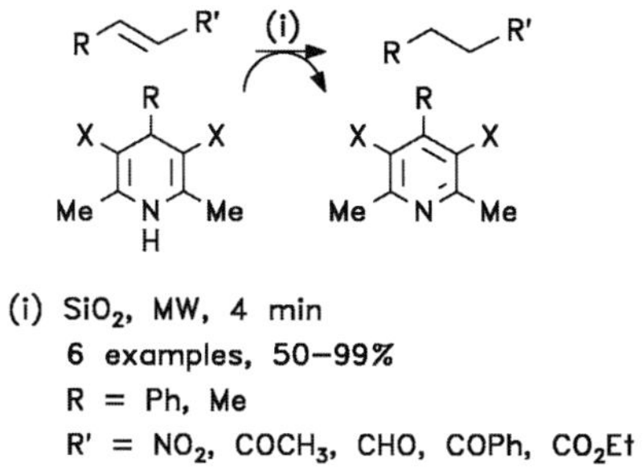

(i) SiO$_2$, MW, 4 min
6 examples, 50–99%
R = Ph, Me
R' = NO$_2$, COCH$_3$, CHO, COPh, CO$_2$Et

Scheme 4.6

4.2.2. *C—C multiple bond reduction using other methods*

Borohydrides normally do not attack carbon–carbon multiple bonds, and thus, α,β-unsaturated imines (1-aza-1,3-butadienes) are reduced only at their C=N bond, under both thermal and microwave conditions. However, the corresponding (1-aza-1,3-butadiene)tricarbonyliron(0) complexes show a totally different reactivity under the same conditions, and a simultaneous reduction of both C=N and C=C takes place if microwave irradiation is applied[25]. When the reaction was performed with sodium borodeuterid, 1,2,3-trideutero, secondary amines were obtained. In contrast to their behaviour under microwave conditions, these complexes were totally inert to reduction by NaBH$_4$ under thermal conditions (Scheme 4.7)[25].

In this context, it is worthwhile to mention that reduction of C=C bonds in α,β-unsaturated carbonyl compounds can also occur as a side reaction in the reduction of

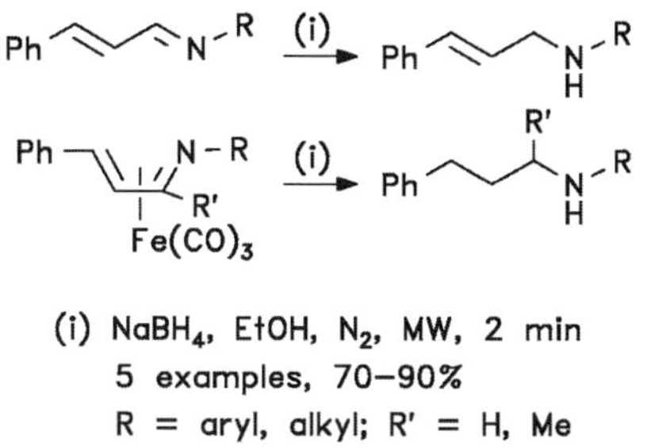

(i) NaBH$_4$, EtOH, N$_2$, MW, 2 min
5 examples, 70–90%
R = aryl, alkyl; R' = H, Me

Scheme 4.7

carbonyl groups with sodium borohydride, and mixtures of saturated and unsaturated alcohols may be obtained in some cases (Schemes 4.9 and 4.11)[26,27].

Microwave irradiation also shows a beneficial effect in the preparation of solid-supported palladium catalysts for hydrogenation reactions. Thus, alumina- and silica-supported palladium catalysts were synthesised by conventional and microwave heating, and their physical properties and catalytic activity in the hydrogenation of benzene were compared. The alumina-based system prepared under microwave conditions showed turnover numbers an order of magnitude higher than the conventionally prepared catalysts[28].

Hydroacylation of alkenes was achieved in the presence of Wilkinson's catalyst and microwave irradiation under solvent-free conditions. As an example, benzaldehyde was reacted with dec-1-ene to give 1-phenylundecan-1-one in 83% yield within 30 min. Both domestic microwave ovens and single-mode reactors have been used for this reaction. The presence of an amine such as 2-amino-3-picoline or aniline and a carboxylic acid is crucial for the success of the reaction, showing that the formation of an imine plays an important role as an intermediate in the mechanism of this reaction[29].

The copper-catalysed hydrosilylation of vinylpyridines was significantly improved by microwave irradiation (Scheme 4.8). With 2-vinylpyridine, the reaction times were decreased by a factor of 360, and the yield of the product was enhanced from 5% under thermal conditions to 75% in the microwave-assisted reaction. Due to the much cleaner reaction under microwave conditions, the work-up procedure was considerably simplified[30].

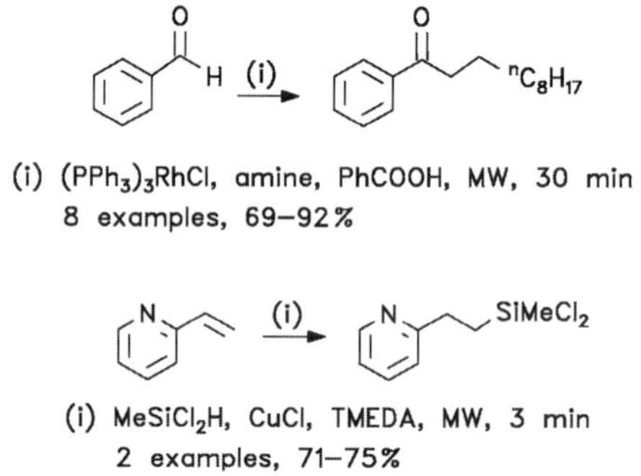

(i) $(PPh_3)_3RhCl$, amine, PhCOOH, MW, 30 min
8 examples, 69–92%

(i) $MeSiCl_2H$, CuCl, TMEDA, MW, 3 min
2 examples, 71–75%

Scheme 4.8

4.3. Reduction of carbonyl groups

Alcohols are easily accessible by reduction of carbonyl compounds, such as aldehydes, ketones or carboxylic acid derivatives. While aldehydes, ketones and esters have been frequently used in microwave-assisted reductions, there have been no reports about the use of microwave technology in the reduction of nitriles or amides.

Borohydrides and reagents derived therefrom are preferred over the use of aluminium hydrides, for economic and safety considerations and the much higher ease to handle

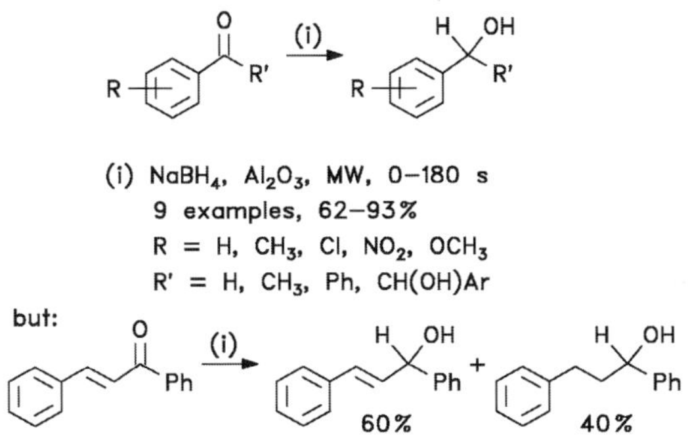

(i) NaBH$_4$, Al$_2$O$_3$, MW, 0–180 s
   9 examples, 62–93%
   R = H, CH$_3$, Cl, NO$_2$, OCH$_3$
   R' = H, CH$_3$, Ph, CH(OH)Ar

but:

60%   40%

Scheme 4.9

borohydrides. Other methods of carbonyl reduction use low-boiling alcohols or formates as a hydrogen source.

4.3.1. *Carbonyl reduction using borohydrides*

Varma reported a facile and rapid method for the reduction of aldehydes and ketones to
the respective alcohols, using alumina-supported sodium borohydride and microwave
irradiation under solvent-free conditions. Aldehydes tend to react at room temperature,
while for the reduction of ketones, short microwave irradiation of 30–180 s was applied
to produce the corresponding alcohols in 62–92% yield. With unsaturated carbonyl
compounds, reduction at the conjugated C=C bond might occur as a side reaction
under these conditions (Scheme 4.9)[26].

   Using a similar reaction protocol, β-trimethylsilyl carbonyl compounds were successfully reduced to their corresponding alcohols in short reaction times and good yields
(Scheme 4.10)[31].

   A study of the influence of the nature of the solid support showed that silica, celite,
cellulose or magnesium sulphate in combination with borohydride can also be used
successfully in the microwave-assisted reduction of carbonyl compounds. The choice of
the solid support has been reported to influence the chemoselectivity of the reduction
of chalcone. Under optimised conditions the reduction of the alkene can be suppressed
using borohydride on silica, whereas the use of cellulose as solid support seems to favour
C=C reduction (Scheme 4.11)[27].

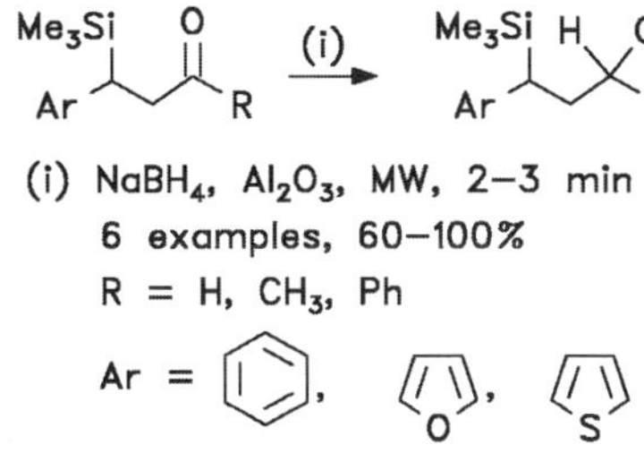

(i) NaBH$_4$, Al$_2$O$_3$, MW, 2–3 min
   6 examples, 60–100%
   R = H, CH$_3$, Ph

Ar =

Scheme 4.10

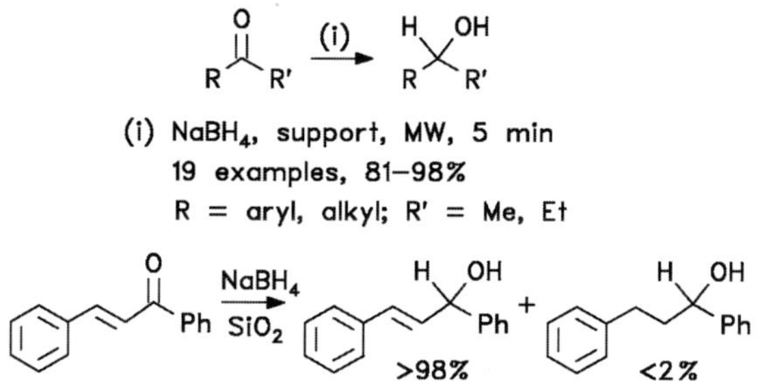

(i) NaBH$_4$, support, MW, 5 min
19 examples, 81–98%
R = aryl, alkyl; R' = Me, Et

Scheme 4.11

Application of alumina-supported sodium borodeuteride under the reaction conditions described by Varma[26] provided deuterated alcohols from aldehydes and ketones
with a high degree of deuterium incorporation. The method is thus suitable for isotopic
labelling procedures (Scheme 4.12)[32].

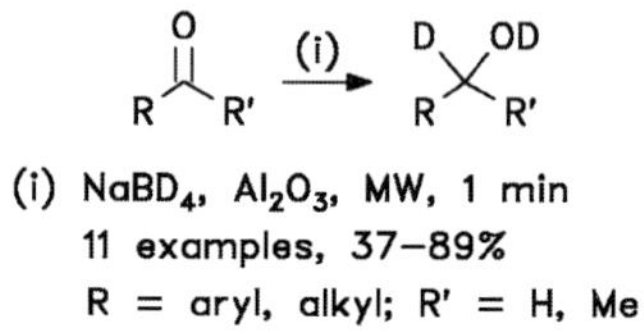

(i) NaBD$_4$, Al$_2$O$_3$, MW, 1 min
11 examples, 37–89%
R = aryl, alkyl; R' = H, Me

Scheme 4.12

Esters are more difficult to reduce, and usually, no reaction takes place with sodium
borohydride. However, the potassium borohydride/lithium chloride system was found
to reduce esters under microwave conditions in a solvent-free reaction[33]. The reactions
are generally completed in 2–8 min and provide the corresponding alcohols in 55–95%
yield (Scheme 4.13).

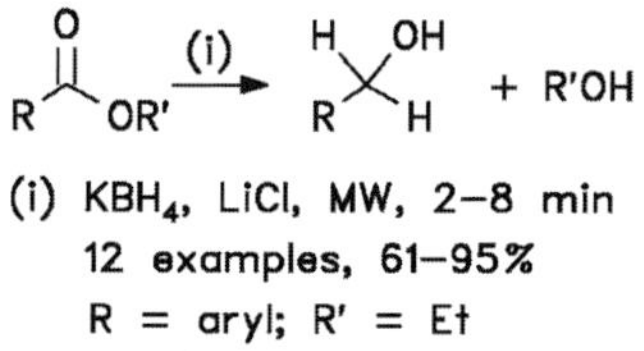

(i) KBH$_4$, LiCl, MW, 2–8 min
12 examples, 61–95%
R = aryl; R' = Et

Scheme 4.13

4.3.2. Carbonyl reduction under Meerwein–Ponndorf–Verley conditions

The Meerwein–Ponndorf–Verley reaction is a useful method for the reduction of carbonyl groups to alcohols. Most typically, aluminium isopropoxide is used as a reducing
agent. The acetone produced can be easily removed by distillation, thus driving the equilibrium reaction in the desired direction. When the carbonyl compound was refluxed

with aluminium isopropoxide in isopropanol with microwave heating, higher yields were obtained compared to the thermal reaction (Scheme 4.14). The effect was attributed to superheating under microwave irradiation leading to enhanced reaction rates[34].

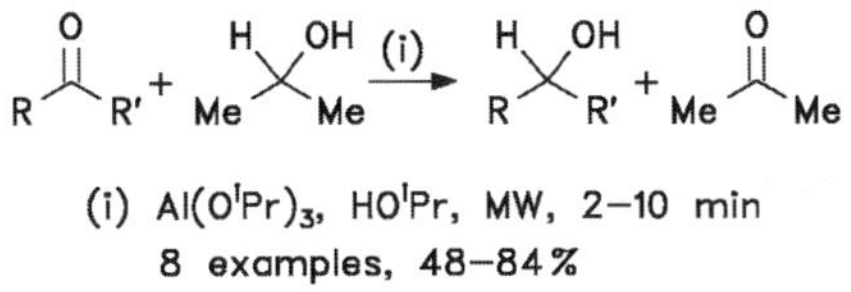

(i) Al(O^{i}Pr)$_3$, HO^{i}Pr, MW, 2–10 min
8 examples, 48–84%

Scheme 4.14

A further improvement of this reaction was achieved by substituting aluminium iso-propoxide by alumina-supported KOH (Scheme 4.15). This method allows for selective 1,2 reduction of α,β-unsaturated carbonyl compounds and the reduction of acyloins to their respective diols[35].

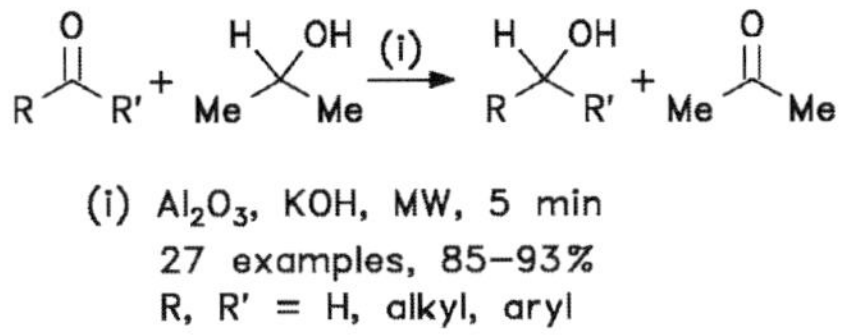

(i) Al$_2$O$_3$, KOH, MW, 5 min
27 examples, 85–93%
R, R' = H, alkyl, aryl

Scheme 4.15

4.3.3. Carbonyl reduction by transfer hydrogenation

(Carbonyl)chlorohydridotris(triphenylphosphine)ruthenium(II) was used as a catalyst in the transfer hydrogenation of benzaldehyde with formic acid as a hydrogen source. Under these conditions, the reduction of benzaldehyde to benzyl alcohol is accompanied by esterification of the alcohol with the excess of formic acid to provide benzyl formate (Scheme 4.16). In this microwave-assisted reaction, the catalyst displayed improved turnover rates compared to the thermal reaction (280 vs. 6700 turnovers/h), thus leading to shorter reaction times[36].

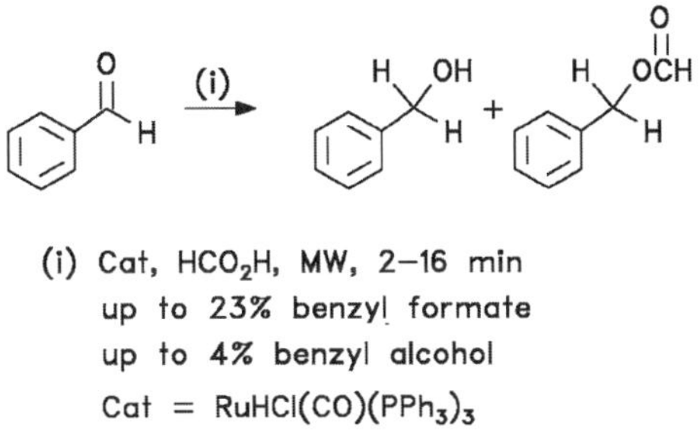

(i) Cat, HCO$_2$H, MW, 2–16 min
up to 23% benzyl formate
up to 4% benzyl alcohol
Cat = RuHCl(CO)(PPh$_3$)$_3$

Scheme 4.16

Enantioselective reduction of acetophenone was achieved in a ruthenium-catalysed hydrogen transfer reaction using isopropanol as the hydrogen source in the presence of mono-tosylated (R,R)-diphenylethylenediamine, ephedrine or norephedrine as chiral auxiliary ligands. Under optimised conditions, (R)-1-phenylethanol was obtained in 90% yield and 82% enantiomeric excess (ee) within 9 min. t-Butylphenylketone was reduced under similar conditions in almost quantitative yield but in moderate ee (Scheme 4.17)[37].

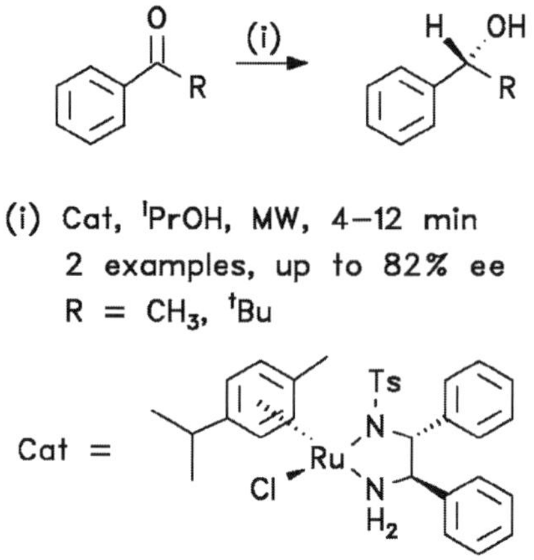

(i) Cat, ^{i}PrOH, MW, 4–12 min
     2 examples, up to 82% ee
     R = CH$_3$, ^{t}Bu

Cat =

Scheme 4.17

4.3.4. Carbonyl reduction by the Cannizzaro reaction

The Cannizzaro reaction, that is, the base-catalysed disproportionation of a carbonyl compound to an alcohol and a carboxylic acid, has gained some importance as an eco-nomically viable alternative to the reduction with borohydrides. However, the reaction is restricted to carbonyl compounds without any α-hydrogen, which do not undergo competing aldol reactions. Thus, mainly aromatic aldehydes are used for this kind of transformation. The protocols developed for microwave applications typically involve solvent-free conditions using alumina as the solid support. Under these conditions, a significant acceleration of the reaction was achieved.

Microwave irradiation of only 15 s was reported to give almost quantitative yields of both the alcohol and the carboxylic acid in the sodium hydroxide catalysed Canniz-zaro reaction of substituted benzaldehydes and other aromatic heterocyclic aldehydes (Scheme 4.18). Basic alumina was used as the solid support. The alcohol was selectively

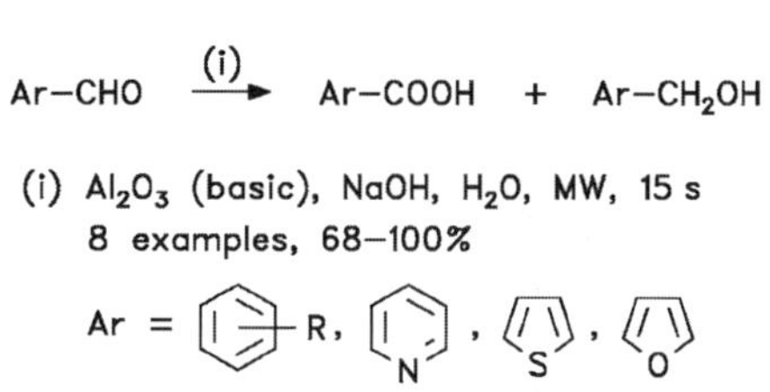

$$\text{Ar–CHO} \xrightarrow{\text{(i)}} \text{Ar–COOH} + \text{Ar–CH}_2\text{OH}$$

(i) Al$_2$O$_3$ (basic), NaOH, H$_2$O, MW, 15 s
     8 examples, 68–100%

Ar =

Scheme 4.18

extracted from the solid support with an organic solvent. The alcohol was selectively extracted from the solid support with an organic solvent. Subsequently, the carboxylic acid was released from the alumina with water and isolated after acidic work-up[38].

γ-Alumina was found to catalyse the Cannizzaro reaction of aromatic aldehydes even in the absence of any additional base to give the corresponding alcohols and carboxylic acids in high yield. The reaction was once more carried out under solvent-free conditions, and both the alcohol and the carboxylic acid were isolated by stepwise elution from the solid support as described earlier. With terephthalaldehyde, a selective Cannizzaro reaction at only one of the two reactive sites was achieved (Scheme 4.19)[39].

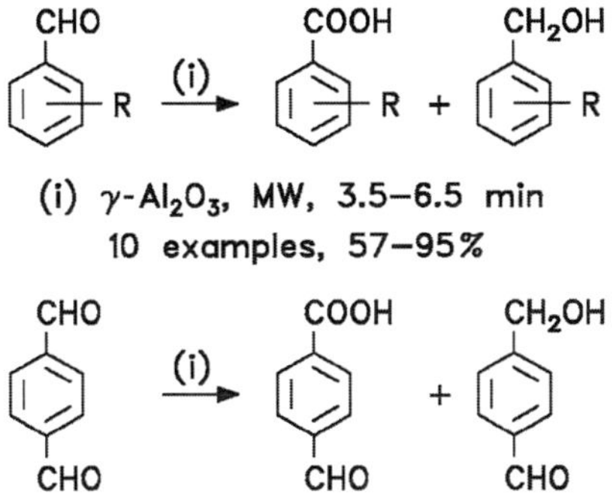

(i) γ-Al$_2$O$_3$, MW, 3.5–6.5 min
10 examples, 57–95%

Scheme 4.19

In an attempt to couple halobenzaldehydes with amines, Al$_2$O$_3$ was pre-absorbed with the substituted benzaldehydes and imidazole or piperidine as a base and irradiated with microwaves. However, the corresponding benzylic alcohols and benzoic acids were unexpectedly obtained by the Cannizzaro route (Scheme 4.20). The products of Cannizzaro reactions were also obtained as the main products, when microwave-assisted condensation reactions of benzaldehydes with vinyl acetate using barium hydroxide as the catalyst were attempted[40].

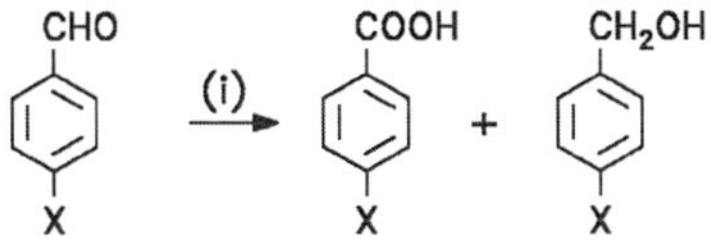

(i) Al$_2$O$_3$, imidazole or piperidine, MW, 5 min
11 examples, 80–95% (X = F, Cl, Br)

Scheme 4.20

The so-called 'crossed' Cannizzaro reaction is synthetically more useful than the Cannizzaro reaction itself, as it can be applied for the preparation of alcohols in high yields, without loss of 50% of the product in the formation of the corresponding carboxylic acid. Typically, paraformaldehyde is used as a sacrificial reducing agent, together with the carbonyl compound which is to be transformed into the alcohol. The reaction thus serves as an alternative method to the use of complex hydrides for the reduction of aromatic aldehydes.

Varma[41] was the first to explore the application of microwave irradiation to a 'crossed' Cannizzaro reaction (Scheme 4.21). A mixture of the aldehyde with 2 equiv. of paraformaldehyde and 2 equiv. of $Ba(OH)_2 \cdot 8H_2O$ was irradiated in a domestic microwave oven for 0.25–2 min at 900 W. Yields of the alcohols ranged from 80 to 99%, whereas the production of the corresponding carboxylic acid as the by-product could be suppressed to 1–20%. Under thermal conditions in an oil bath at 100–110°C, similar results were obtained although longer reaction times were required. The same reactions attempted with calcium hydroxide failed to provide the Cannizzaro products[41].

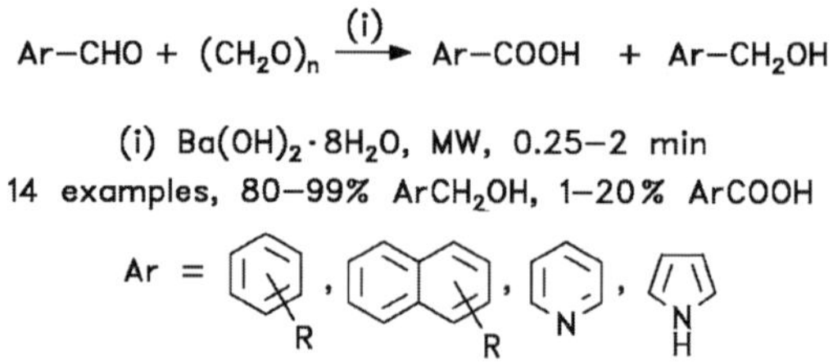

$$\text{Ar–CHO} + (\text{CH}_2\text{O})_n \xrightarrow{\text{(i)}} \text{Ar–COOH} + \text{Ar–CH}_2\text{OH}$$

(i) $Ba(OH)_2 \cdot 8H_2O$, MW, 0.25–2 min
14 examples, 80–99% $ArCH_2OH$, 1–20% $ArCOOH$

$$Ar = $$

Scheme 4.21

Aromatic aldehydes can also be converted selectively into the corresponding benzylic alcohols in a 'crossed' Cannizzaro reaction, if NaOH is used as a base. Similarly, the reaction is performed under solvent-free conditions by mixing the aldehyde with the base and an excess of paraformaldehyde and irradiating with microwave for 20–25 s. An alternative protocol uses 40% formalin solution and basic alumina to obtain comparable yields. The thermal reaction in refluxing methanol was found to require 12 h, providing considerably lower yields of the benzylic alcohols (Scheme 4.22)[42].

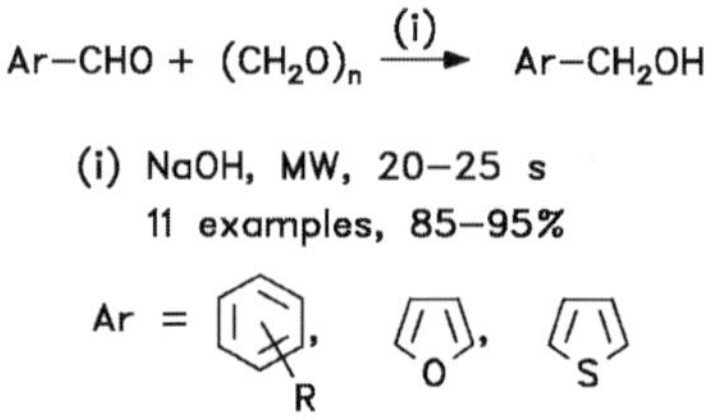

$$\text{Ar–CHO} + (\text{CH}_2\text{O})_n \xrightarrow{\text{(i)}} \text{Ar–CH}_2\text{OH}$$

(i) NaOH, MW, 20–25 s
11 examples, 85–95%

$$Ar = $$

Scheme 4.22

To avoid strongly basic conditions that might lead to the formation of by-products with sensitive substrates, a more recent report quotes the use of $KF\text{-}Al_2O_3$ in combination with paraformaldehyde for the 'crossed' Cannizzaro reaction of aromatic aldehydes. Similarly, microwave irradiation completed the solvent-free reaction within a few minutes, whereas the thermal reaction in dioxane required reaction times of up to 3 h (Scheme 4.23)[43].

4.3.5. Carbonyl reduction using other methods

The reductive coupling of carbonyl compounds to pinacols (i.e., 1,2-diols) is usually performed in the presence of a low-valent metal such as Li(0), Sm(II) or Ti(III). Under

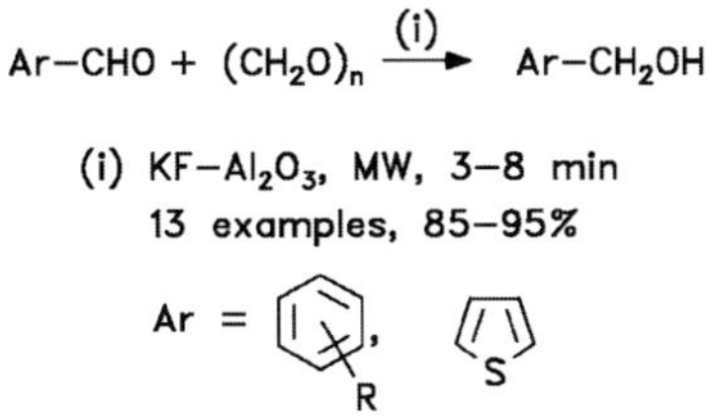

$$\text{Ar–CHO} + \text{(CH}_2\text{O)}_n \xrightarrow{\text{(i)}} \text{Ar–CH}_2\text{OH}$$

(i) KF–Al$_2$O$_3$, MW, 3–8 min
13 examples, 85–95%

Ar =

Scheme 4.23

microwave conditions, the solvent-free synthesis of bis(trimethylsilyl)pinacols from aldehydes or ketones and trimethylsilyl chloride was achieved using montmorillonite K10 clay as the solid support. The products were obtained as diastereomeric mixtures in 56–90% yield (Scheme 4.24)[44].

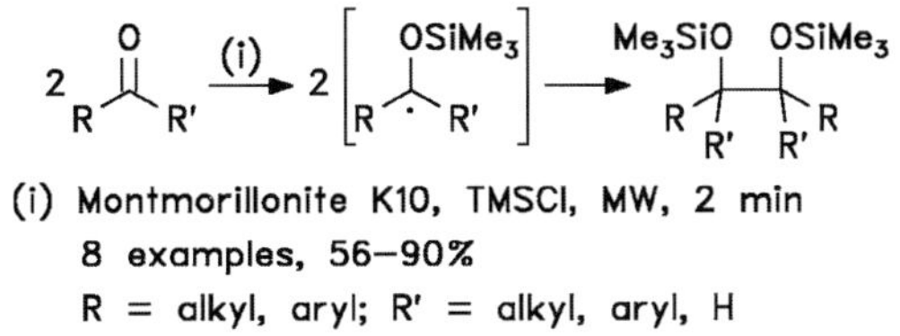

(i) Montmorillonite K10, TMSCl, MW, 2 min
8 examples, 56–90%
R = alkyl, aryl; R' = alkyl, aryl, H

Scheme 4.24

4.4. Reduction of nitrogen functional groups

Amines are a very important class of compounds in chemistry. A convenient method for their synthesis involves reduction of appropriate nitrogen functional groups, for example, imines, nitro groups or hydrazones. Surprisingly, no microwave-assisted reduction of azides has been reported in the literature, although azides have been used in cycloaddition reactions without any reported major hazards[45,46]. Also, to our knowledge, other nitrogen-containing functional groups such as nitroso compounds, oximes or hydroxylamines have not been involved in microwave-assisted reductions up to now.

The reduction of hydrazones can also be used as an indirect method of reducing, carbonyl compounds to the corresponding alkyl compounds. In addition, hydrazides have been used in the reductive synthesis of aldehydes.

4.4.1. Reduction of imines

4.4.1.1. Reduction of imines using borohydrides In analogy to the reduction of carbonyl groups, imines or enamines can be reduced with borohydrides to give the corresponding amines. The imino species is often synthesised *in situ* and subjected to reduction without isolation. A microwave-assisted two-step reaction for solvent-free reductive amination of carbonyl compounds using wet montmorillonite K10 clay supported sodium borohydride has been developed. Secondary and tertiary amines were thus obtained in high yield within a few minutes (Scheme 4.25)[47].

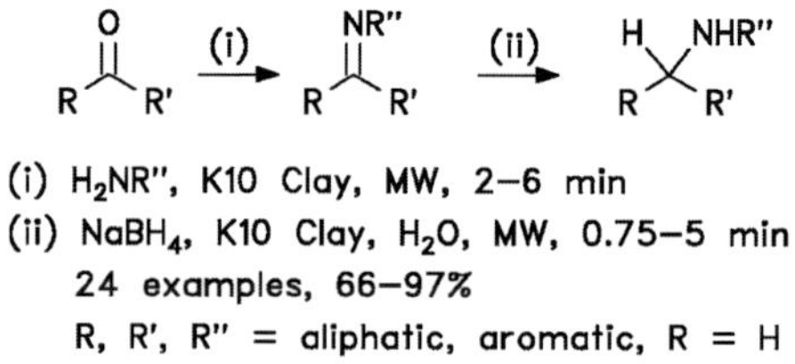

(i) H_2NR'', K10 Clay, MW, 2–6 min
(ii) $NaBH_4$, K10 Clay, H_2O, MW, 0.75–5 min
     24 examples, 66–97%
     R, R', R" = aliphatic, aromatic, R = H

Scheme 4.25

Microwave-induced imine formation, subsequent reduction with $NaBH(OAc)_3$ and cyclisation of the resulting amino acid with isothiocyanates was used in an efficient 'one-pot' multi-step synthesis of thiohydantoins (Scheme 4.26). The reductive amination was conducted as a two-step procedure to avoid direct reduction of the aldehyde at high temperatures[48].

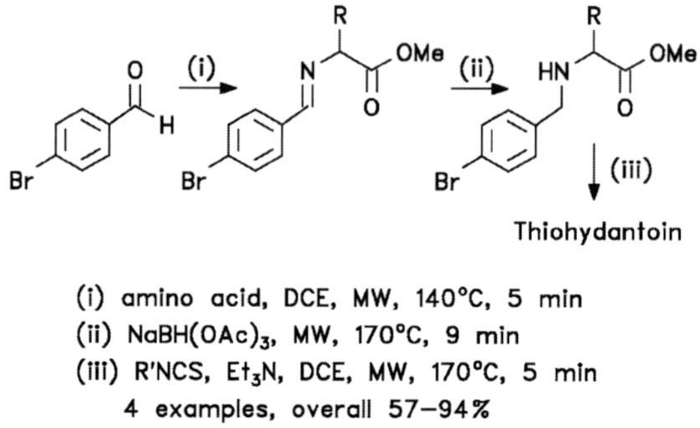

Thiohydantoin

(i) amino acid, DCE, MW, 140°C, 5 min
(ii) $NaBH(OAc)_3$, MW, 170°C, 9 min
(iii) R'NCS, Et_3N, DCE, MW, 170°C, 5 min
     4 examples, overall 57–94%

Scheme 4.26

Microwave-assisted stepwise imine formation and subsequent reduction with $NaBH_4$ were also used as the key steps in the synthesis of ephedrine from L-phenylacetylcarbinol. The reactions were performed on a multigram scale in a domestic microwave oven to provide the product in satisfactory yield within a total reaction time of 19 min (Scheme 4.27)[49].

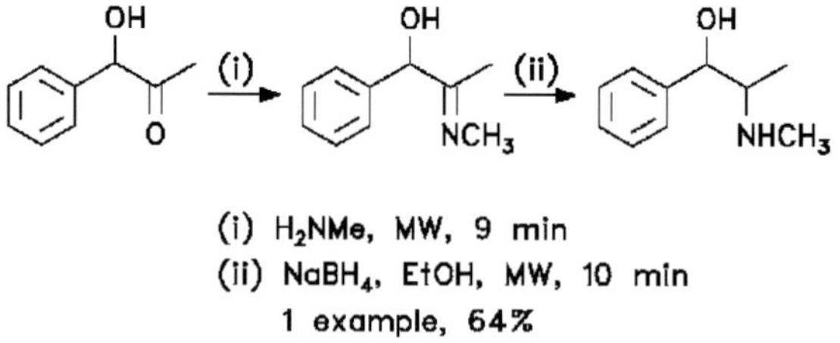

(i) H_2NMe, MW, 9 min
(ii) $NaBH_4$, EtOH, MW, 10 min
     1 example, 64%

Scheme 4.27

The microwave synthesis of optically pure imines for subsequent diastereoselective boronate reduction at room temperature has been described[50]. The reduction of (1-azabuta-1,3-diene)tricarbonyliron(0) complexes and their free ligands using sodium borohydride or borodeuteride under thermal or microwave conditions has been discussed in Section 4.2.2[25].

4.4.1.2. *Reduction of imines using formates*   Ammonium formates and formic acid have been employed as reducing agents in the synthesis of secondary amines from imines. By simple mixing of the reagents and microwave irradiation without solvent, the amines were produced in good yields within 2.5–10 min (Scheme 4.28)[51].

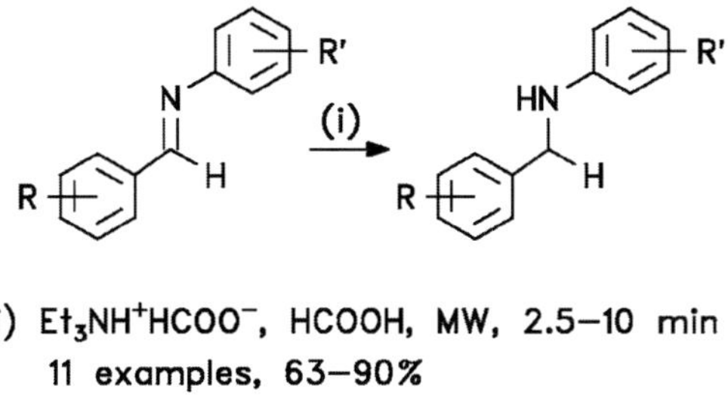

(i) Et$_3$NH$^+$HCOO$^-$, HCOOH, MW, 2.5–10 min
11 examples, 63–90%

Scheme 4.28

Formation of the imine and subsequent reduction can often be achieved in 'one pot'. Thus, a microwave-assisted reductive amination-cyclisation domino reaction was used as the key step in the synthesis of perhydrocyclo-penta[*ij*]quinolizines from 1,5,9-triketones. This type of heterocycle is an important structural element in a series of alkaloids. The reaction of the triketone with ammonium formate in PEG-200 was performed within 1 min using microwave irradiation of 370 W in a domestic microwave oven. A mixture of two of three possible stereoisomers was obtained in 87% overall yield (Scheme 4.29)[52].

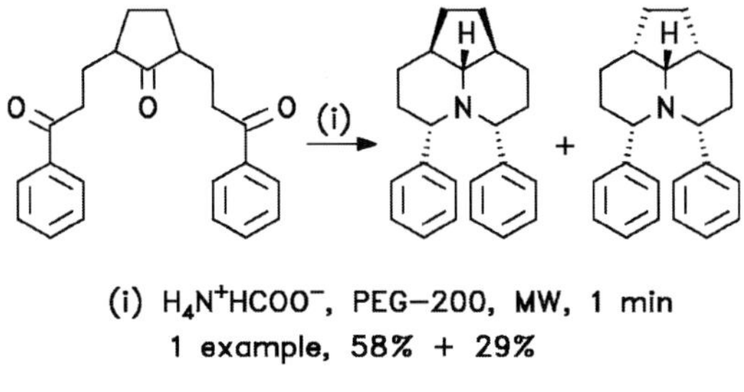

(i) H$_4$N$^+$HCOO$^-$, PEG–200, MW, 1 min
1 example, 58% + 29%

Scheme 4.29

A modified protocol of the Eschweiler–Clarke reaction, a reductive transamination, was also used for an efficient *N*-alkylation of hexahydroazepine and benzylamine in the presence of formic acid and aldehydes or ketones[53].

The Leuckart reductive amination of carbonyl compounds with ammonium formate or formamide was found to benefit strongly, when the reaction was carried out under solvent free conditions with microwave irradiation. Yields of *N*-alkylated formamides of up to 97% were produced in reaction times of about 30 min, as compared to thermal conditions where temperatures of about 240°C were used and low yields were obtained (Scheme 4.30)[54].

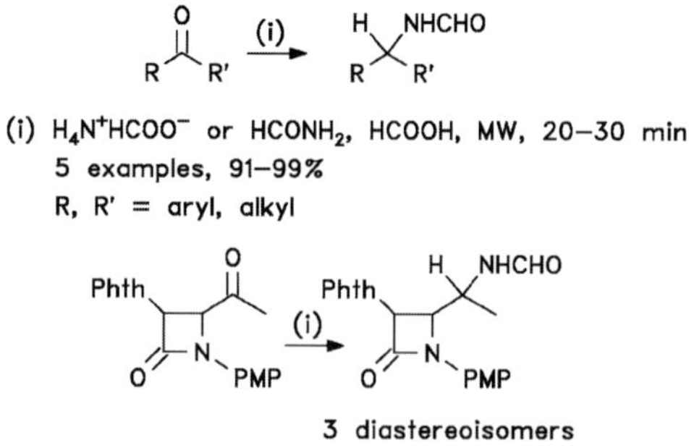

(i) $H_4N^+HCOO^-$ or $HCONH_2$, HCOOH, MW, 20–30 min
5 examples, 91–99%
R, R' = aryl, alkyl

3 diastereoisomers

Scheme 4.30

The same methodology was applied in the reaction of a 4-acetyl β-lactam with a mixture of formamide and formic acid as the aminoformylating agent, to give the diastereomeric products within 20 min in 73% yield (Scheme 4.30)[55].

4.4.1.3. Reduction of imines using other reducing agents   The syntheses of homoallylic hydroxylamines and homoallylic hydrazides were achieved by reductive coupling of aldonitrones or hydrazones with allyl bromide. The microwave-assisted reaction with gallium or bismuth in the presence of 0.1 equiv. of NH_4Cl or Bu_4NBr is complete within 4–5 min, as compared to 6–12 h under classical conditions (Scheme 4.31). It is worth noting that the nitrone derived from 3-nitrobenzaldehyde was selectively allylated, without accompanying reduction of the nitro group by the low-valent metal[56].

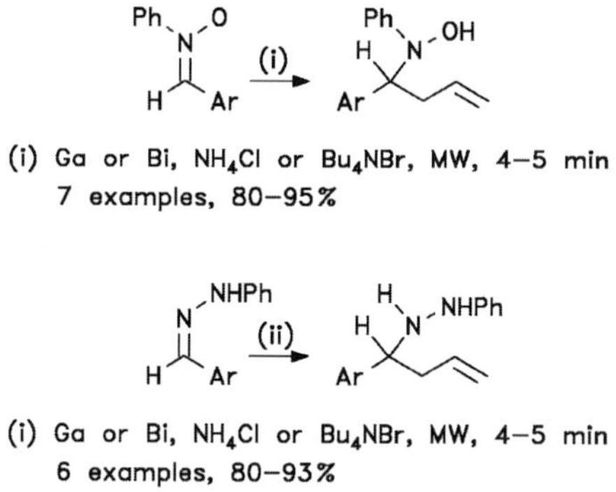

(i) Ga or Bi, NH_4Cl or Bu_4NBr, MW, 4–5 min
7 examples, 80–95%

(i) Ga or Bi, NH_4Cl or Bu_4NBr, MW, 4–5 min
6 examples, 80–93%

Scheme 4.31

4.4.2. Reduction of nitro groups

Aromatic nitro compounds are often straightforward to synthesise, and their reduction gives easy access to aromatic amines. A variety of reducing agents for conversion of nitro aromatics to amines have been used in microwave conditions, for example, metal powders (e.g., Zn, Fe, Sn) or other compounds that are easy to oxidise (e.g., $SnCl_2$, hydrazine hydrate, sulphide or hypophosphite). All reductions lead selectively to the amines; no method for a microwave-assisted selective synthesis of intermediate reduction products (e.g., hydroxylamines, oximes, nitroso compounds) has so far been reported.

Rapid reduction of aromatic nitro compounds into amines has been described using sodium hypophosphite and $FeSO_4 \cdot 7H_2O$. The reactions showed best results in terms of yields and purity, when the substrates were pre-absorbed on alumina and irradiated by microwaves under solvent-free conditions. The reaction is chemoselective and does not affect functional groups such as CN, OH, COOH, $CONH_2$ or halogens. In addition, oximes were not reduced under the given reaction conditions, but were dehydrated to the corresponding nitriles instead (Scheme 4.32)[57].

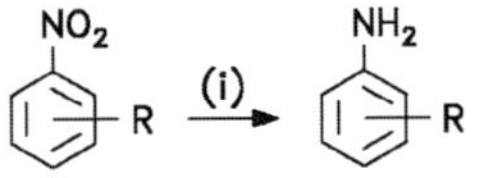

(i) NaH_2PO_2, $FeSO_4 \cdot 7H_2O$, MW, 50–100 s
   15 examples, 69–88%
   R = H, CH_3, OH, $CONH_2$, Ph, COOH, CN, NH_2

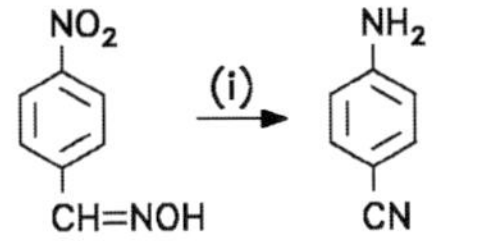

Scheme 4.32

A combination of iron(III) salts with solid-supported hydrazine hydrate (pre-absorbed on Al_2O_3) were successfully used in reductions of aromatic nitro groups (Scheme 4.33). The nature of the solid support plays a crucial role in this reaction and alumina was found to be most effective. In the absence of a solid support, the reaction

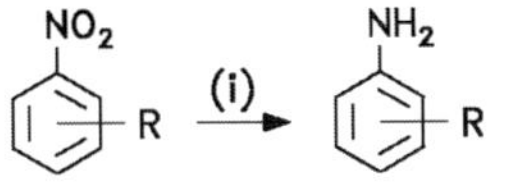

(i) $H_2NNH_2 \cdot H_2O$, $FeCl_3 \cdot 6H_2O$, Al_2O_3 MW, 6–10 min
   11 examples, 83–97%   (Ref. 58)

(i) NaHS, Al_2O_3, MW, 2–4 min
   8 examples, 70–95%   (Ref. 59)

(i) $SnCl_2 \cdot 2H_2O$, EtOH, MW
   8 examples, 55–99%   (Ref. 1)

(i) Sn powder, HCl, MW, 50 min
   1 example, 94%, R = 2-CH_3 (Ref. 60)

(i) Zn, NH_4Cl, MeOH, MW, 2 min   (Ref. 61)

(i) $R_4N^+HCOO^-$, Pd/C, $[BMIM]^+[PF_6]^-$, MW, 90 min
   3 examples, 67–92%   (Ref. 23)

Scheme 4.33

suffers from poor miscibility of the reagents and intensive gas formation that renders the reaction uncontrollable[58].

As an alternative to the above reaction conditions, aromatic nitro compounds can be reduced on Al_2O_3 support using sodium hydrogen sulphide as the reducing agent (Scheme 4.33). Reaction times of 2–4 min are required under microwave conditions (600 W), as compared to 10 h in the thermal reaction (no temperature given). The effect of electron-withdrawing and -donating substituents on the aromatic ring was studied and found to have little influence on the reaction rates or yields[59].

The microwave-assisted reduction of aromatic nitro compounds to amines has also been achieved with $SnCl_2·2H_2O$/EtOH as the reducing agent (Scheme 4.33). Alkenes and esters are not affected by this reagent[1].

The use of metal powders, especially when applied in high excess rather than as a catalyst, is often regarded incompatible with microwave heating because of the strong coupling of metals with microwaves and the risk of destructive arcing. This problem was addressed in the microwave-assisted reduction of an aromatic nitro compound with tin powder (Scheme 4.33). Although this example does not show any appreciable advantage over the thermal reaction, in terms of reaction rates or yields, the study shows that metal powders can be safely used in microwave reactions. However, appropriate conditions to prevent arcing (e.g., high pressure, use of high-boiling-point solvents, low-power irradiation) might be necessary[60].

As part of an efficient, facile and practical liquid-phase combinatorial synthesis of benzimidazoles under microwave irradiation, the reduction of a polymer bound *o*-nitroaniline with zinc metal in methanol was completed in a very short time (2 min)[61] (Scheme 4.33). However, no detailed reaction conditions were given.

Reduction of aromatic nitro compounds can also be achieved under transfer hydrogenation conditions. Nitrophenyl-dihydropyrimidones were thus reduced to the corresponding amines using ammonium formate and 5% Pd/C as a catalyst at 120°C (Scheme 4.34). A significant rise in pressure was observed during the reaction, apparently due to the production of a sufficient partial pressure of hydrogen for the reduction to occur. It is worthwhile highlighting that under these and even under more drastic reaction conditions, the C=C bond in the heterocycle is not hydrogenated, presumably due to steric hindrance and/or electronic effects[62].

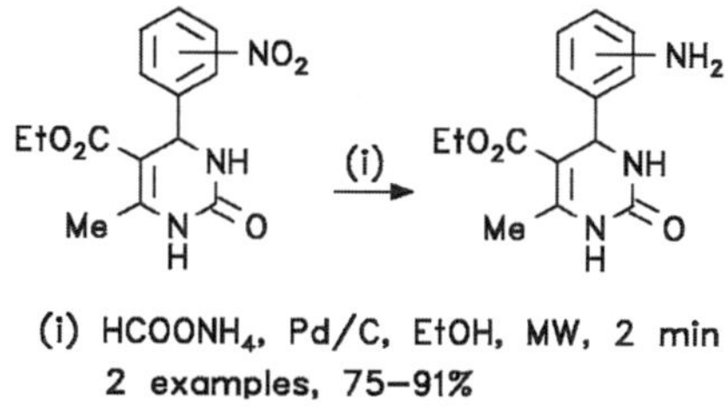

(i) HCOONH₄, Pd/C, EtOH, MW, 2 min
2 examples, 75–91%

Scheme 4.34

The same combination of reagents in an ionic liquid were reported to reduce nitro groups in the presence of nitriles or carbonyl-containing functional groups in high yields at temperatures of 150°C (Scheme 4.33)[23].

4.4.3. Reduction of hydrazones and hydrazides

Carbonyl groups can be transformed into amines *via* formation of their hydrazones followed by transfer hydrogenation. Thus the benzoyl formate shown in Scheme 4.35 was converted into the phenylhydrazone under microwave irradiation in ethylene glycol as solvent. Subsequent reduction using ammonium formate and 10% Pd/C as a catalyst provides the amine in an overall reaction time of 10 min and a total yield of 83%.[14]

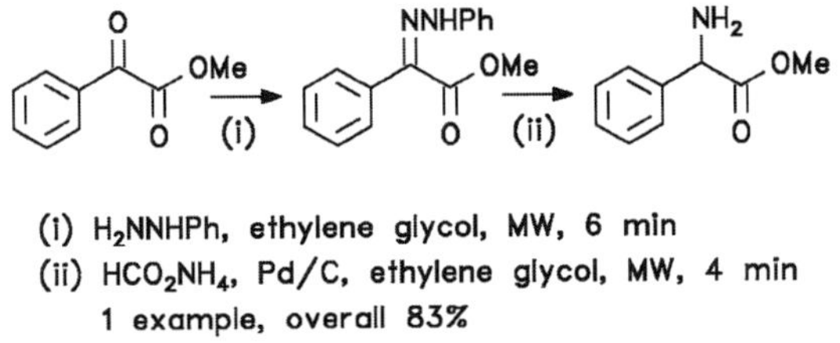

(i) H₂NNHPh, ethylene glycol, MW, 6 min
(ii) HCO₂NH₄, Pd/C, ethylene glycol, MW, 4 min
1 example, overall 83%

Scheme 4.35

The Wolff–Kishner reaction is commonly used for the reduction of carbonyl groups to the corresponding hydrocarbons *via* the hydrazones. Conventional methods usually require harsh conditions, for example, temperatures up to 200°C, reaction times in a range of several hours and the use of strong bases in high concentrations. Application of microwave irradiation to achieve this reaction under milder conditions would appear to be an obvious solution. Although initial attempts used very cautious conditions because of the assumed risks connected with the use of a potential explosive such as hydrazine. However, none of the reports cited in this section refer to any hazards, and the first report that appeared in the literature even recommends the reaction as a student's experiment for undergraduate teaching.

The first microwave-assisted Wolff–Kishner reduction was described by Parquet and Lin in 1997[63]. The transformation of isatin to oxindole was performed on a small scale in a domestic microwave oven in two steps with a total reaction time of 40 s, as compared to 3–4 h if classical heating was utilised (Scheme 4.36). The first step involved the transformation of the carbonyl group into the hydrazone with 55% hydrazine in ethylene glycol and medium power microwave irradiation for 30 s. In the subsequent reduction step, KOH in ethylene glycol was used to substitute the more hazardous sodium ethoxide. The reaction mixture was irradiated for 10 s and the product was obtained in a yield of 32%.

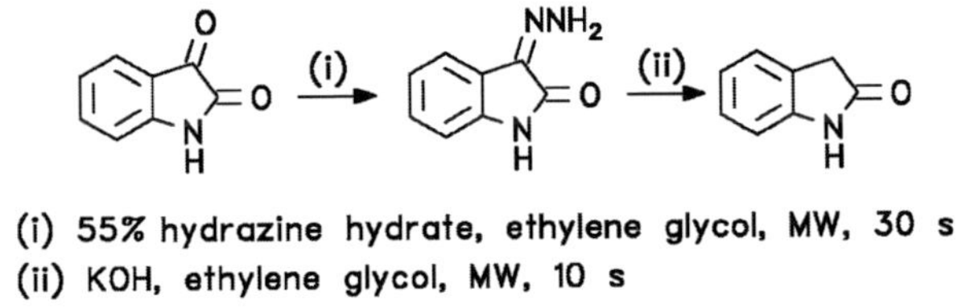

(i) 55% hydrazine hydrate, ethylene glycol, MW, 30 s
(ii) KOH, ethylene glycol, MW, 10 s

Scheme 4.36

Compared to conventional methods, the microwave-assisted approach has a major advantage in requiring a very short reaction time. The method is technically simple and allows for the use of less hazardous reagents (e.g., KOH substituting the conventionally used NaOEt). Still, the reaction lacks from a drawback of being performed in a two-step sequence, with necessary isolation of the intermediate hydrazone.

In a more general approach, eight examples of the Wolff–Kishner reduction of aromatic aldehydes and ketones are described using 80% hydrazine hydrate in toluene[64] (Scheme 4.37). The reaction times are longer than described in the previous paper because less reactive substrates were used. Still, both the formation of the hydrazone and the reduction step are considerably faster than under thermal conditions; the reduction proceeds at ambient pressure and in the absence of a solvent. The microwave reduction is compatible with other reducible functional groups such as aromatic OMe, Me, Cl or COOMe, which can otherwise cause problems under conventional reaction conditions[64].

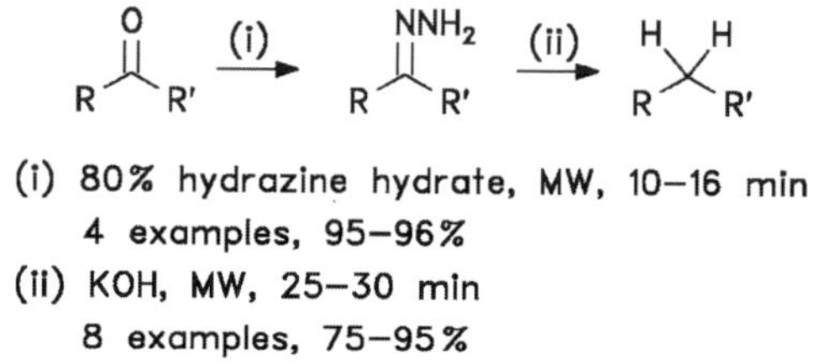

(i) 80% hydrazine hydrate, MW, 10–16 min
    4 examples, 95–96%
(ii) KOH, MW, 25–30 min
    8 examples, 75–95%

Scheme 4.37

In the same study, a 'one-pot' variant avoiding the isolation of the intermediate hydrazone was attempted. However, reduction of the crude hydrazone leads to the formation of by-products; for example, in the reaction of benzophenone, a mixture of diphenylmethane and benzophenone azine was found (Scheme 4.38)[64].

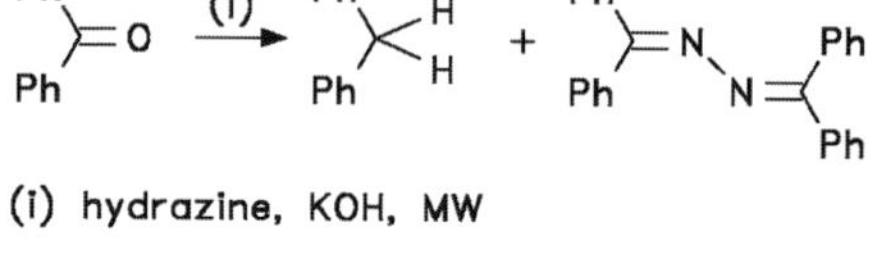

(i) hydrazine, KOH, MW

Scheme 4.38

The first report on a successful microwave-assisted one-step reduction of ketones to their respective hydrocarbons *via* the hydrazones appeared in 2002[65]. This so called Huang–Minlon variant of the Wolff–Kishner reduction was successfully applied to some aromatic and aliphatic aldehydes and ketones, including intermediates in the synthesis of the alkaloid flavopereirine. The reactions were performed by mixing the carbonyl compound with 2 equiv of hydrazine hydrate and an excess of powdered KOH in a commercial microwave oven. The mixtures were irradiated at 150 W for a few minutes before 250–350 W irradiations were applied (Scheme 4.39). The reaction was shown to be of limited use in the case of 3-acetyl pyrrole. This compound is deacylated under Huang–Minlon conditions, irrespective of whether conventional heating or microwave irradiation is applied.

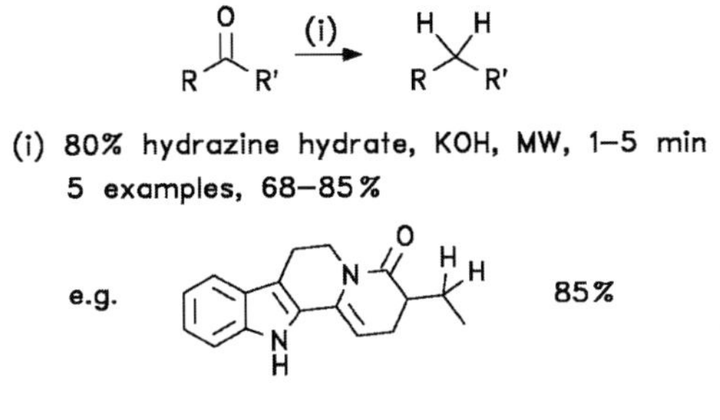

(i) 80% hydrazine hydrate, KOH, MW, 1–5 min
    5 examples, 68–85%

e.g.   85%

Scheme 4.39

A further extension of the Huang–Minlon modification of the Wolff–Kishner reduction using similar conditions has also been reported[66].

In the McFadyen–Stevens reaction, microwave irradiation has also been employed to convert carboxylic acids to aldehydes *via* the *p*-toluenesulphonyl hydrazides[65]. The *p*-toluenesulphonyl hydrazide is mixed with sodium carbonate, glass powder and ethylene glycol, and pre-irradiated at 150 W for a few minutes before the actual reaction conditions were applied. The reaction was successfully used in the synthesis of the alkaloid Nauclefidine (Scheme 4.40).

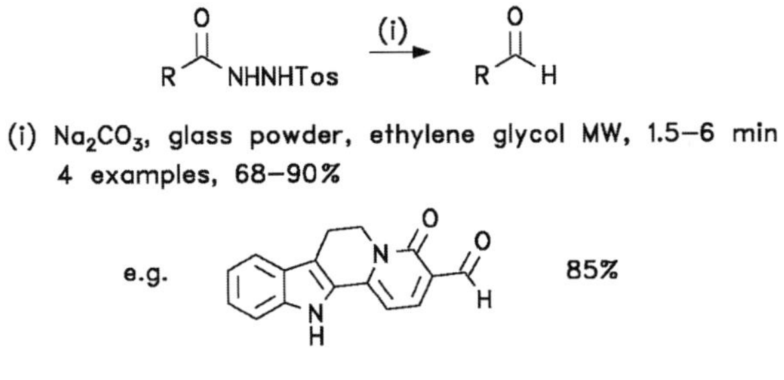

(i) Na$_2$CO$_3$, glass powder, ethylene glycol MW, 1.5–6 min
    4 examples, 68–90%

e.g.   85%

Scheme 4.40

4.5. Hydrodehalogenation

The dehalogenation of aromatic compounds has become an increasingly important method in organic chemistry, as it allows for the use of halogens as a sort of protecting group of reactive sites in the aromatic ring or allows for specific isotopic labelling of aromatic compounds. A further promising application is in the detoxification of chlorinated organic compounds or their transformation into value-added products.

Bose reported the dehalogenation of bromoanthracene, bromonaphthalene and several bromobenzenes under microwave-assisted hydrogen transfer conditions (ammonium formate, 10% Pd/C, ethylene glycol). An application of this reaction is in the synthesis of several β-lactams and isoquinoline derivatives[14].

Aryl halides can be dehalogenated with triethylsilane in the presence of a palladium catalyst. The method is versatile and can also be used for the reduction of acyl chlorides to aldehydes, or benzylic bromides to the corresponding hydrocarbons. If different types of halides are present in the molecule, selective dehalogenation takes place. Thus, an aryl iodide can be reduced in the presence of a chloride, and benzylic bromide is reduced more easily than an aryl bromide. Finally, the method is even able to distinguish between two aryl bromides in the same molecule (Scheme 4.41)[67].

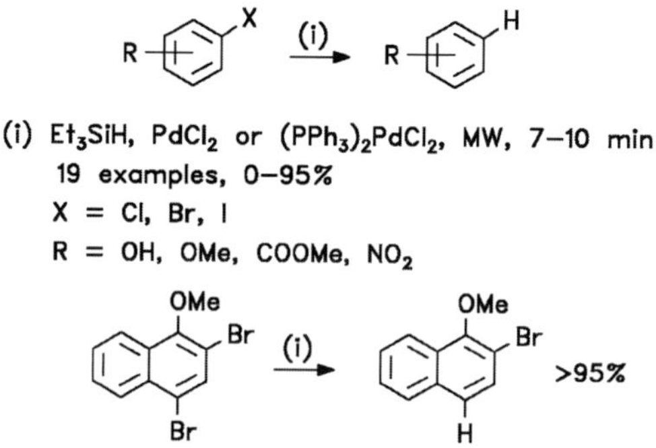

(i) Et$_3$SiH, PdCl$_2$ or (PPh$_3$)$_2$PdCl$_2$, MW, 7–10 min
19 examples, 0–95%
X = Cl, Br, I
R = OH, OMe, COOMe, NO$_2$

>95%

Scheme 4.41

Tin hydrides bearing highly fluorinated substituents (fluorous chemistry) were used in the radical-mediated reduction of 1-bromoadamantane. The reaction was complete within 3 min under 35 W microwave irradiation (Scheme 4.42)[68].

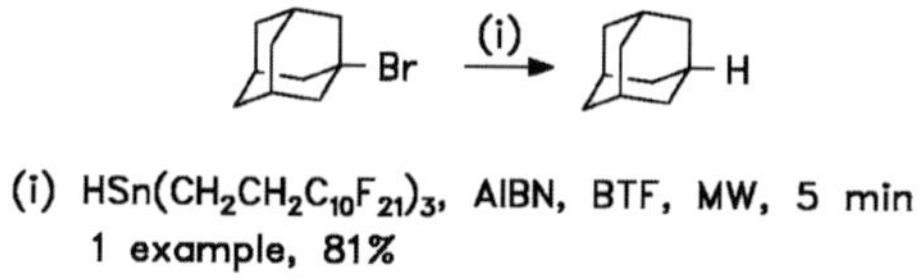

(i) HSn(CH$_2$CH$_2$C$_{10}$F$_{21}$)$_3$, AIBN, BTF, MW, 5 min
1 example, 81%

Scheme 4.42

A sodium hydroxide/polyethylene glycol system was used in the reductive decyanation of diphenylacetonitriles to give diphenylmethanes in high yield. Thermal reaction times of 60 min could be reduced to 2 min under microwave irradiation (Scheme 4.43)[69].

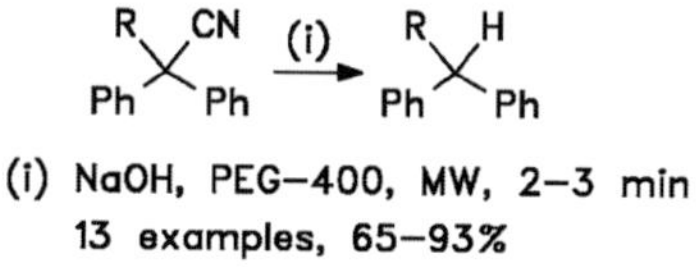

(i) NaOH, PEG–400, MW, 2–3 min
13 examples, 65–93%

Scheme 4.43

Rapid and specific isotopic labelling of aromatic compounds was achieved through microwave-assisted dehalogenation with deuterated formates and Pd/C or Pd(OAc)$_2$ as catalysts (Scheme 4.44)[70]. A combination of catalytic transfer hydrogenation with an aromatic dehalogenation was used to prepare several deuterated compounds from p-bromocinnamic acid. The reactions required only 40–90 s under microwave irradiation, as compared to 2 h under thermal conditions. The choice of the catalyst has a crucial effect on the outcome of the reaction. Changing from RhCl$_3$ to Pd(OAc)$_2$ switches the selectivity from hydrogenation of the C=C bond to dehalogenation. Use of the combined catalyst system and microwave irradiation leads to simultaneous hydrogenation and dehalogenation (Scheme 4.44)[20].

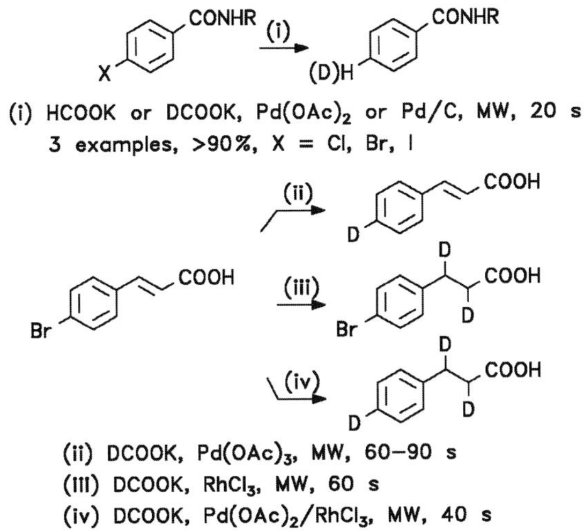

(i) HCOOK or DCOOK, Pd(OAc)$_2$ or Pd/C, MW, 20 s
3 examples, >90%, X = Cl, Br, I

(ii) DCOOK, Pd(OAc)$_3$, MW, 60–90 s
(iii) DCOOK, RhCl$_3$, MW, 60 s
(iv) DCOOK, Pd(OAc)$_2$/RhCl$_3$, MW, 40 s

Scheme 4.44

Reductive dehalogenation of chlorinated phenols to phenol, cyclohexanol and other chlorine-free compounds takes place rapidly with hydrogen gas and Pd/C in an aqueous system or under solvent-free conditions. Thus, pentachloro phenol was able to be completely dechlorinated within 20 min (Scheme 4.45). This methodology enables a facile route for rapid and complete detoxification of highly toxic polychlorinated aromatic hydrocarbons and environmental remediation[71,72].

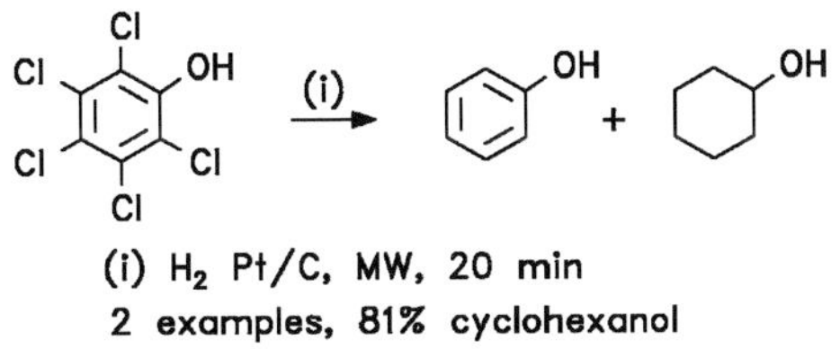

(i) H$_2$ Pt/C, MW, 20 min
2 examples, 81% cyclohexanol

Scheme 4.45

Microwave heating was also successfully used in the preparation of more efficient catalysts for hydrodehalogenation reactions, although in these cases, the actual dehalogenations with hydrogen gas were performed in a flow reactor under thermal conditions.

Silica- or alumina-supported Pd-Fe mono- or bimetallic dehalogenation catalysts were prepared by conventional and microwave-assisted calcination. The catalysts synthesised under microwave conditions gave a higher conversion in the dechlorination of chlorobenzene. The effects were attributed to an enhanced crystallite size, a lower susceptibility to alloy formation and differences in the Pd morphology[73,74].

The catalytic hydrodechlorination of chlorobenzene has also been studied on Pd/Nb$_2$O$_5$ catalysts. Also in this case, the catalysts prepared under microwave irradiation are more resistant towards deactivation than the corresponding catalysts prepared under thermal conditions[75].

Similarly, a Pd/Al$_2$O$_3$ catalyst system for the hydrodechlorination of CCl$_2$F$_2$ to CH$_2$F$_2$ was developed. Microwave irradiation produced the catalyst in a much shorter reaction times, and a twofold increase in catalytic activity was observed[76].

4.6. Conclusions

Microwave irradiation has been demonstrated to facilitate the reduction of many useful functional groups in organic chemistry. In many cases, mild reagents and reaction conditions have been used, increasing the potential for the technique to be applied to the reduction of sensitive substrates. In other examples, the use of supported reagents has allowed the development of clean and environmentally acceptable processes where the need for solvent and long reaction times has been considerably reduced. Additionally, technically simpler and less time consuming work-up procedures can often be used due to higher yields of products and reduced amounts of side products. This also makes many of the microwave-assisted reduction schemes well suited for incorporation into multi-step reaction sequences without the need for intermediate purification, or in combinatorial synthesis for high-throughput drug discovery.

It is notable that so far the reduction of only a relatively small number of organic functional groups has been reported. Thus, this area has still a high potential for further research and development.

4.7. References

1. Lidström, P., Tierney, J., Wathey, B. and Westman, J., Microwave assisted organic synthesis—a review, *Tetrahedron*, 2001, **57**, 9225–9283.
2. Wathey, B., Tierney, J., Lidström, P. and Westman, J., The impact of microwave-assisted organic chemistry on drug discovery, *Drug Discov. Today*, 2002, **7**, 373–380.
3. Mingos, D.M.P. and Baghurst, D.R., Applications of microwave dielectric heating effects to synthetic problems in chemistry, *Chem. Soc. Rev.*, 1991, **20**, 1–47.
4. Perreux, L. and Loupy, A., A tentative rationalization of microwave effects in organic synthesis according to the reaction medium, and mechanistic considerations, *Tetrahedron*, 2001, **57**, 9199–9223.
5. Strauss, C.R. and Trainor, R.W., Developments in microwave-assisted organic chemistry, *Aust. J. Chem.*, 1995, **48**, 1665–1692.

6. Strauss, C.R., A combinatorial approach to the development of environmentally benign organic chemical preparations, *Aust. J. Chem.*, 1999, **52**, 83–96.
7. Bose, A.K. and Manhas, M.S., Banik, B.K. and Robb, E.W., Microwave-induced organic-reaction enhancement (MORE) chemistry – techniques for rapid, safe and inexpensive synthesis, *Res. Chem. Intermed.*, 1994, **20**, 1–11.
8. Loupy, A., Petit, A., Hamelin, J., Texier-Boullet, F., Jacquault, P. and Mathé, D., New solvent free organic synthesis using focused microwaves, *Synthesis*, 1998, 1213–1234.
9. Varma, R.S., Solvent-free organic synthesis, *Green Chem.*, 1999, **1**, 43–55.
10. Varma, R.S., Solvent-free accelerated organic syntheses using microwaves, *Pure Appl. Chem.*, 2001, **73**, 193–198.
11. Pillai, U.R., Sahle-Demessie, E. and Varma, R.S., Environmentally friendlier organic transformations on mineral supports under non-traditional conditions, *J. Mater. Chem.*, 2002, **12**, 3199–3207.
12. Banik, B.K., Manhas, M.S., Newaz, S.N. and Bose, A.K., Facile preparation of carbapenem synthons via microwave-induced rapid reaction, *Bioorg. Med. Chem. Lett.*, 1993, **3**, 2363–2368.
13. Bose, A.K., Banik, B.K., Barakat, K.J. and Manhas, M.S., Microwave-induced organic-reaction enhancement (more) chemistry. 5. Simplified rapid hydrogenation under microwave irradiation-selective transformations of beta-lactams, *Synlett*, 1993, 575–576.
14. Banik, B.K., Barakat, K.J., Wagle, D.R., Manhas, M.S. and Bose, A.K., Microwave-induced organic reaction enhancement (MORE) chemistry, Part 13. Microwave-assisted rapid and simplified hydrogenation, *J. Org. Chem.*, 1999, **64**, 5746–5753.
15. Danks, T.N., Microwave assisted synthesis of pyrroles, *Tetrahedron Lett.*, 1999, **40**, 3957–3960.
16. Rao, H.S.P. and Jothilingam, S., One-pot synthesis of pyrrole derivatives from (*E*)-1,4-diaryl-2-butene-1,4-diones, *Tetrahedron Lett.*, 2001, **42**, 6595–6597.
17. Dayal, B., Ertel, N.H., Rapole, K.R., Asgaonkar, A. and Salen, G., Rapid hydrogenation of unsaturated sterols and bile alcohols using microwaves, *Steroids*, 1997, **62**, 451–454.
18. Leskovšek, S., Šmidovnik, A. and Koloini, T., Kinetics of catalytic transfer hydrogenation of soybean oil in microwave and thermal field, *J. Org. Chem.*, 1994, **59**, 7433–7436.
19. Al-Qahtani, M.H., Cleator, N., Danks, T.N., Garman, R.N., Jones, J.R., Stefaniak, S., Morgan, A.D. and Simmonds, A.J., Microwave enhanced hydrogenation reactions using solid hydrogen, deuterium and tritium donors, *J. Chem. Res. (S)*, 1998, 400–401.
20. Chappelle, M.R., Kent, B.B., Jones, J.R., Lu, S.Y. and Morgan, A.D., Development of combined microwave-enhanced labelling procedures for maximising deuterium incorporation, *Tetrahedron Lett.*, 2002, **43**, 5117–5118.
21. Desai, B. and Danks, T.N., Thermal- and microwave-assisted hydrogenation of electron-deficient alkenes using a polymer-supported hydrogen donor, *Tetrahedron Lett.*, 2001, **42**, 5963–5965.
22. Danks, T.N. and Desai, B., Alumina-supported formate for the hydrogenation of alkenes, *Green Chem.*, 2002, **4**, 179–180.
23. Berthold, H., Schotten, T. and Hönig, H., Transfer hydrogenation in ionic liquids under microwave irradiation, *Synthesis*, 2002, 1607–1610.
24. Torchy, S., Cordonnier, G., Barbry, D. and Van den Eynde, J.J., Hydrogen transfer from Hantzsch 1,4-dihydropyridines to carbon–carbon double bonds under microwave irradiation, *Molecules*, 2002, **7**, 528–533.
25. Akisanya, J., Danks, T.N. and Garman, R.N., Reaction of (1-azabuta-1,3-diene)tricarbonyliron(0) complexes with sodium borohydride under microwave conditions, *J. Organomet. Chem.*, 2000, **603**, 240–243.
26. Varma, R.S. and Saini, R.K., Microwave-assisted reduction of carbonyl compounds in solid state using sodium borohydride supported on alumina, *Tetrahedron Lett.*, 1997, **38**, 4337–4338.
27. Chen, S.T., Yu, H.M., Chen, S.T. and Wang, K.T., Microwave-assisted solid reaction: reduction of ketones using sodium borohydride, *J. Chin. Chem. Soc.*, 1999, **46**, 509–511.
28. Prasad, P.S.S., Lingaiah, N., Rao, P.K., Berry, F.J. and Smart, L.E., The influence of microwave-heating on the morphology and benzene hydrogenation activity of alumina-supported and silica-supported palladium catalysts, *Catal. Lett.*, 1995, **35**, 345–351.
29. Loupy, A., Chatti, S., Delamare, S., Lee, D.Y., Chung, J.H. and Jun, G.H., Solvent-free chelation-assisted hydroacylation of olefin by rhodium(I) catalyst under microwave irradiation, *J. Chem. Soc., Perkin Trans. 1*, 2002, 1280–1285.
30. Abramovitch, R.A., Abramovitch, D.A., Iyanar, K. and Tamareselvy, K., Transfer hydrogenation, atom economy, activation, palladium, chemistry. Application of microwave-energy to organic-synthesis - improved technology, *Tetrahedron Lett.*, 1991, **32**, 5251–5254.
31. Zadmard, R., Saidi, M.R., Bolourtchian, M. and Nakhshab, L., Microwave-assisted reduction of beta-trimethylsilyl carbonyl compounds by sodium borohydride, *Phosphorous, Sulphur Silicon Relat. Elem.*, 1998, **143**, 63–66.
32. Erb, W.T., Jones, J.R. and Lu, S.Y., Microwave enhanced deuterations in the solid state using alumina doped sodium borodeuteride, *J. Chem. Res. (S)*, 1999, 728–729.

33. Feng, J.C., Liu, B., Dai, L., Yang, X.T. and Tu, S.J., Microwave assisted solid reaction: reduction of esters to alcohols by potassium borohydride-lithium chloride, *Synth. Commun.*, 2001, **31**, 1875–1877.
34. Barbry, D. and Torchy, S., Accelerated reduction of carbonyl compounds under microwave irradiation, *Tetrahedron Lett.*, 1997, **38**, 2959–2960.
35. Kazemi, F. and Kiasat, A.R., Reduction of carbonyl compounds to the corresponding alcohols with isopropanol on dehydrated alumina under microwave irradiation, *Synth. Commun.*, 2002, **32**, 2255–2260.
36. Gordon, E.M., Gaba, D.C., Jebber, K.A. and Zacharias, D.M., Catalytic transfer hydrogenation of benzaldehyde in a microwave-oven, *Organometallics*, 1993, **12**, 5020–5022.
37. Lutsenko, S. and Moberg, C., Microwave-mediated ruthenium-catalysed asymmetric hydrogen transfer, *Tetrahedron: Asymmetry*, 2001, **12**, 2529–2532.
38. Sharifi, A., Mojtahedi, M.M. and Saidi, M.R., Microwave irradiation techniques for the Cannizzaro reaction, *Tetrahedron Lett.*, 1999, **40**, 1179–1180.
39. Pourjavadi, A., Soleimanzadeh, B. and Malandi, G.B., Microwave-induced Cannizzaro reaction over neutral gamma-alumina as a polymeric catalyst, *React. Funct. Polymers*, 2002, **51**, 49–53.
40. Mečiarová, M., Poláčková, V. and Toma, Š., The effect of microwave and ultrasonic irradiation on the reactivity of benzaldehydes under Al_2O_3, $Ba(OH)_2$, and K_2CO_3 catalysis, *Chem. Pap. (Chemicke Zvesti)*, 2002, **56**, 208–213.
41. Varma, R.S., Naicker, K.P. and Liesen, P.J., Microwave-accelerated crossed Cannizzaro reaction using barium hydroxide under solvent-free conditions, *Tetrahedron Lett.*, 1998, **39**, 8437–8440.
42. Thakuria, J.A., Baruah, M. and Sandhu, J.S., Microwave induced an efficient synthesis of alcohols via cross-Cannizzaro reaction, *Chem. Lett.*, 1999, 995–996.
43. Reddy, B.V.S., Srinivas, R., Yadav, J.S. and Ramalingam, T., $KF-Al_2O_3$ mediated cross-Cannizzaro reaction under microwave irradiation, *Synth. Commun.*, 2002, **32**, 219–223.
44. Bolourtchian, M., Zadmard, R. and Saidi, M.R., Microwave promoted reductive coupling of carbonyl compounds to bis(trimethylsilyl) pinacols under solvent-free conditions, *Synth. Commun.*, 1998, **28**, 2017–2020.
45. Louërat, F., Bougrin, K., Loupy, A., Ochoa de Retana, A.M., Pagalday, J. and Palacios, F., Cycloaddition reactions of azidomethyl phosphonate with acetylenes and enamines. Synthesis of triazoles, *Heterocycles*, 1998, **48**, 161–170.
46. Alterman, M. and Hallberg, A., Fast microwave-assisted preparation of aryl and vinyl nitriles and the corresponding tetrazoles from organo-halides, *J. Org. Chem.*, 2000, **65**, 7984–7989.
47. Varma, R.S. and Dahiya, R., Sodium borohydride on wet clay: solvent-free reductive amination of carbonyl compounds using microwaves, *Tetrahedron*, 1998, **54**, 6293–6298.
48. Öhberg, L. and Westman, J., One-pot three-step solution phase syntheses of thiohydantoins using microwave heating, *Synlett*, 2001, 1893–1896.
49. Shukla, V.B., Madyar, V.R., Khadilkar, B.M. and Kulkarni, P.R., Biotransformation of benzaldehyde to L-phenylacetylcarbinol (L-PAC) by Torulaspora delbrueckii and conversion to ephedrine by microwave radiation, *J. Chem. Technol. Biotechnol.*, 2002, **77**, 137–140.
50. Ridha, T., Ben Hassine, B. and Genet, J.P., Synthesis of optically pure imines using microwave activation and application to the preparation of new secondary amines, *Compt. Rend. Acad. Sci. Ser. II Fas. C*, 2000, **3**, 35–42.
51. Moghaddam, F.M., Khakshoor, O. and Ghaffarzadeh, M., Microwave assisted reduction of Schiff bases by triethylammonium formate/formic acid system, *J. Chem. Res. (S)*, 2001, 525–527.
52. Rao, H.S.P., Jeyalakshmi, K. and Senthilkumar, S.P., Novel domino products from the reaction of phenyl vinyl ketone and its derivatives with cyclic ketones, *Tetrahedron*, 2002, **58**, 2189–2199.
53. Torchy, S. and Barbry, D., N-alkylation of amines under microwave irradiation: modified Eschweiler–Clarke reaction, *J. Chem. Res. (S)*, 2001, 292–293.
54. Loupy, A., Monteux, D., Petit, A., Aizpurua, J.M., Dominguez, E. and Palomo, C., Towards the rehabilitation of the Leuckart reductive amination reaction using microwave technology, *Tetrahedron Lett.*, 1996, **37**, 8177–8180.
55. Loupy, A., Monteux, D., Petit, A., Mérienne, C., Aizpurua, J.M. and Palomo, C., Leuckart reductive amination of a 4-acetylazetidinone using microwave technology, *J. Chem. Res. (S)*, 1998, 187–187.
56. Laskar, D.D., Gohain, M., Prajapati, D. and Sandhu, J.S., Microwave-induced organometallic reactions in aqueous media. Use of Ga and Bi for the allylation of aromatic *N*-oxides and hydrazones, *New J. Chem.*, 2002, **26**, 193–195.
57. Meshram, H.M., Ganesh, Y.S.S., Sekhar, K.C. and Yadav, J.S., Microwave thermolysis VII: selective diversity in the reduction using sodium hypophosphite under microwave irradiation, *Synlett*, 2000, 993–994.
58. Vass, A., Dudás, J., Tóth, J. and Varma, R.S., Solvent-free reduction of aromatic nitro compounds with alumina-supported hydrazine under microwave irradiation, *Tetrahedron Lett.*, 2001, **42**, 5347–5349.
59. Kanth, S.R., Reddy, G.V., Rao, V.V.V.N.S.R., Maitraie, D., Narsaiah, B. and Rao, P.S., A simple and convenient method for the reduction of nitroarenes, *Synth. Commun.*, 2002, **32**, 2849–2853.

60. Whittaker, A.G. and Mingos, D.M.P., Synthetic reactions using metal powders under microwave irradiation, *J. Chem. Soc., Dalton Trans.,* 2002, 3967–3970.
61. Bendale, P.M. and Sun, C.M., Rapid microwave-assisted liquid-phase combinatorial synthesis of 2-(arylamino)benzimidazoles, *J. Comb. Chem.,* 2002, **4**, 359–361.
62. Stiasni, N. and Kappe, C.O., A tandem intramolecular Michael-addition/elimination sequence in dihydropyrimidone to quinoline rearrangements, *ARKIVOC,* 2002, 71–79.
63. Parquet, E. and Lin, Q., Microwave-assisted Wolff–Kishner reduction reaction, *J. Chem. Educ.,* 1997, **74**, 1225–1225.
64. Gadhwal, S., Baruah, M. and Sandhu, J.S., Microwave induced synthesis of hydrazones and Wolff–Kishner reduction of carbonyl compounds, *Synlett,* 1999, 1573–1574.
65. Jaisankar, P., Pal, B. and Giri, V.S., Microwave assisted McFadyen–Stevens and Huang–Minlon reactions, *Synth. Commun.,* 2002, **32**, 2569–2573.
66. Chattopadhyay, S., Banerjee, S.K. and Mitra, A.K., The Huang–Minlon modification of Wolff–Kishner reduction in rapid and simple way using microwave technology, *J. Indian Chem. Soc.,* 2002, **79**, 906–907.
67. Villemin, D. and Nechab, B., Rapid and efficient palladium catalysed reduction of aryl halides by triethylsilane under microwave irradiation, *J. Chem. Res. (S),* 2000, 432–434.
68. Olofson, K., Kim, S.-Y., Larhed, M., Curran, D. and Hallberg, A., High-speed, highly fluorous organic reactions, *J. Org. Chem.,* 1999, **64**, 4539–4541.
69. Bendale, P.M., Chowdhury, B.R. and Khadilkar, B.M., Polyethylene glycol mediated reductive decyanation of diphenylacetonitrile moderately enhanced by microwave heating, *Indian J. Chem., Sect. B,* 2001, **40**, 433–435.
70. Jones, J.R., Lockley, W.J.S., Lu, S.Y. and Thompson, S.P., Microwave-enhanced aromatic dehalogenation studies: a rapid deuterium-labelling procedure, *Tetrahedron Lett.,* 2001, **42**, 331–332.
71. Wada, Y., Yin, H.B., Kitamura, T. and Yanagida, S., Microwave-assisted hydrogenation of chlorinated phenols for complete dechlorination, *Chem. Lett.,* 2000, 632–633.
72. Wada, Y., Yin, H.B. and Yanagida, S., Environmental remediation using catalysis driven under electromagnetic irradiation, *Catal. Surveys Jpn. ,*2002, **5**, 127–138.
73. Berry, F.J., Smart, L.E., Prasad, P.S.S., Lingaiah, N. and Rao, P.K., Microwave heating during catalyst preparation: influence on the hydrodechlorination activity of alumina-supported palladium-iron bimetallic catalysts, *Appl. Catal., A,* 2000, **204**, 191–201.
74. Lingaiah, N., Prasad, P.S.S., Rao, P.K., Berry, F.J. and Smart, L.E., Structure and activity of microwave irradiated silica supported Pd-Fe bimetallic catalysts in the hydrodechlorination of chlorobenzene, *Catal. Commun.,* 2002, **3**, 391–397.
75. Gopinatha, R., Rao, K.N., Prasad, P.P.S., Madhavendra, S.S., Narayanan, S. and Vivekanandan, G., Hydrodechlorination of chlorobenzene on Nb_2O_5-supported Pd catalysts influence of microwave irradiation during preparation on the stability of the catalyst, *J. Mol. Cat. A,* 2002, **181**, 215–220.
76. Prasad, P.S.S., Lingaiah, N., Chandrasekhar, S., Rao, K.S.R., Rao, P.K., Raghavan, K.V., Berry, F.J. and Smart, L.E., Microwave irradiation: an effective method for the preparation of low dispersed Pd/Al_2O_3 catalysts used in the hydrodechlorination of CCl_2F_2 to CH_2F_2, *Catal. Lett.,* 2000, **66**, 201–204.

5 Speed and efficiency in the production of diverse structures: microwave-assisted multi-component reactions

JACOB WESTMAN

5.1. Background

5.1.1. *Introduction*

During the last 10 years pharmaceutical companies have invested significant resources in developing robotics and miniaturisation for biological screening purposes. As a result of these investments, the capacity of biologists to perform *in vitro* high-throughput screening has dramatically improved. The main limitation of this new screening technology, except from only being a model of reality with all advantages and disadvantages that this implies, lies in the capacity of the chemists to furnish biologists with the proper molecules of adequate diversity and in adequate numbers.

Traditionally, drug discovery involved the optimisation of lead structures, most likely derived from biological sources, through a multi-step process of serial synthesis and screening. This approach is extremely costly, as each compound will have to be individually synthesised by a synthetic chemist. The need to find more cost effective methods of drug development has led pharmaceutical companies to examine combinatorial chemistry. Combinatorial chemistry is based on the simple principle that the more compounds there are and the greater the diversity of these compounds is, the better is the chance of finding one that can be developed into a drug. Unfortunately, because of a number of issues associated with combinatorial strategies, which are not discussed here, combinatorial chemistry has so far not fully lived up to its promises.

5.1.2. *Designing the method*

Since the introduction of microwave-assisted organic synthesis in 1986[1], the number of publications covering microwave chemistry has consistently increased, with over 1600 papers currently published and the number rapidly increasing. There are numerous advantages in employing microwave synthesis, for example, generally higher yields, environmentally benign methods and novel reaction routes. However, the main benefit is the dramatic reduction in reaction times.

The technology has been around for 16 years and has now established itself as an accepted method in organic synthesis, and therefore, it is now time to look at microwave synthesis in a broader perspective. Since the reaction time can be dramatically reduced by using microwave dielectric heating, the task to optimise the reaction conditions for a set of diverse substrates is now a rapid process. By adopting the advantages of microwave technology in all steps of the process of synthesising compounds, methods can be developed where both the preparation of intermediates, synthesis, work-up and need for purification can be taken into consideration when designing the method.

As a result, one can reduce the total time for compound production, as well as reduce the total cost per compound. The development of 'green chemistry' methods is also of great relevance and involves the use of benign reagents (environmentally friendly, low reactivity), low excess of reagents, non-hazardous solvents, etc. This strategy fits well with the microwave approaches where often milder reagents, smaller quantities and higher concentrations are employed. This furthermore reduces the costs for destruction of hazardous waste.

5.1.3. *Benefits with multi-component reactions*

A major challenge of modern drug discovery is the design of highly efficient chemical reaction sequences, which allows maximisation of structural complexity and diversity with just a minimum number of synthetic steps. Multi-component reactions are therefore becoming more and more popular in the area of combinatorial chemistry, since they provide the possibility to introduce a large degree of chemical diversity in only one reaction step[2]. The products of multi-component reactions are formed simply by mixing and processing the starting materials; thus, these reactions require a minimum of preparative work in their set-up. Multi-component reactions including as many as seven or more components in 'one-pot' have been described[3], but the most common number of components is three or four. Several of the well-known multi-component reactions have a long history, beginning with Strecker's synthesis of amino acids in 1850[4]. The multi-component reaction strategy has then been utilised successfully in reactions, such as Robinson's synthesis of the alkaloid tropinone and Hantzsch's synthesis of 1, 4-dihydropyridines. Many of the multi-component reactions described herein involve the reaction between an amine and aldehydes that *in situ* forms imines, which react with the next reagent in a subsequent step. These *one-pot multi-step* or *domino* reactions eliminate the need for work-up or purification between the different reaction steps, making the reaction more eco-friendly and faster since less chemical waste is produced by reducing the number of purification steps of intermediates. In the past decade, there has been tremendous development in three- and four-component reactions involving Passerini-, Ugi- and Mannich-type reactions and the number of multi-component reactions are rapidly increasing (Fig. 5.1).

5.1.4. *Multi-component versus one-pot synthesis*

Cascade reactions signify reactions involving two or more bond forming transformations that take place under unchanged reaction conditions and in which the subsequent reactions result from the functionality formed in the previous step[5].

This can be achieved either in a 'true' multi-component mode, where all the substrates and reagents are mixed together initially and subsequently processed, or in a 'one-pot' multi-step procedure, where one or several of the reaction components are added to the reaction at a later stage. Both approaches eliminate the need for work-up or purification between the different reaction steps making the reaction both more eco-friendly and faster.

It is sometimes hard to distinguish between a 'true' multi-component reaction and a multi-step 'one-pot' reaction; however, it might be important to the outcome of the

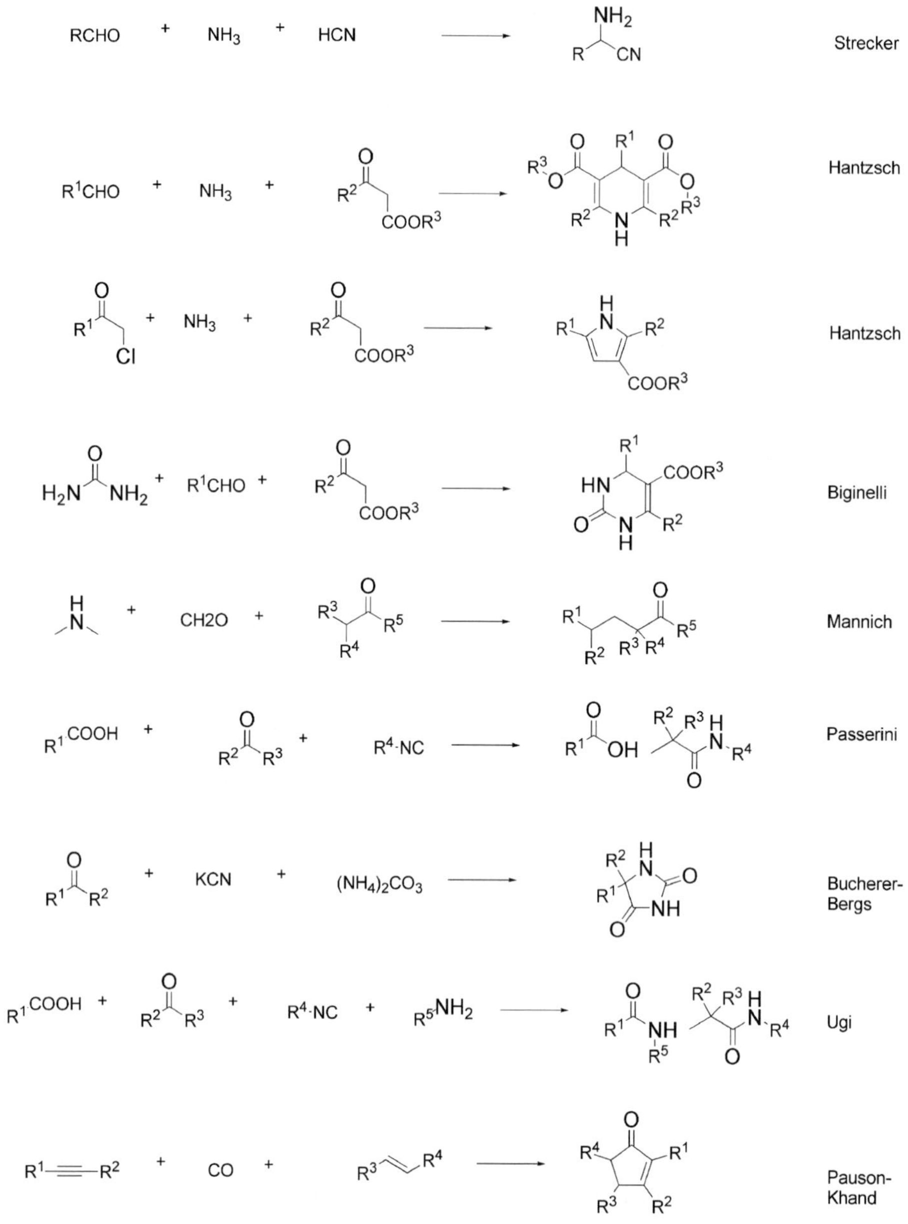

Figure 5.1 Examples of multi-component reactions.

reaction. In several cases, a typical multi-component reaction works much better in a multi-step 'one-pot' set-up or *vice versa*. In the next section, both 'true' multi-component reactions and two-step 'one-pot' reactions will be presented. However, to reduce the number of reactions presented, only 'one-pot' reactions that will work as 'true' multi-component reactions are presented.

5.2. Multi-component reactions

In this section, a number of interesting reactions in the area of multi-component reactions under microwave conditions are presented. The section is intended as an overview of the subject.

5.2.1. *Hantzsch reaction*

The classical Hantzsch reaction, the formation of dihydropyridines from an aldehyde, a β-keto ester and an amine, was first described in 1882[6]. In the 1940s, the interest for this substance class increased due to its pharmacological activity, for example, 4-aryl-1,4-dihydropyrdines form an important class of calcium channel antagonists such as Nifedipin.

The Hantzsch reaction has also been performed several times under microwave conditions. Alajarin and co-workers[7] presented the first example in 1992 followed by Khadilkar and co-workers[8] a few years later. Domestic microwave ovens were used in both examples. Recently Öhberg and Westman[9] reported a further refinement on the subject. By using a commercial microwave designed for organic synthesis, incorporating both temperature and pressure control, they were able to synthesise a small 24-compound library. The results illustrated that temperature control is essential for the development of a reproducible high-yielding protocol. In their study, they improved the yields by 20–40% compared to both conventional heating as well as the use of domestic microwave ovens. Water was used as the solvent instead of ethanol, which is commonly the preferred solvent, and a reaction time of 10–15 min at 140°C was employed (see Scheme 5.1). Six different aldehydes and four different β-keto esters or 1,3-dicarbonyl substrates were used for the library synthesis. On completion of the reactions, the samples were recrystallised from ethanol and water.

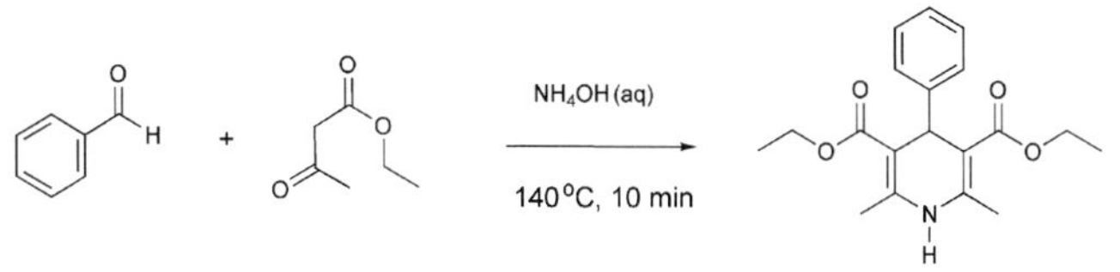

Scheme 5.1

N-Substituted 1,4-dihydropyridines without substituents in positions 2 and 6 also exhibit many pharmaceutical activities and are highly light sensitive in the solid state. It has been shown that these compounds could be dimerised and that the dimers are of interest both as potential inhibitors of HIV-1 protease and as anti-cancer agents[10]. For the preparation of 2,6-unsubstituted 1,4-dihydropyridines, propiolates are used instead of β-dicarbonyl compounds. Cyclocondensation of aldehydes, propiolates and ammonium acetate in refluxing acetic acid followed by N-acylation, or alternatively, cyclocondensation of propiolates, aromatic aldehydes and primary amines under reflux results in N-substituted 1,4-dihydropyridines. In 2001, Balalaie and co-workers[11] reported the latter variant of this reaction under microwave conditions in a domestic oven (Scheme 5.2). They also performed this acid-catalysed reaction under solvent-free conditions; they investigated the best solid support for this reaction using the condensation of benzaldehyde, ethyl propiolate and benzylamine as a model reaction. From studies employing silica gel, montmorillonite K10, acidic alumina and zeolite HY, silica gel was found to be the preferred solid support (62–90% yields with different substrates). Under neat conditions, in the absence of solid support, the reactants and products adhered to the vessel and led to irreproducible results. The cyclocondensation was also carried out in acetic acid alone, but the yields and reproducibility were poor and the work-up was difficult. In all experiments, the optimised time of irradiation was 4 min. This observation of low reproducibility is fairly common when using domestic ovens without temperature control under open vessel conditions. From our own experience (unpublished results), we have on the contrary found that acetic acid is often a very suitable solvent for acid-catalysed reactions. Balalaie and co-workers[11] claim that their methods are clean and environmentally friendly, but since the product was extracted three times with chloroform and filtered in order to separate the silica gel from the product prior to purification, significant amounts of solvents are still used. Hence, these reaction conditions and work-up cannot be considered to be truely eco-friendly.

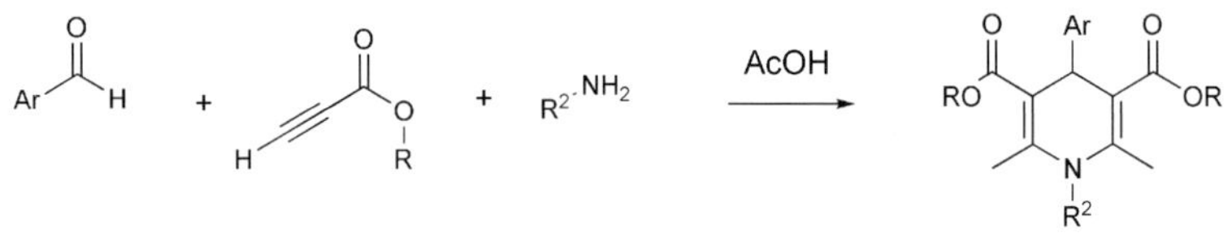

Scheme 5.2

In 1998, Cotterill and co-workers published one of the first papers describing the use of microwave heating in the context of combinatorial chemistry[12]. They performed a three-component Hantzsch synthesis of substituted pyridines in a 96-deep well plate format under solvent-free conditions. The reactions were performed on bentonite clay, using NH_4NO_3 as the amine source. NH_4NO_3 also acts as an *in situ* oxidising agent to oxidise the initially formed dihydropyridines to their pyridine counterparts (Scheme 5.3). The products were formed in over 70% purity but no yields were reported. Unfortunately, with the use of microtitre plates and domestic microwave ovens, the reaction temperature cannot be accurately controlled and the energy distribution over the microtitre plate will be uneven; this in turn results in different wells having different temperatures as described earlier by Combs and co-workers[13] (see also Chapter 8). Thus, the optimisation of the reaction conditions in microtitre plates will be quite problematic.

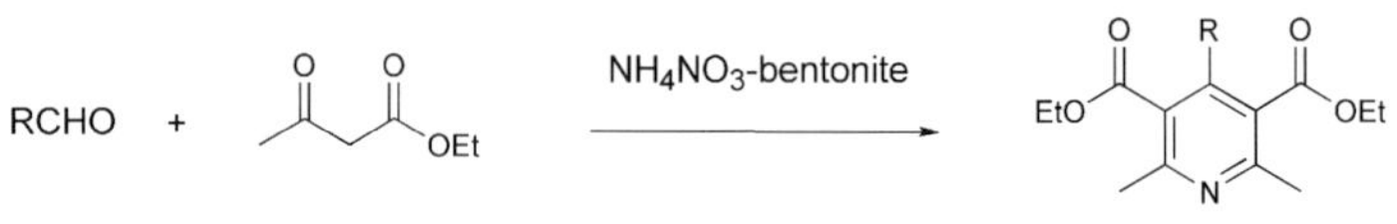

Scheme 5.3

5.2.2. Biginelli reaction

A reaction closely related to the Hantzsch reaction is the Biginelli three-component dihydropyrimidine synthesis (Scheme 5.4). The Biginelli reaction has been described several times under microwave conditions, both in solution and under solvent-free conditions[14,15]. However, as Stadler and co-workers[16] have reported, two issues that have so far been somewhat neglected in microwave-assisted processes are throughput and automation. They, therefore, performed the Biginelli reaction in an automated system that had a microwave cavity integrated with a liquid handler and a reaction vessel gripper allowing the reactions to be performed in a sequential continuous process. They reported a 48-member library synthesis in solution using acetic acid/ethanol as the main solvent system. By using a microwave-assisted Biginelli reaction, they achieved higher yields and an improved purity profile compared to conventional methods. The authors also reported that sequential processing presented distinct advantages as compared to its parallel counterpart. It allowed for better control of the reaction parameters, as well as rapid optimisation of reaction conditions. This allowed some reactions to be performed with alternative conditions during the production of the library, which cannot be carried out using a parallel approach.

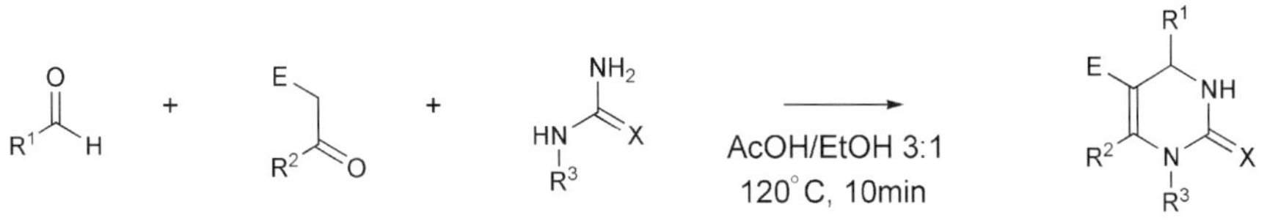

Scheme 5.4

5.2.3. Ugi reaction

The famous Ugi reaction, the 'one-pot' condensation of a carboxylic acid, an amine, an aldehyde or ketone and an isocyanide to yield an α-acylaminoamide, have recently been used as an efficient method for the synthesis of diverse libraries of small organic molecules such as benzodiazepines, pyrroles, lactams and diketopiperazines[2]. Even though some solution-phase Ugi reactions proceed rapidly, such reactions on solid phase have been found to take between one and several days. In 1999, Hoel and Nielsen[17] performed the first microwave assisted, solid phase Ugi four-component condensation (see Scheme 5.5).

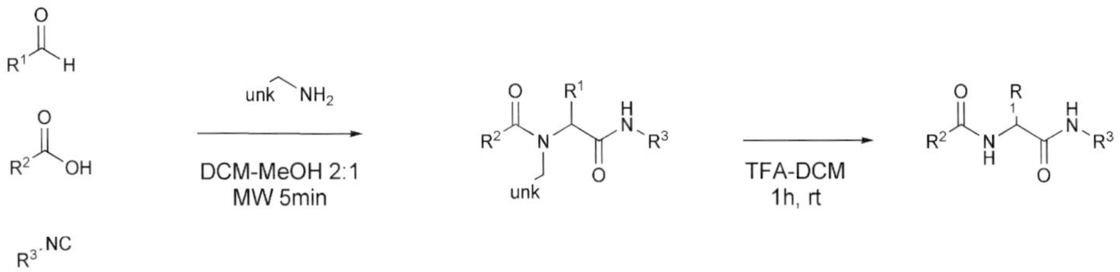

Scheme 5.5

They produced an 18-member library of α-acylamino amides, in acceptable to high yields and purity, from a variety of isocyanides, aldehydes and carboxylic acids by using an amino-functionalised TentaGel resin (TentaGel S RAM). The developed procedure represents a rapid and efficient way of synthesising α-acylamino amides, simplifying the tedious purifications, which can usually accompany multi-component reactions.

Isocyanides are ideal components for multi-component synthesis due to their unique reactivity. However, one drawback is the limited number of commercially available iso-cyanides. Furthermore, the syntheses of isocyanides are not always satisfactory and often use toxic reagents or generate malodorous products. The work-up and purifica-tion of isocyanides are also problematic due to their reactivity. Ley and co-workers[18] have therefore recently developed a new method for the synthesis of isocyanides from isothiocyanates, using polymer-supported reagents. The reagent, 3-methyl-2-phenyl-[1.3.2] oxaza-phopholidine, was first described by Mukaiyama in solution for the trans-formation of isothiocyanates to isonitriles, but despite the mild reaction conditions used, the method was not generally accepted due to the toxicity and instability of the reagent. The separation of the product was also complicated. Ley and co-workers therefore developed a solid-supported version of the reagent (Scheme 5.6). Merri-field resin was treated with 2-aminoethanol in the presence of excess potassium car-bonate, which yielded 2-(polystyrylmethylamino) ethanol (1). Resin 1 was then con-densed with bis(diethylamino)phenylphosine in toluene at 115°C for 10 h to give the 3-polystyrylmethyl-2-phenyl[1.3.2]oxazozphospholidine (2).

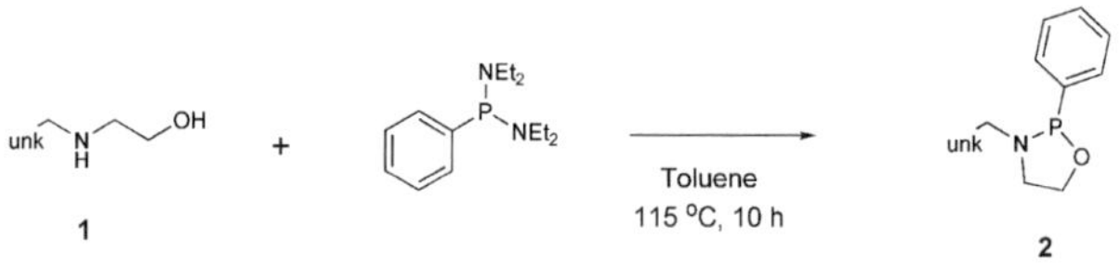

Scheme 5.6

The solid-supported reagent, 2, was subsequently used to convert isothiocyanates to isonitriles (Scheme 5.7). Primary, secondary and tertiary alkyl isocyanides, as well as aromatic analogues were synthesised using this microwave assisted methodology. Isocyamide products were produced in excellent yields and high purity in 30 min to 2 h at 140°C.

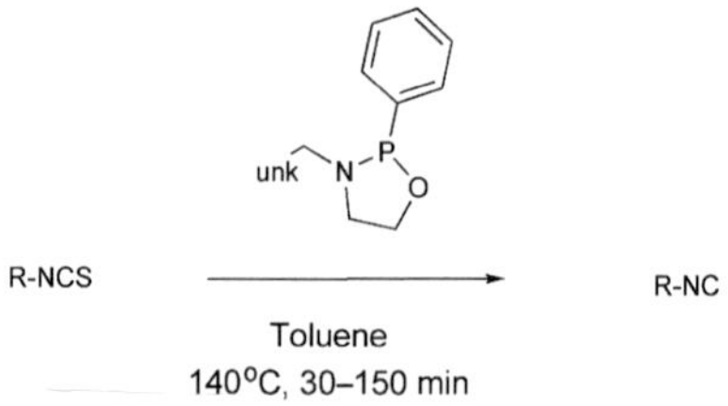

Scheme 5.7

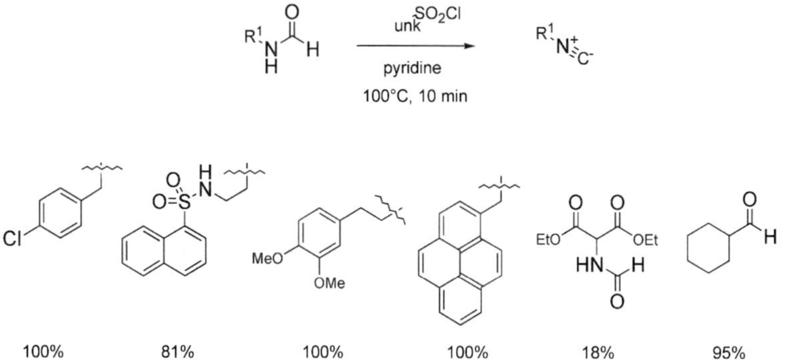

Figure 5.2   Examples of different R-groups used in the formation of isonitriles.

The synthesised isonitriles were afterwards used in 3CC Ugi reactions performed under conventional heating. An alternative microwave-assisted synthesis of isonitriles has recently been reported by Prof. Bradley's group[19]. In order to reduce work-up they also employed a solid supported reagent strategy. Formamides, when treated with solid-supported sulphonyl chloride in pyridine (50 equiv in DCM) at 100°C for 10 min, provide the isonitriles in a quick and convenient fashion. As long as the formamides were not too heavily substituted, the isonitriles were obtained in high yields and purities, Fig. 5.2.

It should also be noted that formamides can easily be formed from both primary and secondary amines under microwave conditions, as described by Danks and co-workers. Primary and secondary amines react with solid-supported ammonium formate under microwave conditions at 100 W for 30 s to provide formamides in good yields after simple filtration and removal of the solvent.

5.2.4. Kindler reaction

Kindler reported the 'one-pot' process for the synthesis of thioamides from elemental sulphur, aldehydes and amines for the first time in 1923 (Scheme 5.8). Thioamides are essential building blocks for the preparation of a number of biologically relevant heterocyclic scaffolds. Among the many different methods to prepare thioamides reported in the literature, the three-component Kindler reaction has so far received comparatively little attention. This might seem surprising considering that a large number of aldehydes and primary/secondary amines are commercially available, potentially offering quick access to a diverse set of synthetically useful thioamides. Two of the main issues are probably the high-reaction temperatures and long-reaction times that are typically required. In addition, the protocols, which use volatile amines or aldehydes, cannot be used without autoclave technology.

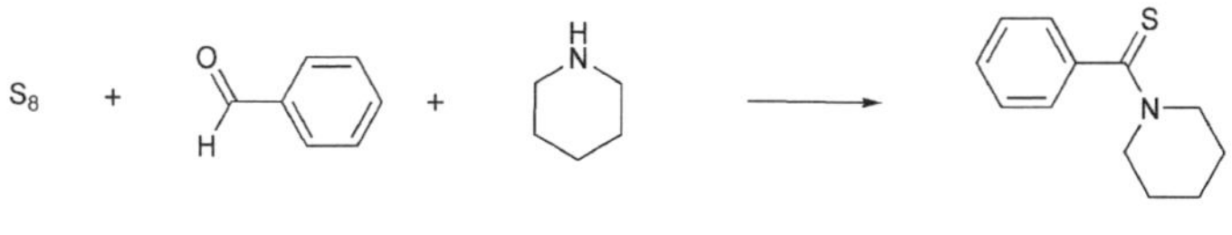

Scheme 5.8

Therefore, Kappe and co-workers[20] performed the Kindler reaction under sealed vessel (autoclave) microwave conditions providing an ideal route for the synthesis of thioamides. This elegant work was preceded by a few earlier reports on microwave-assisted Kindler reactions (and on related Willgerodt–Kindler processes)[21]. These studies were performed in domestic microwave ovens and were restricted to the use of two cyclic secondary amines (morpholine and piperazine). After considerable experimentation involving variation of the molar ratios of the three components, they found that using a 50% excess of amine and a 25% excess of sulphur consistently produced the highest yields of thioamides. This method also allowed the convenient isolation of the product by precipitation in water providing the products in very high purity (96% as determined by ^{1}H NMR and GC/MS), with only trace amounts of sulphur remaining. Doubling the concentration of the reaction mixture further increased the yield; a finding that is fairly common in the area of microwave synthesis. To increase the structural diversity and synthetic value of thioamides that can be accessed using this three-component process, both primary and secondary amines, as well as substituted aromatic and aliphatic aldehydes were used. In fact, even ammonia could be employed under these condensations by using a commercial 7 M solution of ammonia in methanol. In the majority of cases, work-up of the reaction mixtures involved simply pouring the reaction mixture onto ice/water followed by filtration of the solid material.

5.2.5. *Gewald synthesis of 2-acyl amino thiophenes*

In 1961, Gewald and co-workers published the synthesis of poly-substituted thiophenes involving condensation of cyanoacetate and elemental sulphur with ketones or aldehydes in a three-component reaction (Scheme 5.9). Beyond their industrial use in dyes and conducting polymers, 2,5-substituted thiophenes have shown extensive potential in the pharmaceutical industry. Most published Gewald thiophene synthetic procedures require reaction times between 8 and 48 h for the condensation step. Hoener and

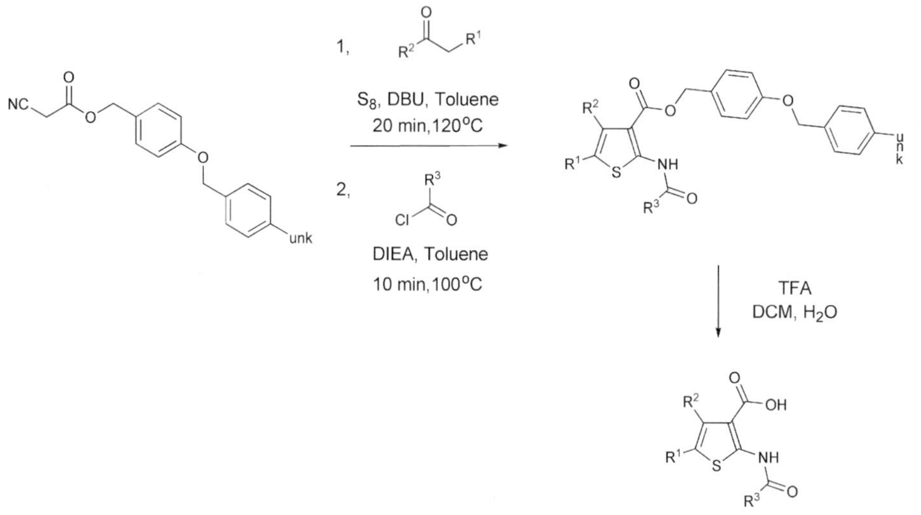

Scheme 5.9

co-workers[22] have developed a solid-phase approach, where the use of microwave heating reduces the reaction time dramatically. The resulting products normally require laborious purification by chromatography and further modifications on their core structure are typically carried out in a second step. In this report, the authors found that the overall procedure could be performed in less than 1 h. The 2-aminothiophene was formed on the commercially available cyanoacetic acid Wang resin in 20 min at 120°C as described in Scheme 5.9. Ethanol, which has been found to be a good solvent for high temperature microwave-assisted solid-phase synthesis[23], is the standard solvent for Gewald transformations in most liquid- or solid-phase protocols. However, the authors wanted a solvent that allowed for direct acylation of the primary amine generated in the first step. Toluene was the solvent of choice and the acylation was performed at 100°C for 10 min. The products were finally cleaved from the resin using TFA in DCM to give the products in 46–99% yield.

5.2.6. *Mannich reaction*

In the Mannich reaction, a carbonyl component usually formaldehyde, a secondary amine and a CH-acidic compound react together to form β-aminoketones. The classical method for the formation of β-aminoketones suffers from many disadvantages such as drastic reaction conditions, formation of undesired side products, little or no stereo- or regioselectivity and low yields. In 2000, Gadhwal and co-workers developed the first microwave-assisted Mannich protocol[24] (Scheme 5.10).

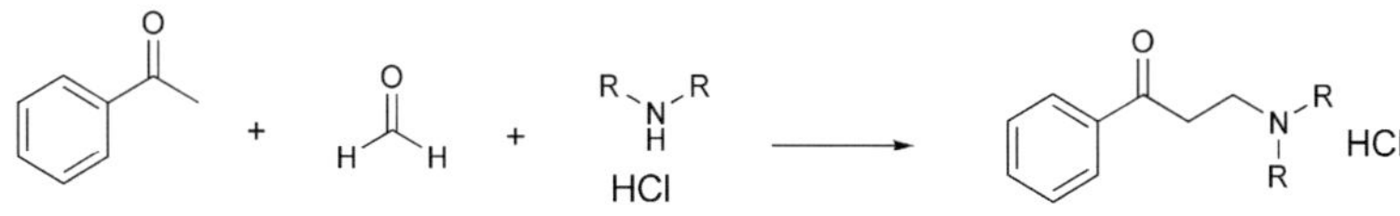

Scheme 5.10

    Primary amines were readily obtained in 60–83% yield after 15 min of microwave irradiation under solvent-free conditions, when using ammonium chloride as the amine source (R=H). When substituted amine hydrochlorides were used, the reaction failed under solvent-free conditions; however, when performed in ethanol once more high yields were obtained (80–83%). In both cases, no traces of side-product formation were found.

5.2.7. *Boronic Mannich reaction*

A further development of the Mannich reaction is the boronic Mannich reaction, which has been described extensively by Petasis and co-workers under conventional heating methods. Gupta *et al.* (Personal Chemistry, Uppsala, Sweden, internal report) have performed this reaction under microwave condition in acetonitrile (Scheme 5.11). A reaction time of 4 min at 120°C afforded the products in yields ranging from 25 to 100%. Alternative amines and boronic acids could be used, but the glyoxylic acid was essential for product formation. The major drawback with this reaction under microwave conditions at present is that the outcome is highly substrate dependent.

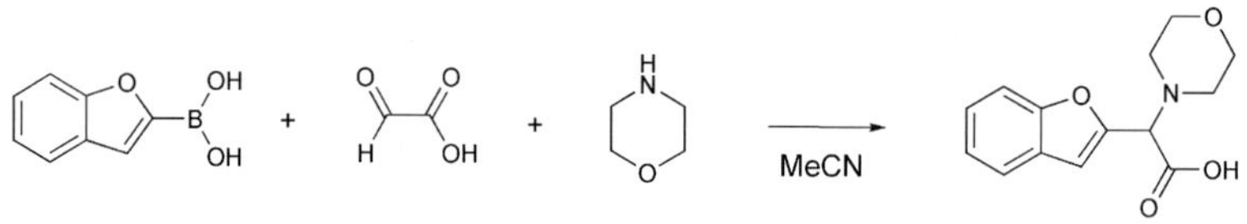

Scheme 5.11

5.2.8. Pauson–Khand reaction

The [2+2+1] cycloaddition of an alkene, an alkyne and carbon monoxide is commonly known as the Pauson–Khand reaction. This transformation has been adopted
many times in the synthesis of complex natural products and related compounds,
which contain a cyclopentenone moiety, for example, prostaglandins. Two independent reports of this reaction appeared almost simultaneously in late 2002 by Iqbal and
co-workers[25] and Fisher and co-workers[26], respectively. They not only used very similar substrate systems in their studies, but they also reached very similar conclusions:
Toluene was found to be the preferred solvent in this reaction, even though it is a very
poor microwave absorber. A reaction time between 5 and 10 min, using dicobaltoctacar-
bonyl or dicobalthexacarbonyl as the carbon monoxide source, and a temperature of
100–120°C resulted in high yields of the products. Fisher and co-workers used 20 mol%
$Co_2(CO)_8$ and cyclohexylamine as an additive (Scheme 5.12), since this system had
been used previously in order to allow a catalytic reaction. Iqbal and co-workers did
not use cyclohexylamine, but instead used 1 equiv. of the carbon monoxide $(Co_2(CO)_6)$
source. In both reports, the products were formed in 40–70% yield.

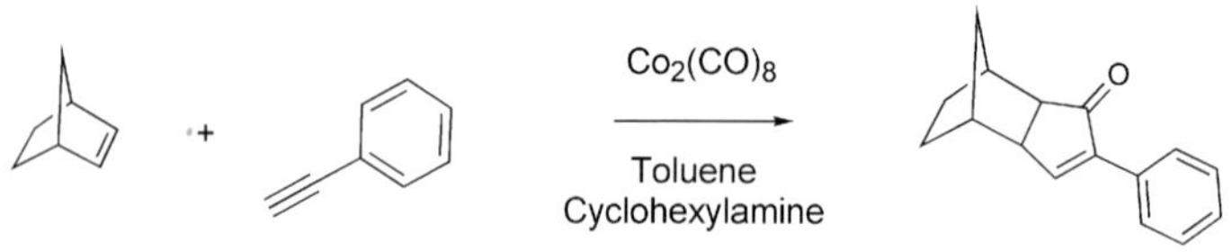

Scheme 5.12

Iqbal and co-workers also used 1,2-dichloromethane as the solvent, which facilitated
purification of the product and led to less of the corresponding *endo*-diastereomer being
formed.

5.2.9. Wittig reaction

Wittig reactions employing microwave heating have been published several times both
in solution and under solvent-free conditions. They have been performed using stable ylides[27], as well as using the pre-formed phosphonium salts, which under basic
conditions forms the ylides *in situ*[28].

One well-known drawback with the Wittig reaction in solution is the formation of
1 equiv. of triphenylphosphine oxide, which is inherently difficult to separate from the
final product. Thus, Westman[23] developed a 'one-pot' multi-step microwave-assisted

method for the Wittig reaction, using solid supported triphenylphosphine, (Scheme 5.13). Here, the formed phosphine oxide remains attached on the solid support and can be swiftly removed by filtration. This was the first report which described results, where all reaction steps of a Wittig sequence, including the initial formation of the ylide, were performed under microwave conditions.

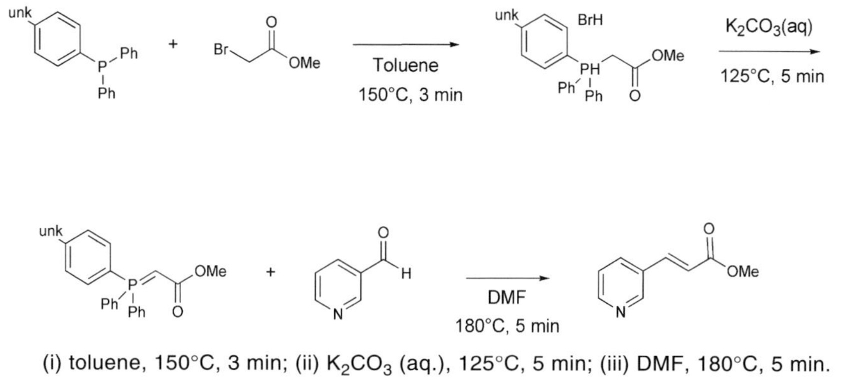
unk
Ph
P
Ph
+
Br
O
OMe
Toluene
150°C, 3 min
unk
BrH
PH
Ph
Ph
OMe
O
K2CO3(aq)
125°C, 5 min
unk
P
Ph
Ph
OMe
O
+
O
H
N
DMF
180°C, 5 min
N
O
OMe
(i) toluene, 150°C, 3 min; (ii) K2CO3 (aq.), 125°C, 5 min; (iii) DMF, 180°C, 5 min.

Scheme 5.13

The Wittig reaction is very rapid under microwave conditions; however, the efficiency of the reaction sequence can be further improved if the ylide can be synthesised with the same speed.

The ultimate simplicity and efficiency was achieved by the development of a multi-component procedure, where both the phosphonium salt and the ylide were formed and reacted *in situ* (Scheme 5.14). By simply mixing all the reagents in methanol and heating the reaction mixture at 150°C for 5 min in the presence of potassium carbonate, the olefins were formed in moderate to excellent yields. These yields from a 'one-pot' procedure were comparable with the results obtained by using a multi-step reaction sequence.

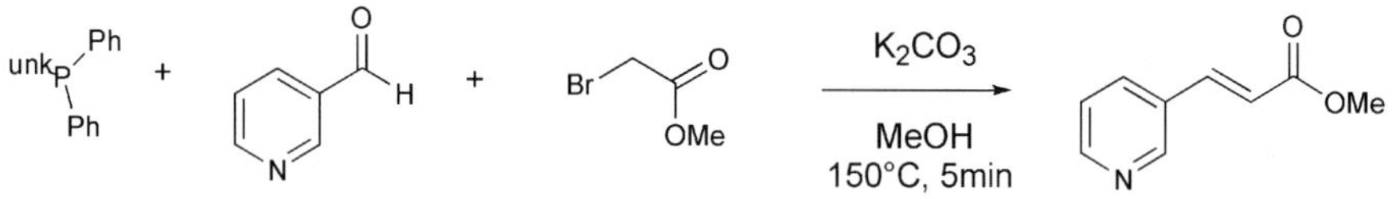
unk
Ph
P
Ph
+
O
H
N
+
Br
O
OMe
K2CO3
MeOH
150°C, 5min
O
OMe
N

Scheme 5.14

It should be noted that methanol is normally not the solvent of choice when performing solid-phase chemistry, because it is a poor swelling solvent for polystyrene resins. However, there have been several recent reports on solid-phase chemistry using unconventional solvents in combination with microwave-assisted organic synthesis (see Chapter 7). Whether the improvements are a sheer consequence of the increased diffusion rates caused by the higher temperatures employed or if it is also a consequence of changed solvent properties is unclear. It is known that solvents change their properties at elevated temperatures and become less polar[29], it is therefore possible that the polarity of methanol in the example above decreases enough to swell the resin and enhance the reaction by making the reaction sites more accessible.

5.2.10. aza-Diels–Alder reaction

The aza-Diels–Alder reaction of Danishefsky's diene with imines provides a convenient method for the synthesis of 2-substituted 2,3-dihydro-4-pyridones, a compound class that has important synthetic applications. Kobayashi and co-workers have studied the reaction in detail using ytterbium (III) triflate as the Lewis acid[30]. Although the reaction is often run at low temperature ($-78°C$ to $0°C$) for a number of hours, we have found that the reaction also worked well at elevated temperatures ($150°C$) in the microwave for a few minutes (J. Westman and A. Hurynowics, unpublished results) (see Scheme 5.15). The imines could either be preformed prior to the addition of the Danishefsky's diene or the reaction could be performed as a multi-component protocol, where all components were added at once.

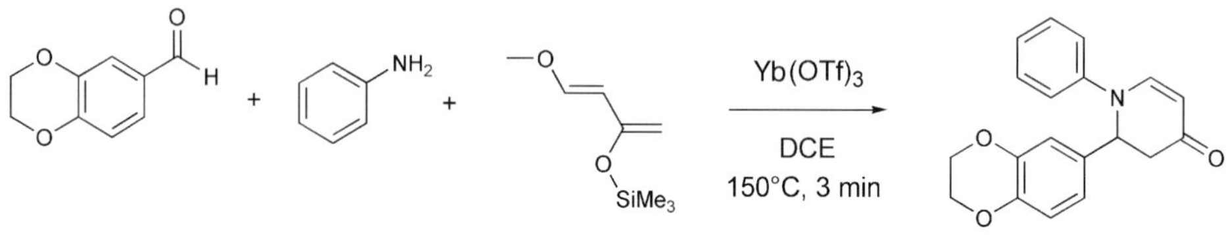

Scheme 5.15

However, the two-step 'one-pot' procedure resulted in purer products and was therefore adopted. The products were formed in acceptable to good yields.

5.3. Versatile reagents in multi-component reactions

5.3.1. (Triphenylphosphoranylidene)ethenone

Westman and co-workers have produced a small library of unsaturated amides, using a three-component reaction of an aldehyde, a secondary amine and (triphenylphosphoranylidene)ethenone[31]. The latter is an example of a multi-purpose reagent that can be used to introduce a carbon–carbonyl building block into the target molecule. This was accomplished by a cascade reaction comprising of an addition reaction and a Wittig olefination reaction. A selection of aldehydes, aromatic as well as aliphatic, and five primary and secondary amines were treated with 1.5 equiv of (triphenylphosphoranylidene)ethenone in DCE. The reaction conditions were varied as indicated in Table 5.1, yielding the products in 18–100% yield. Boc protecting groups are easily cleaved at elevated temperatures, which is why reactions containing N-Boc protected compounds were performed at a slightly lower temperature ($150°C$). If the carbonyl component and the nucleophilic component were present in the same molecule the intermediate product cyclised intramolecularly in a two-component fashion to form various heterocycles, such as tetronic acids, tetronates, coumarins, benzoxepinones and their nitrogen-analogues.

Table 5.1

$$R^1\text{-CHO} + PPh_3C=C=O + R^2(H)N\text{-}R^3 \xrightarrow[\text{5 min}]{180°C} R^1\text{-CH=CH-C(O)-N}R^2R^3$$

	HN-piperazine-Boc	HN-piperazine-(2-pyridyl)	H$_2$N-CH$_2$CH$_2$-Ph	aminoindane	piperidine
1,4-benzodioxane-6-carbaldehyde	100%[b, 16]	90%	59%	84%	90%
cyclohexanecarbaldehyde	74%[b]	56%[a]	n.d[c]	36%	18%[a]
4-(dimethylamino)benzaldehyde	66%[b]	n.d[c]	35%	45%	35%[a]
nicotinaldehyde	74%[b]	n.d[c]	74%[a]	81%	88%
(E)-cinnamaldehyde derivative (MeO, HO, OMe)	99%[b]	n.d[c]	66%[a]	85%[a]	68%[a]

(a) 10 min reaction, time; (b) 150°C reaction temperature; (c) yield not determined due to contamination of triphenylphosphine oxide, but LCMS analysis indicated high yields.

5.3.2. N,N-Dimethylformamide diethyl acetal

N,N-Dimethylformamide diethyl acetal (DMFDEA) is a interesting reagent, since it can be used to form alkylaminopropenones or alkylaminopropenoates when reacted with compounds having activated methylene groups such as β-ketoesters, acetophenone and N-acylglycine. These alkylaminopropenones or alkylaminopropenoates can subsequently be reacted with dinucleophiles to form a variety of heterocycles. Westman and co-workers[32] have used DMFDEA for the synthesis of propenoates and propenones, which were used directly without intermediate purification for the formation of a number of heterocycles in a combinatorial fashion. Some examples are outlined in Scheme 5.16. The reactions were performed in a two-step 'one-pot' procedure, which in this case were more suitable for combinatorial synthesis.

Molteni and co-workers have described a further development of this reaction[33]. Similar to Westman and co-workers, Molteni *et al.* also performed the same type of multi-component reactions in a two-step manner, however they used water as the

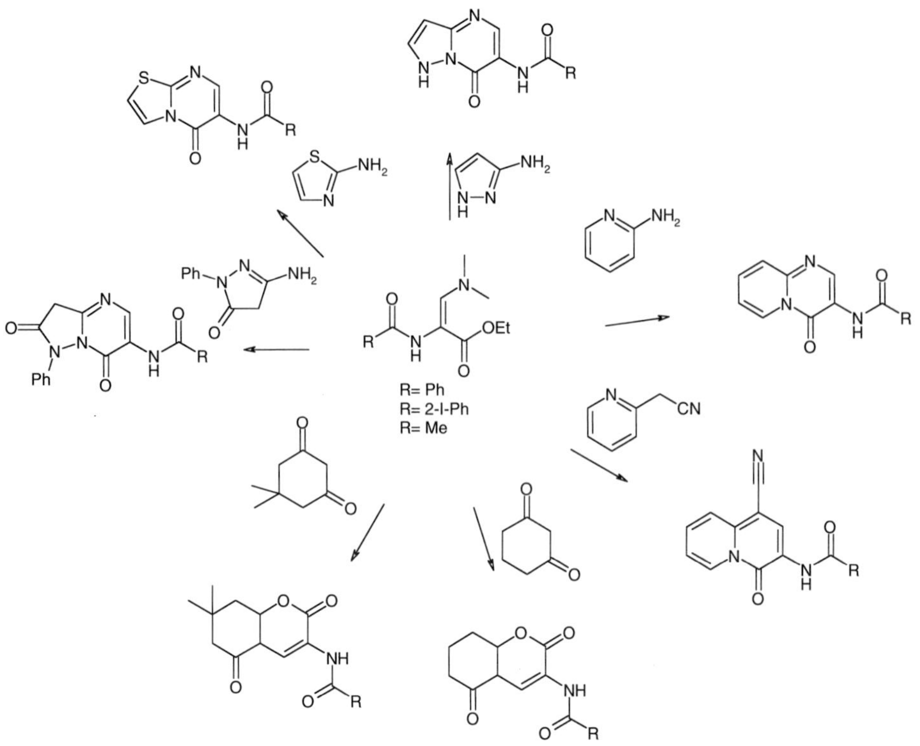

General reaction conditions: 0.2 mmol, 1.2 equiv dinucleophiles, acetic acid/DMF 4:1, 180°C, 5 min.
The LC-purity of the products in the reaction mixture varied from 30 to 90% for this set of reactions.

Scheme 5.16

solvent. The multi-component reaction, Scheme 5.17, gave yields and purities comparable to the two-step procedure.

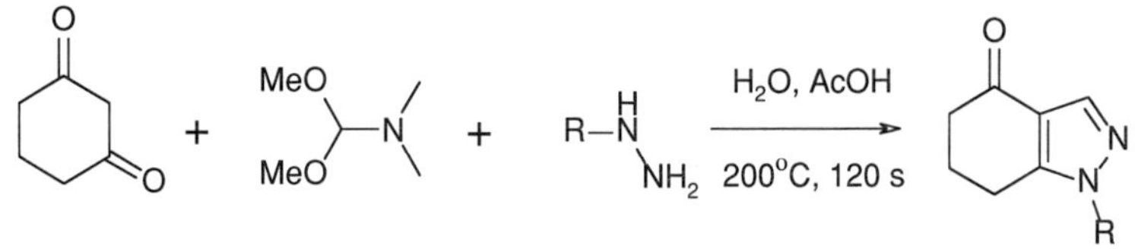

Scheme 5.17

5.3.3. *N,N-Dimethylformamide diethyl acetal on solid support*

Bienaymé[34] reported the formation of substituted isonitriles by treatment of a dialkyl
amine with imidazole diethylacetal and methyl isocyanoacetate. This forms dialky-
lamino propenoates *via* a three-component cascade reaction mechanism (Scheme 5.18).

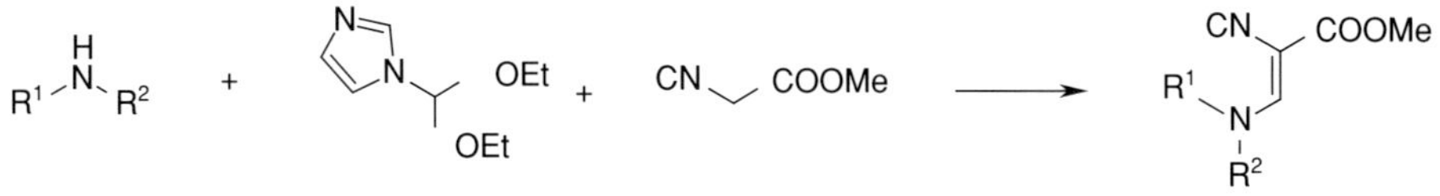

Scheme 5.18

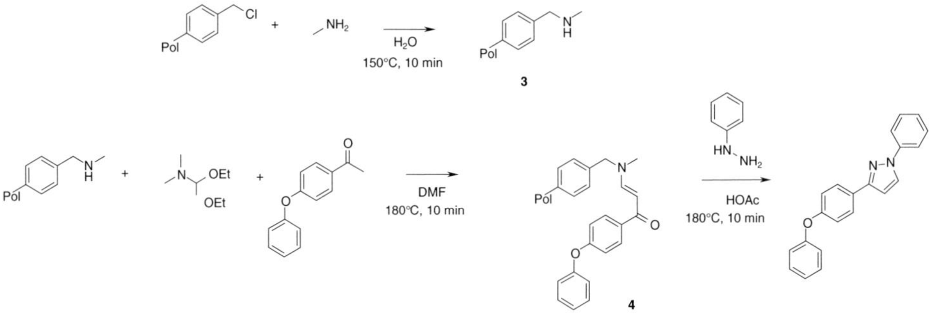

Example of a solid-phase synthesis of 1-phenyl-5-(4-phenoxyphenyl)-pyrazole. Reaction conditions: (i) MeNH₂ in water, 150°C, 10 min; (ii) DMF, 180°C, 10 min; (iii) HOAc, 180°C, 10 min.

Scheme 5.19

The above described methodology was found to be very useful in a solid-supported synthesis of pyrazoles and pyrimidines *via* propenones developed by Westman and co-workers[35] (Scheme 5.19). Merrifield resin was reacted with methylamine in water at 150°C for 10 min to form the solid-supported benzylmethylamine (**3**) in high yield (86% yield, 1.08 mmol/g based on elemental analysis). After washing, the resin was treated with 5 equiv. DMFDEA and 5 equiv of 4-phenoxyacetophenone at 180°C for 10 min in DMF to form the solid supported benzyl methyl aminopropenones (**4**).

After additional washing, the resin was finally treated with 0.5 equiv of the appropriate dinucleophile (the resin-bound intermediate in excess) and reacted in ethanol, DMF or acetic acid at 180°C for 10 min. The resin was filtered off and the filtrate was evaporated to dryness to give the various heterocycles in yields of 88–94% and purities of 85–93%.

5.4. Miscellaneous products

5.4.1. *Imidazoles*

Coleman and co-workers[36] performed a three-component reaction employing 2-oxothioacetamides, aldehydes and alkyl halides in ethanol with Na_2CO_3 as the catalyst to form a 24-member library of substituted 4-sulphonyl-1-H-imidazoles (Scheme 5.20). The products were formed in yields of 40–96% and purities of 20–96%. The library was synthesised in a custom-built reaction block with expandable reaction vessels. The expandable reaction blocks were designed to accommodate the expanding gas volume caused by the elevated temperature without releasing any of it into the cavity. Unfortunately, this approach limited the temperatures used in the reactions to the boiling point

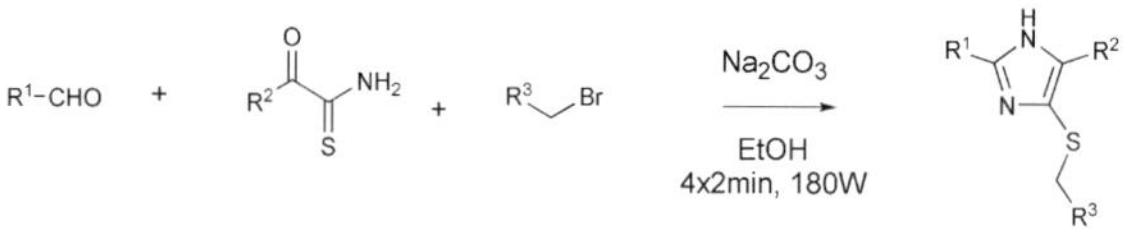

Scheme 5.20

of the solvent, thus loosing the real advantage of employing microwave heating instead of conventional heating.

Using this reaction block in a domestic microwave oven, the optimum conditions for the reaction were found to be four cycles of 2 min irradiation at 180 W followed by 2 min of rest. The cyclic approach was forced by the expandable vessels reaching their full capacity (20 times the reaction volume) after 2 min of irradiation, thereby requiring a rest period to allow the gas to contract before continuing the irradiation. These issues could obviously have been avoided by utilising reaction temperature control. In the above approach, there is no way of assessing the time spent at reflux thereby making reaction optimisation a difficult task due to reproducibility issues.

5.4.2. Substituted imidazoles

The condensation of 1,2-diarylethandienones with aldehydes and amines results in 2,4,5-substituted or 1,2,4,5-substituted imidazoles. Tri-substituted imidazoles are formed when ammonium acetate is used as the amine source, while tetra-substituted imidazoles are formed when both an amine and ammonium acetate are used simultaneously. Triarylimidazoles are used in photography as photosensitive compounds and they are also of interest in medicinal chemistry due to biological activities such as fungicidal, anti-inflammatory and anti-thrombotic effects. In 2000, two groups published their findings regarding the microwave synthesis of substituted imidazoles. Usyatinsky and co-workers[37] and Balalaie and co-workers[38] both synthesised tri- and tetra-substituted imidazoles under solvent-free, conditions, using acidic solid supports such as silica gel, acidic aluminia, Montmorillonite K10, bentonite and Zeolite HY. The classical synthesis of imidazoles requires reflux conditions for several hours in acetic acid.

The synthetic procedure developed by Usyatinsky involves pre-absorbing the mixture of acidic support and ammonium acetate (ammonia source) with an ether solution of the starting materials, evaporating the solvent and heating the solid residue in a domestic microwave oven for 20 min (Scheme 5.21). After irradiation, the reaction mixture was allowed to cool to room temperature and was then washed with a mixture of acetone and triethylamine to extract the product from the solid support. No yields were quoted but the purities of the products were quoted to be 75–85%.

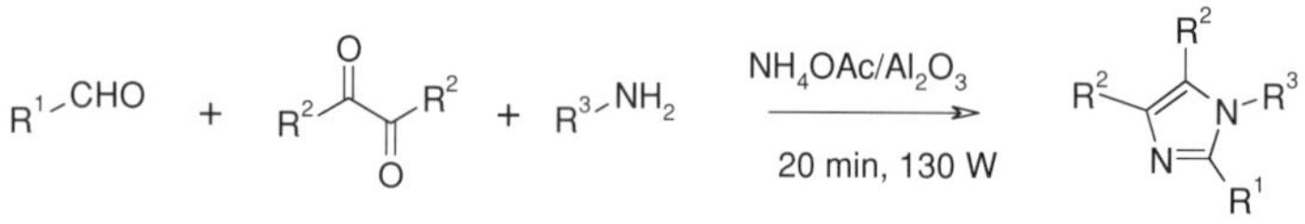

Scheme 5.21

Balalaie and co-workers[38] on the other hand found that Zeolite HY as the acidic catalyst resulted in the highest yields (81–94%) after 6 min of microwave irradiation. When producing large number of substances the solvent-free method suffers from some disadvantages, such as problematic automation and the need for extraction procedures for the separation of the product from the solid support. We therefore performed the reaction in accordance with the classical methods using acetic acid as the solvent (J. Westman, Personal Chemistry, Uppsala, Sweden, unpublished results) (Scheme 5.22). The reaction mixtures were heated at 180°C for 3 min in a closed vessel to give the products in high yields and purities.

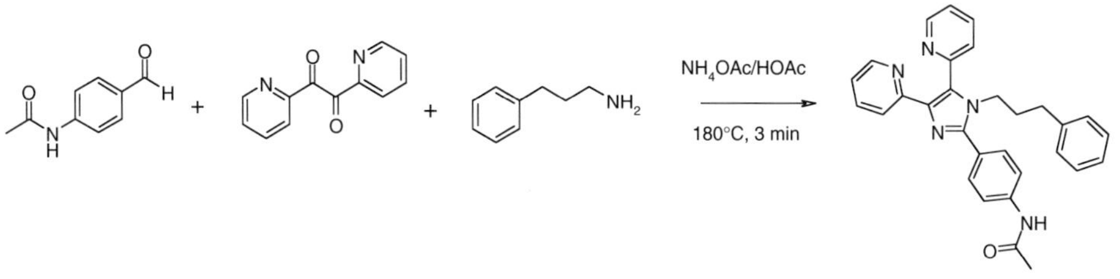

Scheme 5.22

In 2003, Balalaie and co-workers published a new paper in the area of imidazole synthesis[39]. In the above examples reported by Westman, ammonium acetate and primary amines act as the nitrogen source for the imidazole ring. In the Baladie novel 'one-pot' three-component method, ammonium acetate was replaced by benzonitrile derivatives. Thus, benzil, primary amines and benzonitrile derivatives underwent a condensation reaction on acidic silica gel as the solid support, Scheme 5.23. The reaction mixtures were irradiated for 8 min in a domestic oven under solvent-free conditions to give the products in 58–92% yield.

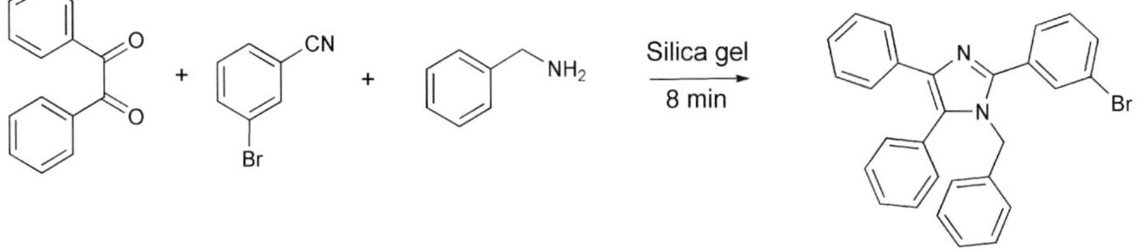

Scheme 5.23

5.4.3. Imidazo-pyridines

In 1998, three different research groups discovered a novel multi-component reaction almost simultaneously[40]. They found (partly by serendipity!) that when using 2-aminopyridines as the amine component in the Ugi four-component reaction (4CR), the formation of imidazo-pyridines was observed. The imidazo-pyridine is the product from the condensation of the 2-aminopyridine with an aldehyde and an isonitrile; the acid, the fourth component of the Ugi 4CR, is required as a catalyst. Lewis acids could also be used. Under conventional conditions, the reaction takes 18–72 h depending on the method used to give the products in high yields. In 1999, Varma and co-workers[41] published this reaction under microwave-assisted solvent-free conditions. Varma and co-workers performed the reaction in a two-step fashion. In step one, they formed the imine by heating the aldehyde and the amino derivative, subsequently the isonitrile was added and the reaction mixture was heated again.

The two-step procedure introduces a limitation in that the alkyl iminium ions can be unstable. The two-step solvent-free procedure is inconvenient in library produc- tion, especially when automated liquid handlers are used. A further complication is the fact that many of the products are highly crystalline, making the isolation from the solid support potentially problematic. Westman and co-worker (J. Westman and A. Franzén, unpublished results) have adopted the method developed by Bienaymé and co-workers[40], in order to find a microwave-assisted solution-phase protocol with short reaction times. The conventional method uses perchloric acid that had been believed to be unsuitable for use under high temperature microwave heating conditions because of its explosive properties. However, the authors found that catalytic amounts of perchloric acid could be used safely in a closed pressurised vial heated at 170°C for 5 min. In a high throughput synthesis set-up, including an automated microwave synthesiser and fully automated RP-HPLC purification, they were able to obtain the products in 20–70% iso- lated yields by heating 0.6 mmol of aldehyde, 1.0 equiv. of an amino derivative, 1.17 equiv. isonitrile and a catalytic amount of perchloric acid (aq) in 2 ml ethanol, at 170°C for 5 min, Scheme 5.24. Both aromatic, aliphatic (even sterically hindered) and heteroaro- matic aldehydes were used and gave the corresponding products in acceptable to good yields. The same is true for the different isonitriles and heteroaromatic amidines used.

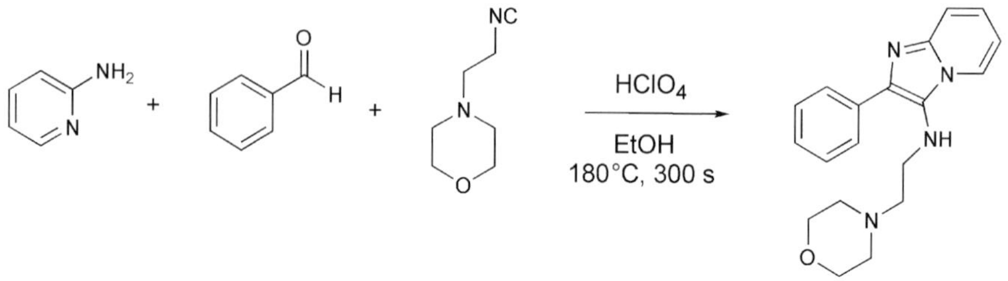

Scheme 5.24

5.4.4. 1,2,4-Triazine

The 1,2,4-substituted triazine core is a versatile scaffold to access a wide range of con- densed heterocyclic ring systems *via* intramolecular Diels–Alder reactions with a vast array of dienophiles. The triazine ring system is also a key component of commer- cial dyes, herbicides, insecticides and also recently appeared in medicinal chemistry. One way to synthesise triazines is to use a three-component reaction that has been described in the literature several times, both under traditional thermal heating and under solvent-free microwave-assisted conditions[42]. However, the previously described methods focussed only on simple aliphatic phenyl and ester substituents.

Zhao and co-workers[43] have described a method which utilises heterocyclic acyl hydrazides and heterocyclic 1,2-diketones to afford 1,2,4 triazines. The resulting triazines were obtained in low yields, even under solvent-free microwave conditions. By adopting the conventional thermal conditions, for example, using acetic acid as the solvent instead under microwave heating conditions at 180°C for 5 min (60°C above the boiling point for acetic acid), high yields (75–92%) of the products were obtained, Scheme 5.25.

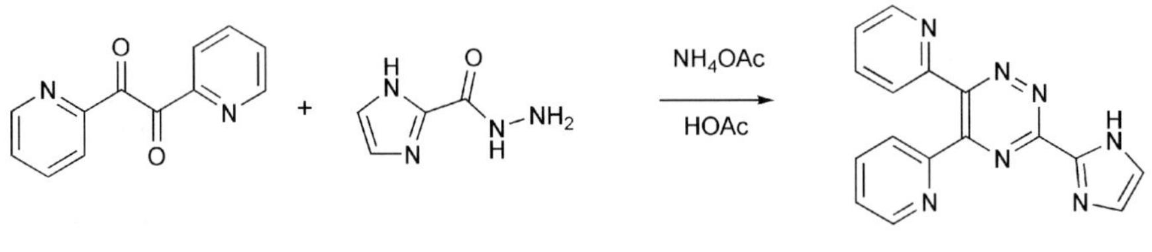

Scheme 5.25

Using this method, the authors synthesised a 48-member library where, upon rapid cooling, ~60% of the desired product precipitated out of the solution making the isolation of the products straightforward.

5.4.5. Indolizines

The indolizines constitute the core structure of many naturally occurring alkaloids, such as (–)-slaframine, (–)- dendroprimine, indalozin 167B and coniceine. There are a number of different routes to the synthesis of indolizines and they are most commonly synthesised by sequential N-quaternisation, intramolecular cyclocondensation reactions or the cycloaddition reaction of N-acyl/alkyl pyridinium salts.

Bora and co-workers[44] have developed a microwave-assisted three-component synthesis of indolizines. The reaction involves a 1,3-dipolar cycloaddition reaction between the in situ generated dipole (from the bromoacetophenone and pyridine) and acetylene, Scheme 5.26. The developed method provides fast access to cycloadducts, which otherwise are accessible only through multi-step synthesis.

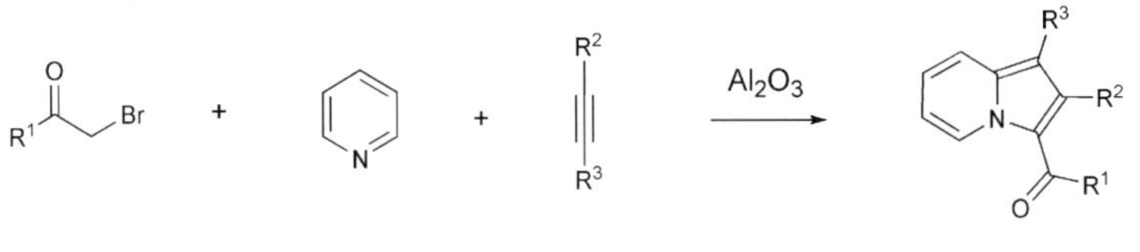

Scheme 5.26

It was found that basic alumina worked well as the basic catalyst for the in situ dipole generation from the N-acyl pyridinium salt. A three-component mixture of phenacyl bromide (1 mmol), pyridine (1.2 equiv.) and the acetylene (1.2 equiv.) was thoroughly mixed in basic alumina (1 g) and then irradiated for 8 min at 80% power in a domestic microwave. The products were formed in 87–94% yields when running the reaction under solvent-free conditions and in 60–71% yields when using anhydrous toluene as the solvent.

5.4.6. *Substituted pralines*

Wilson and co-workers[45] have reported a 'one-pot' two-step synthesis of substituted pralines (see also Chapter 8), Scheme 5.27. The process involves rapid microwave processing of α-amino acid esters and aldehydes at 180°C for 2 min to generate imines, followed by a [3+2] cycloaddition with *N*-substituted maleimides at 180°C for 5 min to give the cycloadducts in >75% yield. In this manner, a library of 800 compounds was produced. Some limitations with this reaction were encountered; aliphatic aldehydes performed poorly, which was believed to be due to poor stabilisation of the intermediate imine or tautomerisation to the undesired enamine. Furthermore, poor solubility of some of the amino acids in DCE limited the diversity of the library. Under conventional heating, most methods required either Lewis or protic acids to catalyse the cycloaddition reaction or reactions on solid support wherein no acid catalysis was used. The drawback with these methods were not only the long reaction times employed, but also the presence of significant decomposition. Contrary to the literature, they found that the use of acid catalysis could be avoided without compromising the quality of the reaction outcome, which simplified the purification of the products.

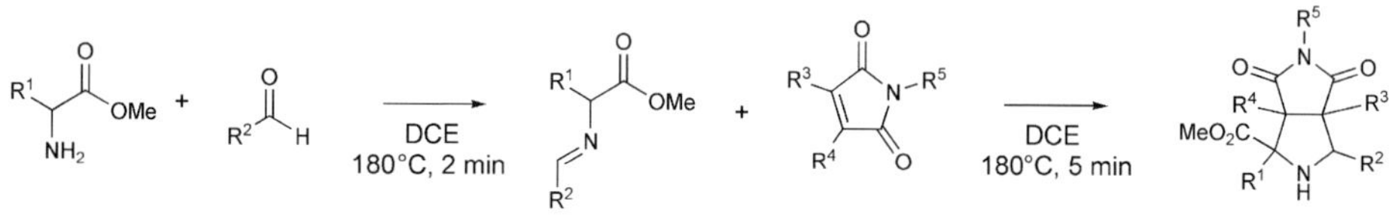

Scheme 5.27

5.4.7. *Quinolines*

The same authors have also reported the synthesis of 2-amino quinolines[46], Scheme 5.28. Secondary amines were reacted with aldehydes to form enamines; subsequent addition of 2-azidobenzophenones initially forms triazoline intermediates, which undergo a thermal rearrangement and intermolecular base-catalysed cyclocondensation to produce 2-amino quinolines. The reactions were run at 180°C for 10 min (the time includes 3 min required for enamine formation) to give the products in 57–100% yields.

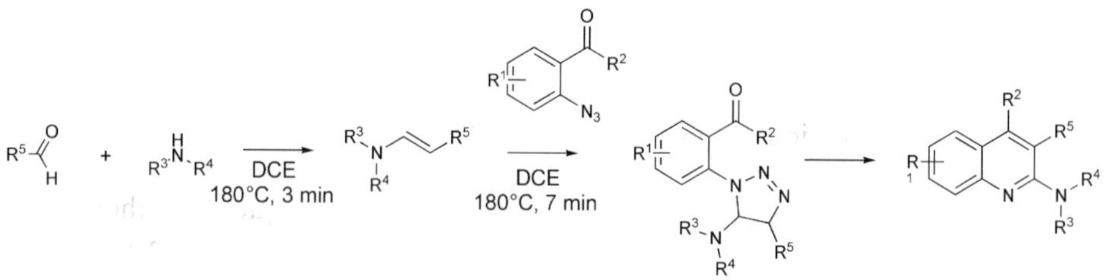

Scheme 5.28

When comparing the results from the corresponding reaction under conventional heating conditions, the author noted that the purity of products resulting from the thermal reactions were reduced due to the presence of decomposed 2-azidobenzophenone.

5.4.8. Quinazolin-4(3H)-ones

In 1998, Rad-Moghadam and Khajavi[47] developed a microwave-assisted three-component cyclocondensation for the synthesis of quinazolin-4($3H$)-ones.

A mixture of the reactants were irradiated in the absence of solvent or any dehydrating agents. In order to control the reaction in the domestic microwave oven, the irradiation was carried out in two stages (t_1 and t_2) with an intermediate cooling period between them. The reaction could proceed *via* two pathways, namely A and B, Scheme 5.29.

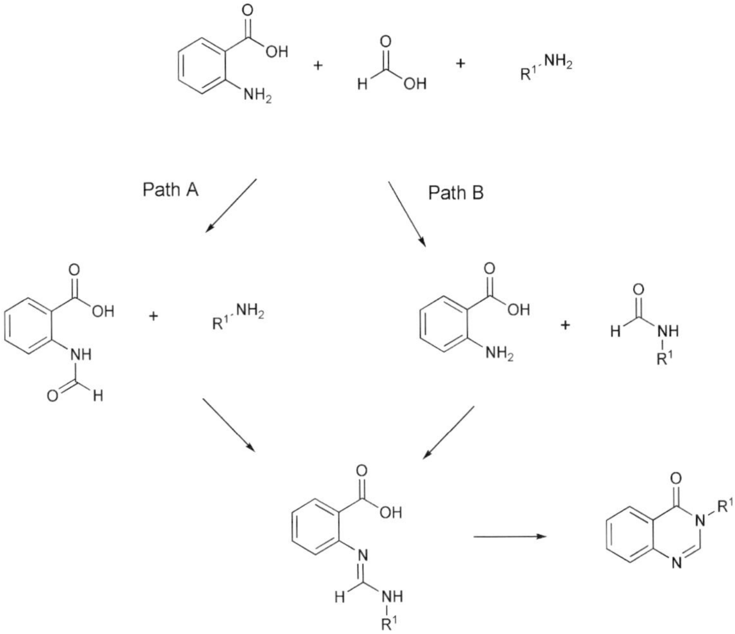

Scheme 5.29

Path A involves N-formylation of anthranilic acid, condensation of the resultant 2-formaminobenzoic acid with the amine followed by intramolecular amidation of the intermediate amidine to form the product. On the other hand, the amine instead of anthranalic acid may be formylated and go through the known Niementowski reaction (path B). When the reaction of 2-formamidobenzoic acid with aniline and the condensation of formanilide with anthralic acid were conducted under microwave irradiation, the desired 3-phenylquinazolin-4($3H$)-one was obtained in both cases in a few minutes in 68–87% yield.

Using ortho esters with a catalytic amount of *p*-toluenesulphonic acid instead of formic acid resulted in the formation of 2,3-disubstituted quinazolin-4($3H$)-ones in 79–89% yield, Scheme 5.30.

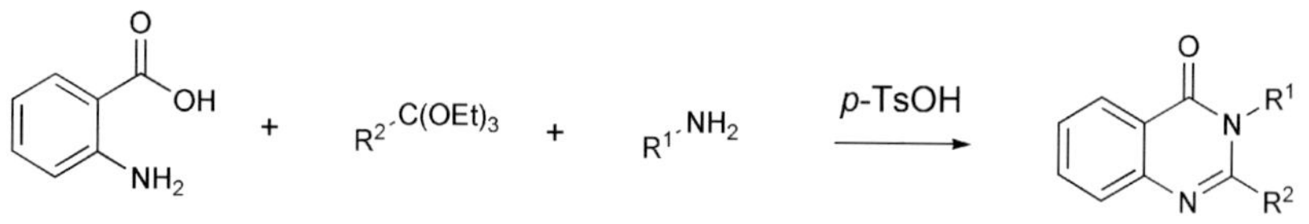

Scheme 5.30

5.4.9. Substituted pyrroles

There are several methods available for the synthesis of pyrroles; most of them involve multi-step synthetic operations. In recent years, a few one-step procedures have been reported both under conventional heating[48] and under microwave conditions[49]. The drawbacks with the conventional heating methods are the long reaction times and the limited scope of substitution available at the ring. Ranu and co-workers[50] have developed a three-component reaction, where an α,β-unsaturated carbonyl compound was treated with an amine and a nitroalkane under solvent-free condition using silica gel as the solid support. Sixteen different tetra-substituted pyrroles were synthesised with reaction times of 5–15 min; the products were obtained in 60–72% yield, Scheme 5.31. This should be compared with a reaction time of 16–18 h under conventional heating conditions in THF; products were obtained in only 30–42% yield. A wide range of structurally varied α,β-unsaturated aldehydes and ketones, including aromatic, aliphatic and heterocyclic units, and a variety of aliphatic and aromatic amines as well as nitroalkanes could be used to produce the corresponding alkyl-substituted pyrroles in a single-step. It should be noted that the reaction with nitromethane does not proceed at all; however, higher homologues like nitroethane and nitropropane underwent smooth reactions.

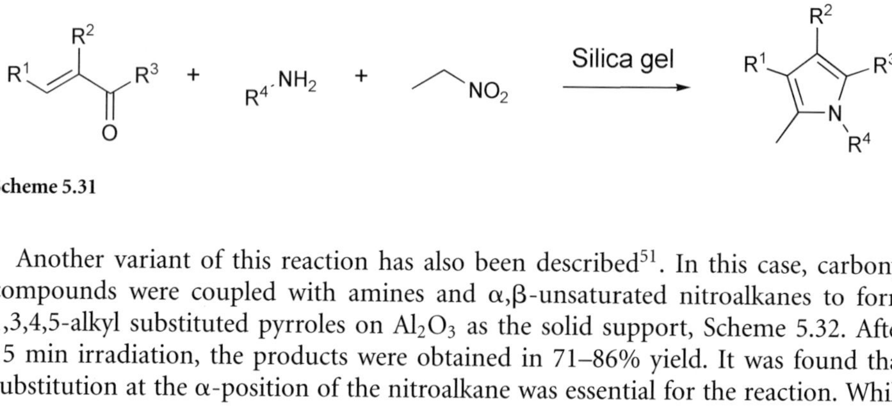

Scheme 5.31

Another variant of this reaction has also been described[51]. In this case, carbonyl compounds were coupled with amines and α,β-unsaturated nitroalkanes to form 1,3,4,5-alkyl substituted pyrroles on Al_2O_3 as the solid support, Scheme 5.32. After 15 min irradiation, the products were obtained in 71–86% yield. It was found that substitution at the α-position of the nitroalkane was essential for the reaction. While

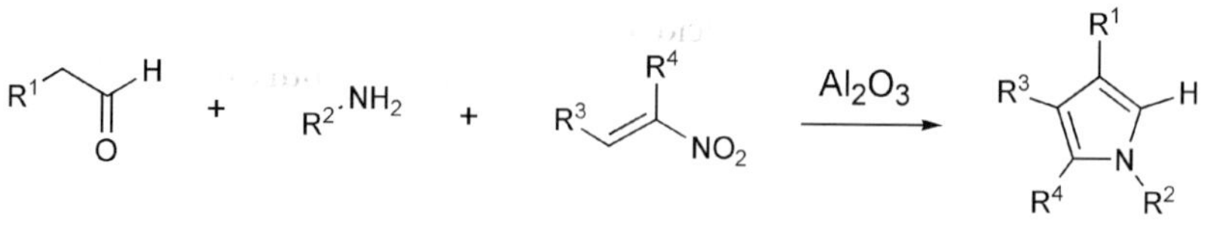

Scheme 5.32

the use of cyclic ketones in place of aldehydes nicely provided the desired fused pyrroles, open-chain ketones gave rise to different products.

5.4.10. *Indoles*

Nemes and co-workers[52] have employed compound **5**, Scheme 5.33, as a key intermediate in the synthesis of non-natural tryptophan derivatives. Several methods for the synthesis of compound **5** have been reported in the literature. When the authors applied the method utilising the condensation of aqueous formaldehyde, substituted indoles and Meldrum's acid, they encountered severe problems regarding the reproducibility of this procedure. To avoid the use of aqueous solution of formaldehyde, the authors turned to the utilisation of paraformaldehyde. The polymer form is easier to handle; however, at ambient or slightly elevated temperatures its depolymerisation is slow and insufficient. To overcome this problem, they used microwave dielectric heating. Indoles bearing both electron donors and electron acceptors in position 5 were allowed to react with Meldrum's acid and paraformaldehyde in the presence of D,L-proline. In all cases, both compound **5** and **6** were formed. When R′ was an electron-withdrawing group, the main product was compound **6**. Compound **6** could, however, be converted to product **5** by heating in acetonitrile in the presence of triethylamine.

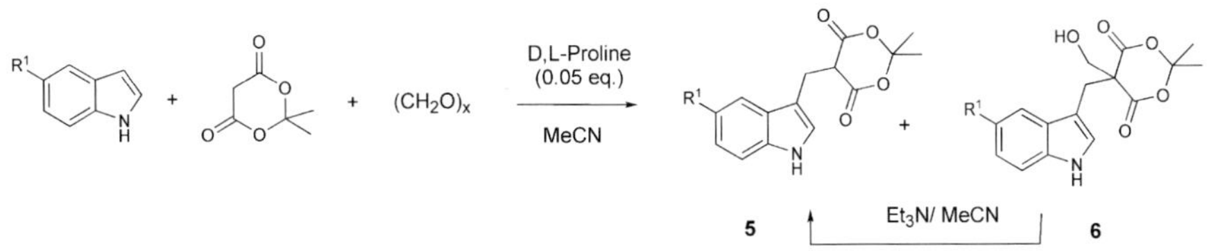

Scheme 5.33

5.4.11. *Spiroindoles*

The great importance of the indole nucleus in the field of medicinal chemistry has attracted the attention of chemists for a long time. Indole derivatives have shown a wide spectrum of biological activities including anti-convulsant and anti-inflammatory activities. Several naturally occurring alkaloids, for example, elegantine, rhynchophylline and surgatoxin, are indole-based compounds with a spiro atom at position 3 of the 2-indoline skeleton. Another example is Strychofline, isolated from Strycnos unsambarensis, which has been found to exhibit anti-mitotic activity in cancer cell cultures. Dandia and co-workers[53] have explored microwave-assisted synthetic routes to various spiro thiazolidine-indoles and spiro-benzothioazine-indoles. These compounds are generally synthesised in a two-step procedure under reflux condition for about 10 h. However, performing a one-step microwave-assisted procedure, the reaction time was reduced to 3–6 min. The reaction was run either in ethanol or under solvent-free conditions using silica gel as the solid support. The condensation of the appropriate indole-2,3-diones with anilines afforded, 3-aryl imino-2H-indol-2-ones which, *in-situ*, were cyclised with mercaptoacetic-, 2-mercapto propionic- or *o*- mercaptobenzoic-acid to give the spiro compounds in 54–89% yield, Scheme 5.34.

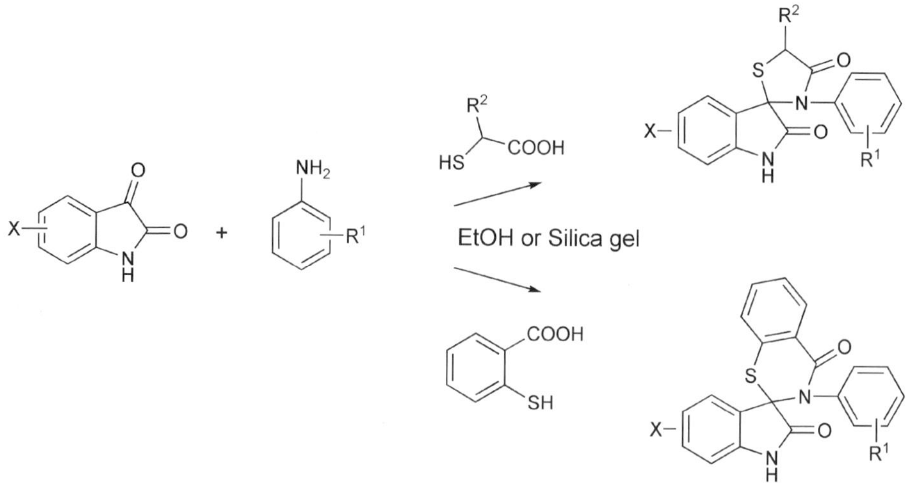

Scheme 5.34

5.4.12. α-Amino phosphonates

During the summer of 2001, two groups reported the development of three-component syntheses of α-amino phosphonates[54]. α-Amino phosphonates are of biological relevance since they act as peptide mimetics. Compounds containing this structural element have been used as enzyme inhibitors, antibiotics, herbicides and pharmacological agents, because of their close structural analogy to α-amino acids.

A number of synthetic methods for the synthesis of aminoalkyl phosphonates have been developed during the past two decades. Generally, they are prepared from the addition of phosphorous nucleophiles to imines in the presence of either base or acid. Standard Lewis acids such as $SnCl_4$, BF_3OEt_2, $ZnCl_3$ and $MgBr_2$ have been used for this transformation. Since the amines and the water that forms during imine formation can decompose or deactivate the Lewis acids, these reactions cannot be carried out in a 'one-pot' operation. However, the use of lanthanide triflates or indium trichloride as catalysts circumvents these problems. These procedures do not require the pre-synthesis of unstable imines, although longer reaction times (10–20 h) are required to obtain the desired products in good yields. Interestingly, metal triflates are found to be effective for the chemoselective reaction with aldehydes without touching the ketones.

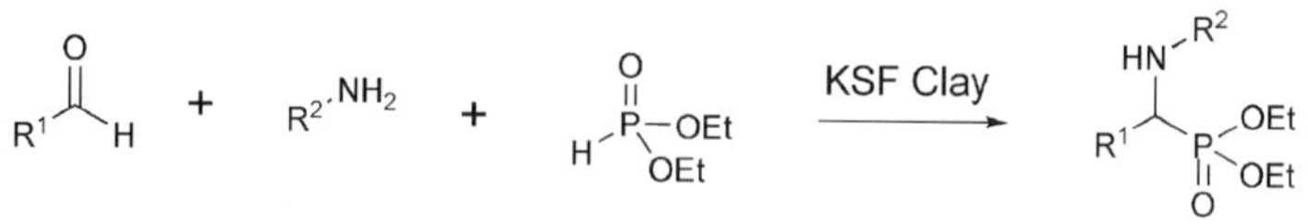

Scheme 5.35

Yadav and co-workers have described a general method for the reaction between benzaldehyde, benzylamine and diethyl phosphite. In the presence of KSF clay, α-amino phosphonates were formed in a few minutes under microwave irradiation, Scheme 5.35.

The reactions proceeded smoothly under solvent-free conditions. Both aromatic and aliphatic aldehydes provided excellent yields of products (80–92%) in a short reaction time, whereas ketones gave phosphonates in good yields (65–80%) after slightly longer irradiation (6–8 min). Several aromatic, α,β-unsaturated, heterocyclic and aliphatic aldehydes also worked well to afford the phosphonates in high yields.

5.4.13. 6-Cyano-5,8-dihydropyrido[2,3-d]pyrimidin-4(3H)-ones

Quiroga and co-workers[55] have described a facile three-component microwave assisted, 'one-pot' synthesis of 5-aryl-6-cyano-7-phenyl-5,8-dihydropyrido[2,3-d]pyrimidin-4 (3H)-ones suitable also in a combinatorial set-up, Scheme 5.36. Equimolar quantities of the starting compounds were placed in open vessels and irradiated in a domestic microwave oven for 15–20 min at 600 W. When irradiation was complete, the resulting solid was treated with ethanol and filtered to give the products in 70–75% yield. Under reflux in ethanol, a much longer reaction time (40–48 h) was required to provide the product in very modest yields (21–25%).

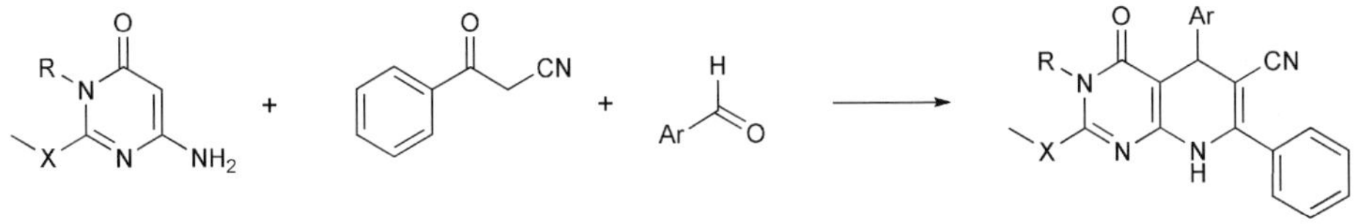

Scheme 5.36

When the substrates were irradiated for just 8–12 min, the reaction in all cases resulted in the formation of stable, hydrated intermediates, Fig. 5.3, which were isolated and characterised. If the intermediate was continuously irradiated (an additional 6–10 min) further, dehydration occured yielding 5-aryl-6-cyano-7-phenyl-5,8-dihydropyrido[2,3-d]pyrimidin-4(3H)-one as the principal product.

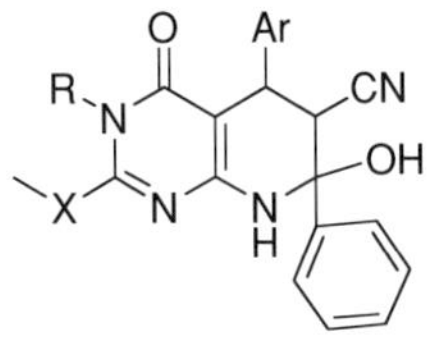

Figure 5.3   5-Aryl-6-cyano-7-phenyl-5,8-dihydropyrido [2,3-d] pyrimidin-4 (3H)-ones.

5.4.14. Multi-component reactions using isatoic anhydride

Isatoic anhydride can be used as a versatile synthon for the synthesis of a diverse set of molecules. The isatoic anhydride can be reacted with amines, amides, hydrazides, isothiocyanates, diketo substrates, aminoacids, amino-, thio or hydroxy anilines, as well as in a three-component reaction with aldehydes and amines to form a large set of diverse pharmacophores.

5.4.14.1. *2,3-Dihydro-2,3-disubstituted benzo(γ)quinazoline* In 1992, Kumar and co-workers[56] described a two-step 'one-pot' synthesis of 2,3-dihydro-2,3-disubstituted benzo(γ)quinazolin-4(1*H*)-ones, by reacting isatoic anhydride with *in-situ* formed imines for approximately 1.5 h. We, however, found that this reaction also worked well as a three-component reaction (J. Westman and K. Orrling, Personal Chemistry, Uppsala, Sweden, unpublished results). By treating isatoic anhydrides with amines and aromatic aldehydes in water at 180–200°C for 15 min, the products were obtained in 22–78% yield, Scheme 5.37. Lower yields were found when electron-withdrawing groups were present in the para position relative to the amine functionality. The reaction also worked poorly, when R^2 ≠ H (i.e N-substituted).

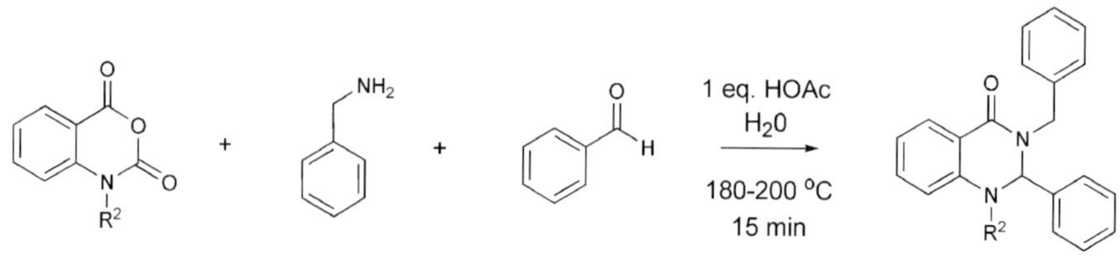

Scheme 5.37

5.4.14.2. *6-Aryl-5,6-dihydrobenzo[4,5]imidazo[1,2-C]-quinazolines* Isatoic anhydride could also be reacted with amino-, hydroxyl- or thiolanilines to form 2-(2-aminophenyl)benz-imidazoles, oxazoles or thiazoles, Scheme 5.38. In the case of 2-(2-aminophenyl)benzimidazoles (X=N), the product was formed after 3 min at 150°C in acetic acid. The products could subsequently be further elaborated; 6-aryl-5,6-dihydrobenzo[4,5]imidazo[1,2-C]quinazolines, a four ring system, was formed by treatment of the 2-(2-aminophenyl) benzimidazole (X=N) with different aldehydes in acetic acid at 150°C for 5 min (J. Westman, and K. Orrling, Personal Chemistry, Uppsala, Sweden, unpublished results). Fifteen compounds were synthesised in 20–75% overall yield. This 3 + 5 min procedure should be compared to the conventional heating protocol developed by Devi and co-workers[57], where each reaction step was run overnight to eventually afford the products in only 30–50% yield.

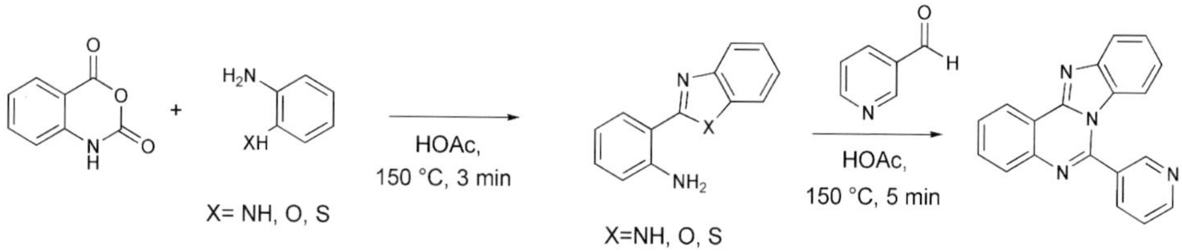

Scheme 5.38

5.4.15. *Pyrido[2,3-d]pyrimidines*

In 1996, Borrell and co-workers published[58] a two-step procedure to form pyrido[2,3-*d*]pyrimidines. Mont and co-workers have recently transformed this reaction to a single-step three-component reaction[59]. α,β-Unsaturated carboxylic acid esters were reacted with guanidine or benzamidine derivatives and malononitrile or methyl cyanoacetate in methanol. When using the guanidine ($R^3 = NH_2$), the products were obtained in 93–99% yield after 10 min at temperatures ranging from 120 to 140°C. When using benzamidines ($R^3 = Ph$), the reactions were performed at 80–120 °C to afford the products in 15–72% yield, Scheme 5.39.

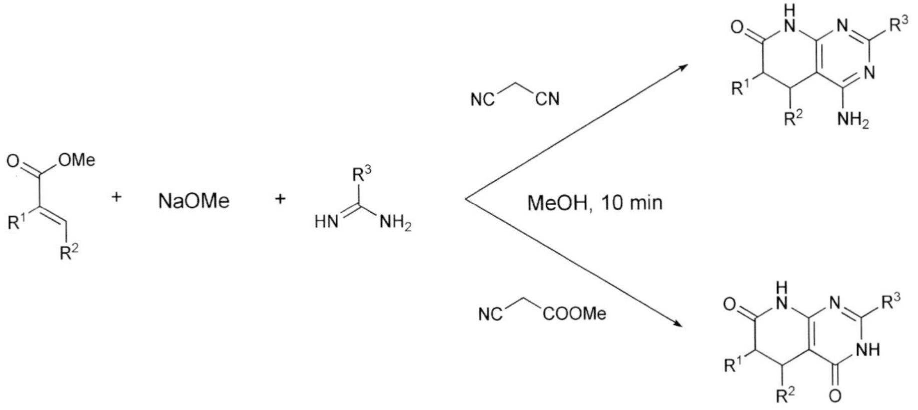

Scheme 5.39

5.5. Summary

The application of microwave technology to speed up chemical reactions is now an accepted and acknowledged tool among chemists. However, combining it with other methodologies such as multi-component reactions can further enhance the benefits offered by the microwaves alone.

This chapter is a non-comprehensive summary of the available papers where microwave dielectric heating has been used in combination with multi-component reactions. The examples highlight the fabulous possibilities that can be found in the cross-section of these technologies. The numbers of papers describing the combination are still limited, but nevertheless microwave heating will have an impact in the area and most certainly, we will see an increased number of publications and increased interest in the near future.

5.6. References

1. Gedye, R., Smith, F., Westaway, K., Ali, H., Baldisera, L., Laberge, L. and Rousell, J., The use of microwave ovens for rapid organic synthesis, *Tetrahedron Lett.*, 1986, **27**, 279–282.
2. Dömling, A. and Ugi, I., Multicomponent reactions with isocyanides, *Angew. Chem., Int. Ed. Engl.*, 2000, **39**, 3168–3210; Weber, L., Illgen, K. and Almsttetter, M., Discovery of new multi component reactions with combinatorial methods, *Synlett*, 1999, **3**, 366–374.
3. Dömling, A., The discovery of new isocyanide-based multi-component reactions, *Curr. Opin. Chem. Biol.*, 2000, **4**, 318–323; Dömling A., and Ugi I.K., A new 5,6-dihydro-2*H*-1,3-oxazine synthesis *via* Asigner-type condensation, *Tetrahedron*, 1993, **49**, 9495–9500.
4. Strecker, A., *Justus Liebigs Ann. Chem.*, 1850, **75**, 27–32.

5. Tietze, L.F., Domino reactions in organic synthesis, *Chem. Rev.*, 1996, 115–136.
6. Hantzsch, A., Uber die synthese pyridinartiger verbindungen aus acetessigather und aldehyammoniak, *Justus Liebigs Ann. Chem.*, 1882, **215**, 1–82.
7. Alajarín, R., Vaquero, J.J., García Navío, J.L. and Alvarez-Builla, J., Synthesis of 1,4-dihydropyridines under microwave irradiation, *Synlett*, 1992, 297–298.
8. Khadilkar, B.M., Gaikar, V.G. and Chitnavis, A.A., Aqueous hydrotrope solution as a safer medium for microwave enhanced Hantzsch dihydropyridine ester synthesis, *Tetrahedron Lett.*, 1995, **36**, 8083–8083; Khadilkar, B.M. and Chitnavis, A.A., Rate enhancement in the synthesis of some 4-aryl-1,4-dihydropyridines using methyl 3-aminocrotonate, under microwave irradiation, *Indian J. Chem., Sect. B*, 1995, **34**, 652–653.
9. Öhberg, L. and Westman, J., An efficient and fast procedure for the Hantzsch dihydropyridine synthesis under microwave conditions, *Synlett*, 2001, 1293–1296.
10. Hilgeroth, A., Baumeister, U. and Heinemann F.W., Solution-dimerization of 4-Aryl-1,4-dihydropyridines, *Eur. J. Org. Chem.*, 2000, 245–250; Hilgeroth, A., Baumeister, U. and Heinemann F.W., Rotameric properties of novel N-acyl and N-acyloxy dimeric 4-phenyl-1,4-dihydropyridines derived from developed solid-state synthesis, *Heterocycles*, 1999, **51**, 2367–2376.
11. Balalaie, S. and Kowsari, E., One-pot synthesis of N-substituted 4-aryl-1,4-dihydropyridines under solvent-free conditions and microwave irradiation, *Monatsh. Chem.*, 2001, **132**, 1551–1555.
12. Cotterill, I.C., Usyatinsky, A.Y., Arnold, J.M., Clark, D.S., Dordick, J.S., Michels, P.C. and Khmelnitsky, Y.L., Microwave assisted combinatorial chemistry. Synthesis of substituted pyridines, *Tetrahedron Lett.*, 1998, **39**, 1117–1120.
13. Glass, B.M. and Combs, A.P., In: *Fifth Intl. Electronic Conf. Synth. Org. Chem.*, http://www.mdpi.net/ecsoc-5/e0027/e0027.htm, 1–30 sept. 2001.
14. Kappe, C.O., Kumar, D. and Varma, R.S., Microwave-assisted high-speed parallel synthesis of 4-aryl-3,4-dihydropyrimidin-2(1H)-ones using a solventless Biginelli condensation protocol, *Synthesis*, 1999, **10**, 1799–1803; Stadler, A. and Kappe, C.O., Microwave-mediated Biginelli reactions revisited. On the nature of rate and yield enhancements, *J. Chem. Soc., Perkin Trans. 2*, 2000, 1363–1368; Dandia, A., Saha, M. and Taneja, H., Synthesis of fluorinated ethyl 4-aryl-6-methyl-1,2,3,4-tetrahydropyrimidine-2-one/thione-5-carboxylates under microwave irradiation, *J. Fluorine Chem.*, 1998, **90**, 17–21; Gupta, R., Gupta, A.K., Paul, S. and Kachroo, P.L., Improved syntheses of some ethyl 4-aryl-6-methyl-1,2,3,4-tetrahydropyrimidin-2-one/thione-5-carboxylates by microwave irradiation, *Indian J. Chem.,Sect. B*, 1995, **34**, 151–152; Krstenansky, J.L. and Khmelnitsky, Y., Biocatalytic combinatorial synthesis, *Bioorg. Med. Chem.*, 1999, **7**, 2157–2162; Stefani, H.A., Gatti, P.M., 3,4-Dihydropyrimidin-2(1H)-ones: fast synthesis under microwave irradiation in solvent free conditions, *Synth. Commun.*, 2000, **30**, 2165–2174; Yadav, J.S., Subba Reddy, B.V., Jagan Reddy, E. and Ramalingam, T., Microwave-assisted efficient synthesis of dihydro pyrimidines: improved high yielding protocol for the Biginelli reaction, *J. Chem. Res., Synop.*, 2000, 354–355.
15. Shaabani, A., Bazgir, A. and Teimouri, F., Ammonium chloride-catalyzed one-pot synthesis of 3,4-dihydropyrimidin-2-(1H)-ones under solvent-free conditions, *Tetrahedron Lett.*, 2003, 857–859.
16. Stadler, A. and Kappe, O.C., Automated library generation using sequential mmicrowave-assisted chemistry. Application toward the Biginelli multicomponent condensation, *J. Comb. Chem.*, 2001, **3**, 624–630.
17. Hoel, A.M.L. and Nielsen, J., Microwave-assisted solid-phase Ugi four-component condensations, *Tetrahedron Lett.*, 1999, **40**, 3941–3944.
18. Ley, S.V. and Taylor, S.J., A polymer-supported [1,3,2]oxazaphospholidine for the conversion of isothiocyanates to isocyanides and their subsequent ude in the Ugi reaction, *Bioorg. Med. Chem. Lett.*, 2002, **12**, 1813–1816.
19. Launay, D., Booth, S., Clemens, I., Merritt, A. and Bradley, M., Solid phase-mediated synthesis of isonitriles, *Tetrahedron Lett.*, 2002, **43**, 7201–7203.
20. Zbruyev, O.I., Stiansi, N. and Kappe, C.O., Preparation of thioamide building blocks via microwave-promoted three-component kindler reaction, *J. Comb. Chem*, 2003, **5**(2), 145–148.
21. Moghaddam, F.M. and Ghaffarzadeh, M., Microwave-assisted rapid hydrolysis and preparation of thioamides by Willgerodt–Kindler reaction, *Synth. Commun.*, 2001, **31**, 317–321; Gupta, M., Paul, S. and Gupta, R., Synthesis of 1,4-dithiocarbonyl piperazines under microwave irradiation in solvent-free conditions, *Synth. Commun.*, 2001, **31**, 53–59; Nooshabadi, M., Aghapoor, K., Reza Darabi, H. and Majid Mojtahedi, M., The rapid synthesis of thiomorpholides by Willgerodt–Kindler reaction under microwave heating, *Tetrahedron Lett.*, 1999, **40**, 7549–7552.
22. Frutos Hoener, A.P., Henkel, B. and Gauvin, J.-C., Novel one-pot microwave assisted Gewald synthesis of 2-acyl amino thiophenes on solid support, *Synlett*, 2003, 63–66.
23. Westman, J., An efficient combination of microwave dielectric heating and the use of solid-supported triphenylphosphine for Wittig Reaction, *Org. Lett.*, 2001, 3745–3747.
24. Gadhwal, S., Baruah, M., Prajapati, D. and Sandhu, J.S., Microwave-assisted regioselective synthesis of β-aminoketones via the mannich reaction, *Synlett*, 2000, **3**, 341–342.

25. Iqbal, M., Vyse, N., Dauvergne, J. and Evans, P., Microwave promoted Pauson–Khand reactions, *Tetrahedron Lett.*, 2002, 43, 7859–7862.
26. Fischer, S., Groth, U., Jung, M. and Schneider, A., Co$_2$(CO)$_8$ catalyzed Pauson–Khand reaction under microwave irradiation, *Tetrahedron Lett.*, 2002, **43**, 2023–2026.
27. Lakhrissi, Y., Taillefumier, C., Lakhrissi, M. and Chapleur, Y., Efficient conditions for the synthesis of *C*-glycosylidene derivatives: a direct and stereoselective route to *C*-glycosyl compounds, *Tetrahedron: Asymmetry,* 2000, 11, 417; Sabitha, G., Reddy, M.M., Srinivas, D. and Yadav, J.S., Microwave irradiation: Wittig olefination of lactones and amides, *Tetrahedron Lett.*, 1999, 40, 165–166; Xu, C., Chen, G., Fu, C. and Huang, X., The Wittig reaction of stable ylide with aldehyde under microwave irradiation: synthesis of ethyl cinnamates, *Synth. Commun.*, 1995, 25, 2229–2234; Fu, C., Xu, C., Huang, Z.-Z. and Huang, X., α,β-Unsaturated sulfones by the Wittig reaction of stable ylide with aldehydes under microwave irradiation, *Org. Prep. Proced. Int.*, 1997, 29, 587; Spinella, A., Fortunati, T. and Soriente, A., Microwave accelerated wittig reactions of stabilized phosphorus ylides with ketones under solvent-free conditions, *Synlett*, 1997, 93–94.
28. Buddrus, J., *Angew. Chem.*, 1974, **94**, 1173–1177.
29. Strauss, C.R. and Trainor, R.W., Invited review. Developments in microwave-assisted organic chemistry, *Aust. J. Chem.*, 1995, **48**, 1665–1692.
30. Kobayashi, S., Araki, M., Ishitani, H., Nagayama, S. and Hachiya, I., Activation of imines by rare earth metal triflates. Ln(OTf)$_3$- or Sc(OTf)$_3$-catalyzed reactions of imines with silyl enolates and Diels–Alder reactions of imines, *Synlett*, 1995, 233–234.
31. Westman, J. and Orrling, K., Cascade synthesis with (triphenylphosphoranylidene)-ethenone as a versatile reagent for fast synthesis of heterocycles and unsaturated amides under microwave dielectric heating, *Comb. Chem. High Throughput Screening*, 2002, **5**, 571–574.
32. Westman, J., Lundin, R., Stålberg, J., Östbye, M., Franzén, A. and Hurynowicz, A., Alkylaminopropenones and alkylamino-propenoates as efficient and versatile synthons in microwave-assisted combinatorial synthesis, *Comb. Chem. High Throughput Screen,* 2002, **5**, 565–570.
33. Molteni, V., Hamilton, M.M., Mao, L., Crane, C.M., Termin, A.P. and Wilson, D.M. Aqueous one-pot synthesis of pyrazoles, pyrimidines and isoxazoles promoted by microwave irradiation, *Synthesis*, 2002, 1669–1674.
34. Bienaymé, H., "Reagent explosion": an efficient method to increase library size and diversity, *Tetrahedron Lett.*, 1998, **39**, 4255–4258.
35. Westman, J. and Lundin, R., Solid phase synthesis of aminopropenones and aminopropenoates. Efficient and versatile synthons for combinatorial synthesis of heterocycles, *Synthesis*, 2003, **7**, 1025–1030.
36. Coleman, C.M., MacElroy, J.M.D., Gallagher, J.F. and O'Shea, D.F., Microwave parallel library generation: comparison of a conventional- and microwave-generated substituted 4(5)-sulfanyl-1*H*-imidazole library, *J. Comb. Chem.*, 2002, **4**, 87–93.
37. Usyatinsky, A.Ya. and Khmelnitsky, Y.L., Microwave-assisted synthesis of substituted imidazoles on a solid support under solvent-free conditions, *Tetrahedron Lett.*, 2000, **41**, 5031–5034.
38. Balalaie, S., Arabanian, A. and Hashtroudi, M.S., Zeolite HY and silica gel as new and efficient heterogenous catalysts for the synthesis of triarylimidazoles under microwave irradiation, *Monatsh. Chem.*, 2000, **131**, 945–948; Balalaie, S. and Arabanian, A., One-pot synthesis of tetrasubstituted imidazoles by zeolite HY and silica gel under microwave irradiation, *Green Chem.*, 2000, **2**, 274–276.
39. Balalaie, S., Hashemi, M.M. and Akhbari, M., A novel one-pot synthesis of tetrasubstituted imidazoles under solvent-free conditions and microwave irradiation, *Tetrahedron Lett.*, 2003, **44**, 1709–1711.
40. Groebke, K., Weber, L. and Mehlin, F., Synthesis of imidazo[1,2-*a*] annulated pyridines, pyrazines and pyrimidines by a novel three-component condensation, *Synlett*, 1998, 661; Blackburn, C., Guan, B., Shiosaki, K. and Tsai, S., Parallel synthesis of 3-aminoimidazo[1,2-*a*]pyridines and pyrazines by a new three-component condensation, *Tetrahedron*, 1998, 3635–3638; Bienaymé, H. and Bouzid, K., A new heterocyclic multicomponent reaction for the combinatorial synthesis of fused 3-aminoimidazoles, *Angew. Chem., Int. Ed. Engl.*, 1998, **37**, 2234–2237.
41. Varma, R.S. and Kumar, D., Microwave-accelerated three-component condensation reaction on clay: solvent-free synthesis of imidazo[1,2-*a*] annulated pyridines, pyrazines and pyrimidines, *Tetrahedron Lett.*, 1999, **40**, 7665–7669.
42. Rostamizadeh, S. and Sadeghi, K., One-pot synthesis of 1,2,4-triazines, *Synth. Commun.*, 2002, **32**(12), 1899–1902.
43. Zhao, Z., Leister, W., Strauss, K., Wisnoski, D.D. and Lindsley, C.W. Broadening the scope of 1,2,4-triazine synthesis by the application of microwave technology, *Tetrahedron Lett.*, 2003, **44**, 1123–1127.
44. Bora, U., Saikia, A. and Boruah, R.C., A novel microwave-mediated one-pot synthesis of indolizines via a three-component reaction, *Org. Lett.*, 2003, **5**, 435–438.
45. Wilson, N.S., Sarko, C.R. and Roth, G.P., Microwave-assisted synthesis of a [3+2] cycloaddition library, *Tetrahedron Lett.*, 2001, **42**, 8939–8941.
46. Wilson, N.S., Sarko, C.R. and Roth, G.P., Microwave-assisted synthesis of 2-aminoquinolines, *Tetrahedron Lett.*, 2002, **43**, 581–583.

47. Rad-Moghadam, K. and Khajavi, M.S., One-pot synthesis of substituted quinazolin-4(3H)-ones under microwave irradiation, *J. Chem. Res.*, 1998, 702–703.
48. Shiraishi, H., Nishitani, T., Sakaguchi, S. and Ishii, Y., Preparation of substituted alkylpyrroles via samarium-catalyzed three-component coupling reaction of aldehydes, amines, and nitroalkanes, *J. Org. Chem.*, 1998, 63, 6234–6238.
49. Danks, D.N., Microwave assisted synthesis of pyrroles, *Tetrahedron Lett.*, 1999, **40**, 3957–3960.
50. Ranu, B.C., Hajra, A. and Jana U., Microwave-assisted synthesis of substituted pyrroles by a three-component coupling of α,β-unsaturated carbonyl compounds, amines and nitroalkanes on the surface of silica gel, *Synlett*, 2000, 1, 75–76.
51. Ranu, B.C. and Hajra, A., Synthesis of alkyl-substituted pyrroles by three-component coupling of carbonyl compound, amine and nitro-alkane/alkene on a solid surface of silica gel/alumina under microwave irradiation, *Tetrahedron*, 2001, **57**, 4767–4773.
52. Nemes, C. and Laronze, J.-Y., Trimolecular condensation of substituted indoles with paraformaldehyde and Meldrum's acid. Convective heating versus microwave irradiation: a comparative study, *Synthesis*, 1999, 254–257.
53. Dandia, A., Saha, M. and Taneja, H., Improved one-pot synthesis of 3-spiro indolines under microwave irradiation, *Phosphorus, Sulfur Silicon*, 1998, 139, 77–85.
54. Kaboudin, B. and Nazari, R., Microwave-assisted synthesis of 1-aminoalkyl phosponates under solvent-free conditions, *Tetrahedron Lett.*, 2001, **42**, 8211–8213; Yadav, J.S., Subba Reddy, B.V. and Madan, Ch., Montmorillonite clay-catalyzed one-pot synthesis of α-amino phosphonates, *Synlett*, 2001, 1131–1133.
55. Quiroga, J., Cisneros, C., Insuassty, B., Abonía, R., Nogueras, M. and Sánchez, A., A regiospecific three-component one-step cyclocondensation to 6-cyano-5,8-dihydropyrido[2,3-d]pyrimidin-4(3H)-ones using microwaves under solvent-free conditions, *Tetrahedron Lett.*, 2001, **42**, 5625–5627.
56. Kumar, R.R. and Reddy, M.S., One-pot synthesis of 2,3-dihydro-2,3-disubstituted benzo[g]quinazolin-4(1H)-ones, *Synth. Commun.*, 1992, **22**, 2499–2508.
57. Devi, R.R. and Reddy M.S., Synthesis of 6,7-dihydro-6-substituted benzimidazo[1,2-c]benzo[g]quinazolines and their heteroaromatic analogues, *Indian J. Chem.* 1994, **33B**, 1013–1016.
58. Borrell, J.I., Teixido, J., Martinez-Teipel, B., Serra, B., Matallana, J.L., Costa, M., and Batllori, X., An uniequivocal synthesis of 4-amino-1,5,6,8-tetrahydropyrido[2,3-d]pyrimidine-2,7-diones and 2-amino-3,5,6,8-tetrahydropyrido[2,3-d]pyrimidine-4,7-diones, *Collect. Czech. Chem. Commun.*, 1996, **61**, 901–909.
59. Mont, N., Teixido, J., Borrell, J.I. and Kappe, C.O., The "Victory" reaction: a three component synthesis of pyrido[2,3-d]pyrimidines, *Poster at CHI's Advancing Library Design and Organic Synthesis Meeting*, Feb. **24–27**, 2003, La Jolla, California.

6 Integrating microwave-assisted synthesis and solid-supported reagents

I.R. BAXENDALE, A.-L. LEE and S.V. LEY

6.1. Introduction

Microwave-assisted chemical processing has received a considerable amount of interest as a technology of the future since its instigation in the late 1980s[1,2] as an alternative heating method for organic reactions. Unfortunately, despite the potential revolutionising impact, microwave chemistry remained for many years more of an academic curiosity than a mainstream scientific tool. The historical reasons for this lack of acceptance are well known and have been extensively reviewed, and hence will not be covered here, except to state that they essentially revolved around the reproducibility of the experiments conducted. As a result, the area of microwave science as applied to organic synthesis remained rather dormant for over a decade. This period ended suddenly with the recent commercialisation of focused multi-mode applicator microwaves that are capable of operating at variable temperature and power settings. These machines have been specifically developed for the synthetic community so as to facilitate the rapid heating of organic reactions with the aim of enhancing productivity and shortening production times. Their introduction has significantly revitalised the general area and permitted microwave chemistry to become a mainstream tool of most bench chemists.

Despite the widespread adoption of this technology, we still do not possess a fully comprehensive knowledge of many of the fundamental aspects of microwave heating and its effects on the promoted chemistries. The area is still very much at an exploratory stage, where significant investigations have yet to be done to augment our understanding. For example, there remains a great deal of contention in the literature surrounding the possibility of rate enhancements due to athermal effects created by microwave irradiation[3-17]. The results and hypotheses from both sides of the discussion are very convincing, although with what amounts to a very small population of verified data it may be premature to draw absolute conclusions, as a number of authors seem apt to do. As with most emerging sciences, the reality will probably be more complex and less generic than the initial investigations seem to imply. This is certainly the case when considering the application of microwave heating to heterogeneous reactions in which significant differences are becoming apparent in the ways these systems can interact with microwave irradiation when compared to their more simplistic homogeneous counterparts. It is however evident, from nearly all the published work, that the application of microwave irradiation for heating chemical reactions is a useful approach for achieving higher reaction kinetics and the formation of cleaner products[18-24].

In addition to our interest in the use of microwaves for general synthetic chemistry[25-32], we have also been particularly intrigued with the potential benefits that could be realised by integrating them into our existing immobilised reagent methodologies and related synthesis platforms[24,33-36]. A significant problem encountered when using supported reagents is the extended reaction times relative to their parent solution-phase equivalent. This reduced reactivity is unfortunately inherent with the immobilisation of the active species on a support material and its resulting heterogeneous nature. This effect therefore represents a significant technological challenge and a potential drawback to the continued development of the area.

In this chapter, we review and discuss some of the applications of microwave-related techniques to the area of accelerated synthesis using polymer-supported reagents. Besides a general description of working examples, we would also like to give a brief discourse on the current understanding and explanations for the reactivity observed. Not all the information contained within this chapter comes from referenced sources; there is an ever-increasing knowledge base of industrial expertise that we are fortunate to be aware of and would like to acknowledge Argonaut, AstraZeneca, GlaxoSmithKline, Millenium Pharamaceuticals, NovaBiochem, Novartis, Organon, Personal Chemistry and Pfizer. Although not peer reviewed by the wider scientific community, this information is still exceedingly valuable and is of a high standard given its economic importance.

6.2. Microwave heating of reactions

There are two distinct synthetic methods of applying microwave heating to reacting samples. The simplest is to conduct the reaction under normal atmospheric pressure in an open reaction vessel. This involves standard laboratory glassware fitted with a modified reflux condenser housed externally to the microwave cavity, using reaction temperatures at or just below the normal reflux point of the solvent. Although not strictly correct, this type of microwave reaction is becoming commonly known as MORE chemistry (microwave-induced organic-reaction enhancement)[37]; however, we feel these types of anagrams are not particularly helpful in defining the area[37–42] though these are frequently used. Even though this method is characteristically closer to standard synthetic methodology, it has been used significantly less than the alternative approach because of certain safety considerations. This alternative approach is microwave flash heating, a technique that utilises rapid heating and cooling of a sample contained in a sealed reaction vessel. In such processes, it is possible to generate high-internal pressures as a result of the elevated reaction temperatures employed, which are usually significantly higher than the standard boiling point of the solvent. This form of rapid thermal transition also seems to be beneficial with regard to the general integrity of the polymer matrix itself (including polymer-supported intermediates), which has increased stability under such short duration, high-pressure thermal exposure. This is certainly in contrast to many other reports that state that significant decomposition of immobilised starting materials and intermediates can occur with the action of prolonged conventional heating cycles at ambient pressure. For this reason, almost all the reactions involving polymer-supported reagents are now conducted using microwave flash heating.

6.2.1. Heating a heterogeneous sample: polymer considerations

When a sample containing a polymer-supported reagent is subjected to electromagnetic irradiation, it is possible for the energy to be directly coupled to polar functional groups. It then takes a defined period of time for the absorbed energy to dissipate to the surrounding environment; this is especially pronounced if the majority of the polymer system has a low-thermal conductivity. Therefore, if the heat distribution time is greater than the time between heating pulses, thermal equilibrium is not reached and significant temperature variations can result. This gives rise to the phenomenon known as superheating[5,43,44]. This type of process has been proposed by Marand and co-workers[45,46] to explain the relative differences in microwave-dependent polymerisation experiments and later by Wei[47,48] to account for the relative rate enhancements in the thermal curing of epoxy resins. Indeed, extensive effort has been invested in the area of microwave curing applications, resulting in a number of interesting observations based on the increased reactivity of functionalised resins[45,46,49,50].

In general, when microwave heating leads to enhanced reaction rates, it is plausible to assume that the reactive sites on the surface of the polymer are subject to selective absorption, which could cause some pathways to predominate. This has certainly been seen in the case of many metal tethered or metal-impregnated catalysts. In these systems, the metal particles can be heated directly without notably heating the support because of their drastically different dielectric properties[51–54]. Roussy and co-workers have also described the effect of structural voids in materials, which can produce field non-uniformities in microwave heated samples resulting in temperature fluctuations[55]. Additional studies aimed at identifying and predicting the temperature variations in porous structures have concentrated on particulate metals immobilised within inorganic supports. These systems, usually comprising of small (1–5 mm) spherical pellets, have been used in continuous flow microwave reactors, and in general indicated only minimal microscopic temperature variations. There was, however, a strong dependence on the dielectric properties of the support material[56]. In cases where the inorganic matrix exhibited a high degree of 'lossy' character[57], only small temperature variations were noticed, but these became more pronounced as the absorption characteristics of the support became lower. This situation would be more representative of a functionalised polymeric resin, although the interaction and dissolution effects of any solvent present would significantly affect the microwave coupling ability. The speculation about the ability to selectively heat active sites to temperatures higher than the bulk surroundings is the subject of extensive research[58]. Such directed or localised superheating to create 'hot spots' by microwave irradiation is highly desirable as it represents a very simple and efficient method of supplying energy to the site of reaction, which otherwise cannot be achieved by conventional methods. Unfortunately, measuring the specific temperatures of individual active sites is beyond the current experimental capabilities and so is severely limiting the progress in this area. This area would offer enormous advantages for synthetic applications, especially for large-scale manufacture, if the required tuning and reproducibility can be obtained. In theory, only minor changes in local temperatures variations would need to be achieved to have quite drastic effects on overall chemical conversions, although these local heating effects can result in exponential reaction kinetics, raising some safety concerns.

6.2.2. *Heating a polymer-solvent: a binary phase system*

The interactive effect of the polymer and the solvent can be easily demonstrated in a very simple but crude microwave heating experiment, where no consideration is given

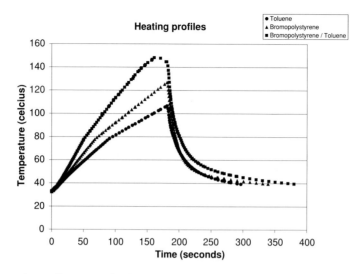

Figure 6.1 Heating profiles associated with the poorly absorbing solvent toluene and then the repeated heating showing the interaction of the solvent with a polymer resin.

to such factors as volume density, sample packing or irradiation penetrating differences. Presented in Fig. 6.1 are the heating profiles for three separate reactions involving the heating of (1) a sample containing neat toluene, (2) dry bromopolystyrene resin and (3) an equivalent weight of pre-swollen bromopolystyrene (toluene).

As can be seen from the data presented in Fig. 6.1, the heating profile of the pre-swollen resin is significantly enhanced in comparison to both the neat toluene and the dry resin samples. Of course, the swelling of the resin is a very important consideration, especially where a polar resin is being used because the charge distribution between the two phases will be enhanced (see later). This causes a difference in the sample's internal electric gradient that will affect the power absorption and ultimately the samples heating characteristics.

Many experimentalists are familiar with this principle of doping a sample with a species that couples better with the microwave irradiation and so can act as a thermal dissipater. What is often less appreciated is the general nature of this process, as not only solid/liquid interfaces but also liquid/liquid biphasic systems such as emulsions show the same effects[59–63]. Figure 6.2 represents the heating profiles of toluene and a perfluorinated solvent first independently and then as an emulsion. A similar trend can be seen in a hexane/acetonitrile mixture, although because of the superior heating capacity of acetonitrile the effect is less pronounced.

To place these results in context it should be reaffirmed that microwave heating is a completely distinct phenomenon operating in an entirely different way from microwave spectroscopy. This latter process involves the direct interaction of photons of a particular energy in order to excite the quantum rotational levels of gas-phase sample. Although in microwave heating, the absorption of microwave irradiation by a sample has been shown to be frequency dependent, it is not a requirement of the system for the energy to be quantised. As a result the heating process does not depend upon the direct absorption of microwave photons, instead the sample is heated via

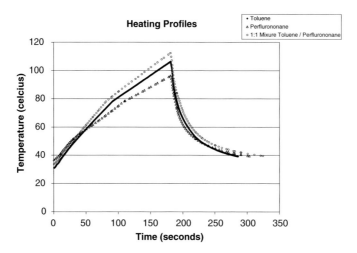

Figure 6.2   Doping effects of solvent combinations.

molecular interactions with the high-frequency electric and magnetic fields[64–67]. It is also understood that in biphasic systems the electrical properties of the two media (i.e., the electrical conductivity and dielectric constants) will differ and therefore the action of an applied electric field will be to induce electrical charges to appear at the boundary between the primary phase and the surrounding continuum. The distortion of the electrical double layer and the creation of Maxwell–Wagner interfacial polarisation will thus generate an electrical dipole moment[68–70]. Interestingly, the magnitudes of these induced dipoles are very large in comparison to those usually associated with molecular species because, although the magnitude of the charges involved may not themselves be large, the distance between their oppositely charged poles is by comparison a significant effect. As a consequence these induced dielectric properties cause enhanced and constructive wave interference within the sample leading to a more effective dielectric coupling and superior sample heating.

In addition the Maxwell–Wagner field theory also predicts that at the interface of two phases a proportion of the total microwave irradiation will be reflected. This can also be explained in terms of Snell's law of total internal reflection that describes the effect of altering the speed of photon movement between two media and is responsible for the bending (reflectance) of incident light. In this case although the emergent light's speed changes, as does the wavelength, the frequency remains constant ($v = f\lambda$). Considering the tremendous surface area and high-conformational diversity encountered throughout a polymer structure, we would expect a large number of interfacial combinations to be created in a system comprising of a polymer mixed with a solvent. Consequently, these phase boundaries would promote multiple internal reflections, thus increasing the available power absorbed before the radiation exits the sample. In effect this is equivalent to increasing the internal path length of the sample[59].

As these two aspects imply, the interaction of two phases can have a significant impact on the heating profile of a sample. It may also be of interest to consider the effect that the other reaction conditions, such as extremes of temperature and pressure, could have upon the development and continuation of such interfacial effects (and the associated polarisation) during the reaction heating cycle. For example, as we will briefly consider later, the chemical nature and physical character of a solvent at non-ambient conditions can be substantially modified, therefore it is worth bearing in mind that as the reaction conditions change, so may the features of the solvent, which will alter the interaction with the secondary system.

6.2.3. *Migration of the reacting species*

The potential for reaction rate enhancement as a consequence of improved transport properties is an important consideration in any heterogeneous reaction. However, in the case of most supported reagent applications, an excess of the immobilised system is used (pseudo-first-order kinetics), making the requirements for diffusion less of a critical factor. Obviously, diffusion still remains a significant consideration for catalytic applications, as the rate of transportation of both reactants and products through the unreactive bulk remains an important issue. One area of interest is the altering of the diffusion dynamics that occur at the elevated operating temperatures and pressures encountered in many of the microwave-assisted reactions. A more intriguing question is: are there any additional advantages to be gained through the use of microwave irradiation? For example, can higher transport numbers be attained through enhanced organisation of the reactants or by selective activation of components within the mixture?

A number of authors have described noticeable improvements in the absorption/desorption and interaction of small organic molecules with polymeric systems under microwave heating conditions[71,72]. Many applications, including the industrial processing of textiles, where microwave irradiation has been shown to increase the fixing rate of dyes in polymeric fabrics[73,74] are encountered in the literature. However, the majority of the effects observed seem to be concerned with rapidly establishing equilibriums or the faster kinetics obtained at higher temperatures, rather than altering the mechanism of transport. This question has been examined in depth by Jeon and Kim[75], who determined that there was no discernible difference in the migration of simple organic solvents such as *n*-hexane, toluene, ethanol or carbon tetrachloride into polymer structures (polyethylene, PVC and silicone rubber), when compared to contemporary approaches based on classical thermal heating. The only differences they observed in absorption were related to the expected solvent swelling compatibility of the resins tested.

It is, however, worth noting that the gelling factors of many polymers are known to be affected under microwave irradiation, especially those systems comprising of very polar monomer units[76–79]. This can have a profound effect on the migration of reactants and products through the polymer structure and is especially associated with binary polymer systems, such as gel grafts, and other highly polar resins, such as polyacrylamide, where it can cause problems with reaction kinetics. In addition, dipole relaxation times can also become relatively large when a viscous or 'mobility'-restricting medium is involved. For example, if a polymer resin is introduced into a system, greater dielectric loss is observed. This has been extensively investigated only for systems containing soluble polymers[80–90], but the concept can be extrapolated to other resin environments as well. The origin of microwave heating lies with the ability of the electric field to induce a polarisation of charges within the heated sample. This induced polarisation due to the imposed restricted freedom of associated species prevents it from following the rapid oscillations of the electromagnetic field, resulting in heating. Therefore, any change in the systems bonding from interactions with the polymeric support (more so in a polar medium) will significantly enhance dielectric loss and microwave coupling. This effect can also have important implications for the generation of hot spots and superheated areas, a concept we discuss later.

6.2.4. Reaction heating: solvent considerations

As previously stated, the selective heating of a single component can occur under microwave conditions; only components that couple with microwaves are directly heated. The non-absorbing components are thus heated indirectly by energy transfer from the other species. As is often the case in chemistry, the most suitable solvents for one aspect of a particular reaction are not usually the most effective for other reasons. Many of the typical solvents used in resin-based reactions are chosen for their effective resin swelling capabilities, but are often very poor microwave absorbers. (For reference there are three main microwave-related solvent characteristics: relative permittivity, dipole moment and the tangent delta—a term that defines the conversion of electromagnetic energy to thermal energy at a given frequency $\delta = \varepsilon''/\varepsilon'$, where ε'' is complex dielectric permittivity and ε' is dielectric loss. These components are related the absorbing ability of the solvent; see Chapter 1 for further details). There are a few exceptions to this statement such as N,N-dimethylformamide, 1,2-dichlorobenzene and acetonitrile, which can be successfully employed. But as already discussed, in a heterogeneous sample, such as a polymer suspension in a poorly absorbing solvent, heating is associated with the phase boundaries or specific interfacial topology. These regions, known as 'heating zones', are characterised by temperature gradients ranging in dimension from the macroscopic to the molecular scale. Furthermore, solvent pockets located deep within the confines of the resin can also experience the equivalent of a thermal squeeze, resulting from the difference in viscosity of the cooler external bulk solution and the expansion restricting nature of the resin channels. These small effects combined with restricted convection can produce temperature variations. This aspect of heating is of particular importance when products show thermal degradation, but require relatively high-activation energies for formation. Thus reaction in the superheated internal structure followed by migration of the product to the more temperate zones of the reaction media can enhance both the yields and the overall purity of the products.

The effect of solvent heating is obviously very important, because the temperature achieved is directly related to the speed of the reaction. It is therefore very significant that higher boiling points can be reached when solvents are subject to microwave irradiation in the absence of any mechanical stirring mechanism, even at atmospheric pressure. This effect, called superheating[5,43,44], is connected to both the chemical and physical properties of the solvent mixture and has been attributed to retardation of nucleation during microwave heating. Temperatures in excess of 20°C above the standard boiling points of certain solvents can be readily achieved (Table 6.1). These raised boiling profiles are created as a result of the propagation of microwave heating throughout the entire reaction volume and are related to microwave power levels[91–94]. As a consequence, a reverse heating gradient is produced in which lower temperatures are recorded at the periphery of the reaction vessel, due to energy dissipation to the surroundings, which is in

Table 6.1 Solvent boiling points (bp) at ambient temperature under normal thermal heating and under microwave conditions[43,91]

Solvent	bp (°C)	MW bp (°C)	Solvent	bp (°C)	MW bp (°C)
Water	100	104	Dichloromethane	40	55
Methanol	65	84	Chlorobenzene	132	150
Ethanol	79	103	1-Chlorobutane	78	100
Butan-1-ol	118	132	Trichloroethylene	87	108
Propan-2-ol	82	100	N,N-Dimethylformamide	153	170
Acetic anhydride	140	155	Acetone	56	81
Diglyme	162	175	Butan-2-one	80	97
Dimethoxyethane	85	106	Tetrahydrofuran	66	81
iso-Pentyl alcohol	130	149	Acetonitrile	81	107
iso-Propyl ether	69	85	Ethyl acetate	78	95

sharp contrast to classical heating procedures. The presence of materials that preferentially absorb microwave irradiation and exhibit 'lossy' character[56] can also lead to better conversions of the reactants and higher reaction rates. The application of highly absorbent materials such as graphite and amorphous carbon have also been used to create hot spot environments, which facilitate greater rates of reaction[95–104].

Under the appropriate conditions the support material can also act as the solvent. For example, a number of publications have described the use of soluble polymers, such as PEG, to serve as both phase-transfer reagents and the solvent for a reaction[105–115]. PEG systems of low-average-molecular weight offer a number of unique attractions because of their relatively low-melting points (MW 400–800) 45–47°C and facile separation from small organic compounds. Under the action of focused microwave irradiation, PEG can function as the solvent environment very effectively, absorbing microwave irradiation and diffusing the heat to the reactants. In addition, this solvent system can also effectively stabilise and support ionic species, especially cations, in a manner similar to more expensive crown ethers[116–118].

If such a chemical enhancement can be realised at ambient pressures, the significance is obviously going to be much greater when sealed reaction vessels at high temperatures are employed. In this case the dimensions of the reaction chamber, the volume of solvent and its specific boiling point and the final pressure reached will all affect the chemistry taking place. In fact, a direct relationship between the observed pressure and temperature and any rate enhancement can usually be made. An issue that needs additional consideration with regards to promoting homogeneous and heterogeneous reactions is the potential new properties of the solvent under these forcing conditions. The specific properties of a given solvent under standard conditions are well documented, but this is not as well defined when elevated temperatures and pressures are applied. Many of the microwave reactors are capable of running reactions at temperatures exceeding 250°C, with associated pressure containment to approximately 20 atm. Under such conditions, the behaviour and attributes of most solvent systems will be significantly different to their normal characteristics. This situation is well illustrated by methanol, which at high temperatures and pressures resembles a solvent more like hexane. For applications involving polymers, this can have a significant effect because of the difference in swelling ability and the solubility of the reactants and products. A

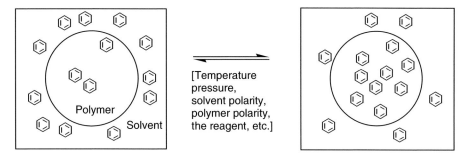

Figure 6.3 A cartoon representation depicting the equilibrium conditions of substrate or solvent molecule interacting with a polymer resin.

biphasic reaction containing a polymeric reagent as one component represents a series of dynamic equilibriums. In addition to any chemical equilibria established due to the interconversion of reactants and products, there are also a series of distribution equilibriums concerned with the varying solubilities and residence of the reacting components in the bulk solvent and the solvent-like polymer species (Fig. 6.3)[119]. Due to the phase boundary, these two environments are in direct competition for the smaller substrate molecules, which will affect both reactivity and product isolation. The solvent will be a critical factor that determines the levels of compartmentalisation. As with all equilibria, temperature and pressure can have a significant effect, but this will be multiplied if the properties of the solvent change (i.e., polarity and solubilising ability). In fact the retention of product within the resin is one of the most common reasons for the isolation of low yields from a resin-based reaction, even under ambient conditions. In addition, increased precipitation and crystallisation of materials involving polymer species is also quite common, because of the selective concentration of compounds into one phase. The situations in the pre-, post- and even during the microwave reaction will be radically different if flash heating techniques are applied. It will therefore often be essential to screen a range of solvents to identify the best combination for a specific reaction and resin type. Although this is a complication to the use of polymers with microwave heating, the possibility arises that these solubility differences and equilibriums can be harnessed to promote novel chemistry, which otherwise would not be possible.

6.3. Microwave reactions with polymer-supported reagents

There are only three partial reviews to date that mention the integrated approach of using polymer-supported reagents with microwave heating, which highlights the novelty and recent establishment of the field[120–122]. It is certainly evident from the growing interest in the two concepts that the complementary nature of their combined use will soon result in significantly more reports. In the following section, we review a number of the existing reactions and supplement them with a selection of related observations to provide a more comprehensive knowledge source for this exciting new field.

6.3.1. *Polymer drying*

The most basic application of microwave heating is in the rapid drying of resins prior to a given reaction. Many polymer reagents are exceedingly hydroscopic, and retained water can affect the properties of the reagent and its resulting reactivity. Extensive use has been made of domestic microwave units to dehydrate resins, especially polystyrene-based material, which are usually very poor microwave absorbers (although this is highly dependant on the loading and type of functionality). This poor absorbency in comparison to the residing water allows for selective heating of the matter resulting in rapid drying (this has also been conducted more effectively on a larger scale under reduced pressure). This approach has also been used for a number of inorganic supports and is a common practice for drying molecular sieves and silica samples.

6.3.2. *Reductive aminations*

It has long been known that imine formation can be rapidly achieved using microwave flash heating[123–125]. In many cases this process is facilitated by using suitable solvents like toluene, which allow the water to be driven from the reaction mixture by either azeotropic distillation or direct evaporation into the vapour phase as a result of selective heating. Chemists at GlaxoSmithKline have extended this coupling sequence to include a rapid reductive amination reaction, using an immobilised cyanoborohydride reagent[126]. In this way a small set of aldehydes and amines were successfully coupled and reduced using an automated microwave procedure (Scheme 6.1). It was noted that under the optimised conditions, a problematic side reaction occurred involving a competing acetylation of the secondary amine products by the acetic acid catalyst. This proved only a minor problem because purification could be easily achieved using an SCX scavenging protocol. This type of direct amide bond formation from an acid and amine has also been described in the literature and has proved to be very facile under microwave heating conditions[127]. One of the other competing reactions that has been identified using both the immobilised and solution-phase cyanoborohydride reagents under the high pressures and temperatures produced using microwave heating is the release of cyanide anion from the borohydride reagent. Although only a minor product, this still obviously indicates some potential health and safety considerations that need to be addressed before doing this transformation.

    Many reductions have also been conducted under similar conditions using the simple immobilised borohydride resin. The majority of these reactions have been conducted because the corresponding room temperature reaction proved to be very slow. Again enhancements in rates and overall yields have been noticed.

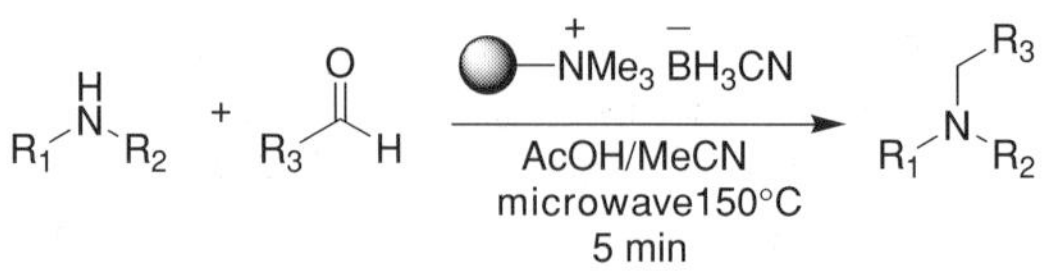

Scheme 6.1

6.3.3. The Henry reaction

The nitro aldol (Henry) reaction is known to be successfully catalysed by many strongly basic ion exchange resins such as those functionalised with chloride or hydroxide ions[128]. More recently, the use of milder bicyclic guanidine bases including the supported TBD variant (1,5,7-triazabicyclo[4.4.0] dec-1-ene) **1** have been shown to be exceedingly effective for the same transformation[129]. In addition, these reactions tended to proceed with higher levels of chemical selectivity, favouring the nitro aldol adducts **2** over the further eliminated alkene by-products **3**. Using an immobilised guanidine reagent, therefore, combines the advantages of increased selectivity with the simplified purification protocols that are particularly important for parallel synthesis applications. Scientists from the pharmaceutical division of Aventis have taken this one step further and investigated this transformation under microwave heating conditions to expedite a more rapid generation of large libraries of intermediates for use as diverse starting materials in subsequent combinatorial programmes (Scheme 6.2)[130]. They reported that in a matter of a few hours, they were able to prepare a reasonably sized array of nitro-alcohols and the corresponding dehydrated β-nitro styrenes. Although no specific details were released regarding the exact distribution of the two possible products, the comment was made that the selectivity of the particular product isolated was very substrate dependent.

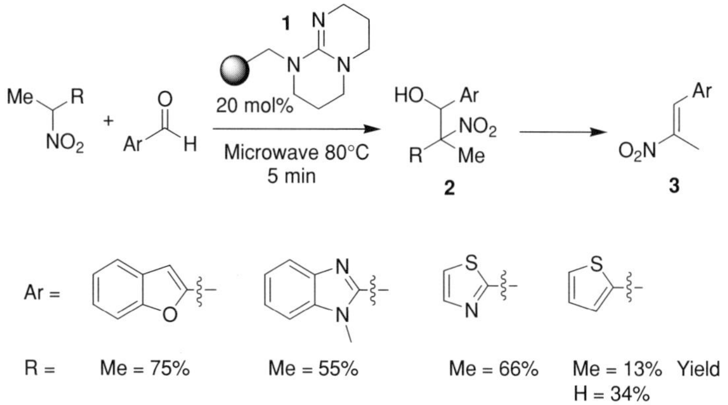

Scheme 6.2

6.3.4. Alkylation reactions

Researchers at Abbott Laboratories have also combined the use of the polymer-supported TBD base **1** and microwave irradiation for the alkylation of various phenolic components[131,132]. These reactions were conducted to evaluate the microwave procedure against the traditional literature conditions, which involved prolonged mixing at ambient temperatures (24–168 h). It was discovered that as expected, elevated temperatures significantly reduced the reaction times, but more significantly, the products in all but a couple of cases could be isolated in higher overall yields and significantly greater purity (Schemes 6.3 and 6.4).

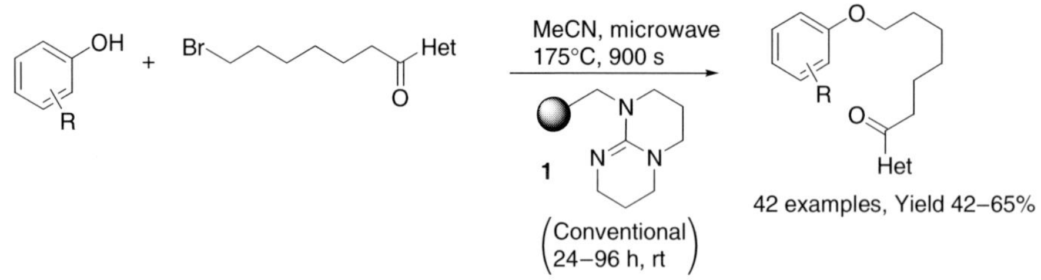

Scheme 6.3

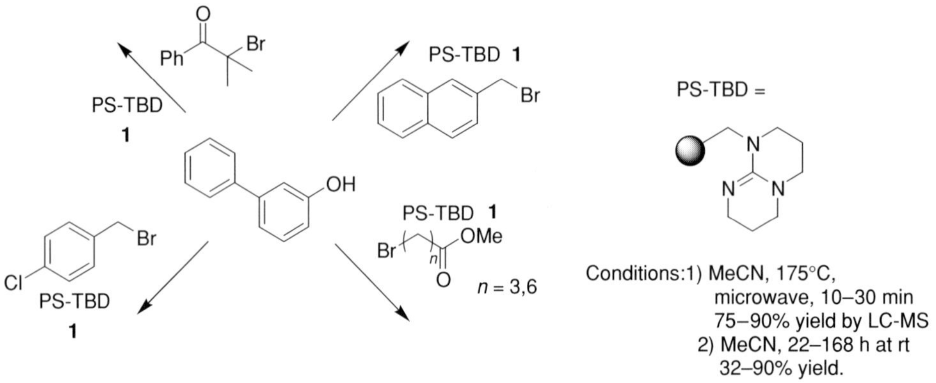

Scheme 6.4

This ability of microwave irradiation to facilitate higher and cleaner conversions seems to be a very common result. Many authors have reported similar improvements in isolated yields and purities in both their own work and in examples where microwave heating procedures have been applied to previous documented transformations. The majority of these improved experimental conditions usually involved microwave flash heating protocols. The principle rationalisation for the observed enhancement is the rapid and homogeneous heating of the sample to the desired reaction temperature, resulting in reduced time periods for alternative pathways to successfully compete. In other words, if one assumes that most purity-reducing reactions are as a result of the decomposition of the starting materials or product at less than optimal reaction temperatures then this would account for the improved product yield and composition. This type of near instantaneous heating and cooling as achieved through microwave heating of the sample is often referred to as the 'square' heating profile (Fig. 6.4). This partial explanation, although important, ignores many additional factors especially when heterogeneous systems are involved (see discussion in the previous sections).

6.3.5. O-Alkylations of carboxylic acids

Flash microwave heating of samples in sealed vessels has also been used by Linclau *et al.* to accelerate the rate of acid esterification reactions using an immobilised DCC derivative

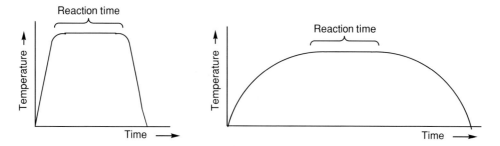

Figure 6.4 Comparison of heating profiles associated with microwave and traditional conductive heating methods.

(Scheme 6.5)[133,134]. In the initial work using this reagent, an overnight reflux (16–20 h) in THF was required to achieve comparable conversions. As previously mentioned, the shorter reaction times provided by this approach are certainly more compatible with the use of resin-type supports. The authors make the comment that under the optimised reaction conditions they identified no significant degradation of the polymer. It was also reported that the acceleration of the reaction did not affect the chemoselectivity of

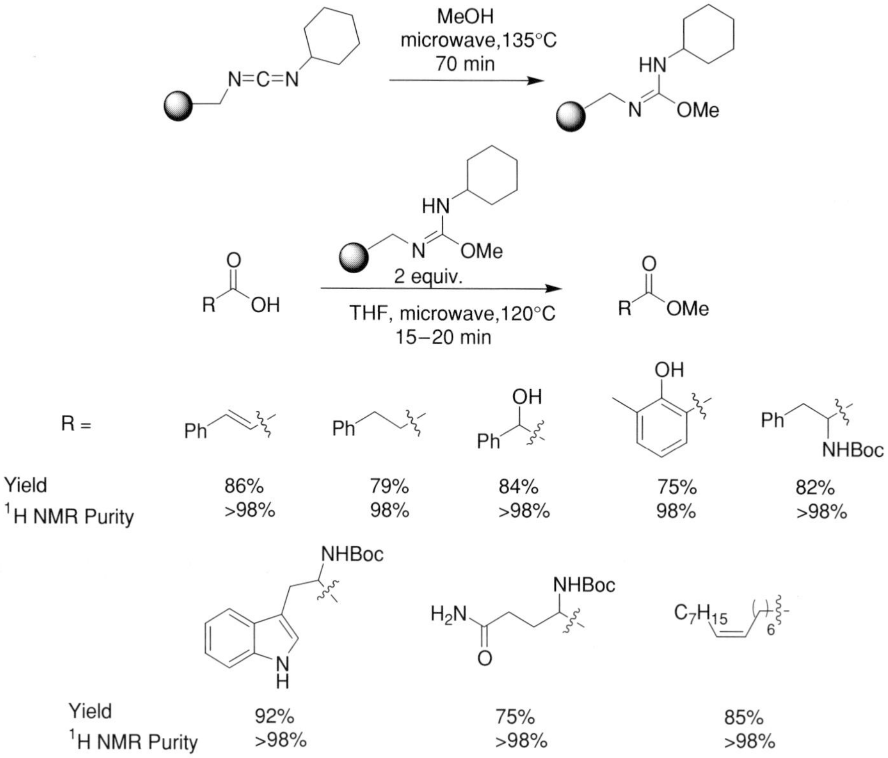

R =	Ph⁀	Ph⁀	Ph (OH)	(OH)	Ph (NHBoc)
Yield	86%	79%	84%	75%	82%
¹H NMR Purity	>98%	98%	>98%	98%	>98%

Yield	92%	75%	85%
¹H NMR Purity	>98%	>98%	>98%

Scheme 6.5

the reagent, with alcohols, phenols and primary amines remaining inert. Besides conducting the esterification reactions under microwave heating, the polymer-supported O-methylisourea reagent was itself formed by irradiation of the carbodiimide resin in dry MeOH. In contrast to the initial unsuccessful attempts at making the reagent using traditional copper(I) catalysis, high pressure thermolysis in a focused microwave oven at 135°C for 70 min allowed complete conversion. This approach to the use of microwaves as a primary heating method really highlights the utility and practicality of such techniques in synthesis.

6.3.6. Wittig reactions

The Wittig reaction, although of high synthetic value, can often be both time consuming and labour intensive to carry out. Both the initial preparation of the phosphonium salt and the ensuing reaction of the ylid species with a suitably functionalised carbonyl are often slow processes. These difficulties are additionally compounded by the relatively low stability of the Wittig intermediates, which often prevent the use of elevated reaction temperatures especially over extended periods of time. Finally, on completion of the reaction, purification can also be troublesome because of problems encountered with the separation of the product from contaminating triphenylphosphine oxide. It would, therefore, be highly advantageous if a method could be found to circumvent these problems. A recent report proposes to have achieved this through the combined use of an immobilised triphenylphosphine equivalent and rapid microwave heating[135]. The initial development phase of the research was directed at procedures for accelerating the three individual steps of the reaction, by employing short but intense microwave heating sequences (Scheme 6.6). This approach proved highly beneficial for shortening the reaction times and simplifying the work-up procedure without jeopardising the yields or purities of the products.

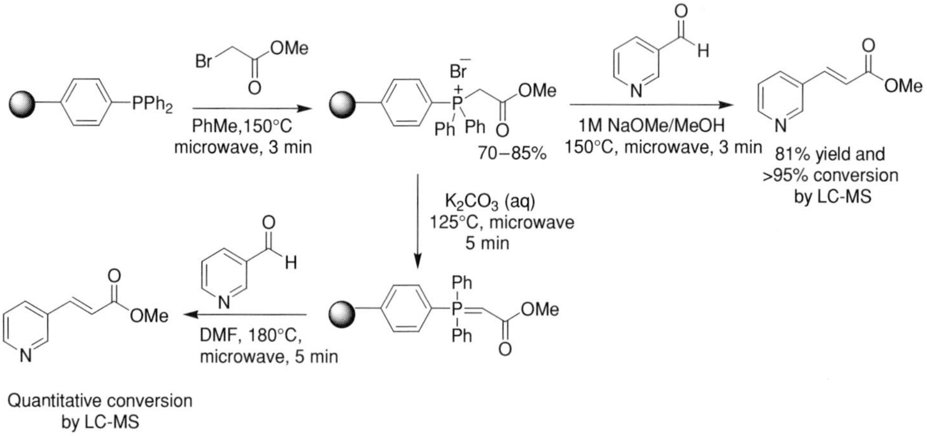

Scheme 6.6

In an extension of this work, the authors devised a more convenient one-pot procedure where all reaction components were heated simultaneously using microwave irradiation (Scheme 6.7). This protocol allowed the generation of a small set of ten compounds in poor to excellent isolated yield following preparative HPLC. This method certainly constitutes a very significant result, especially considering its applicability to high throughput combinatorial synthesis.

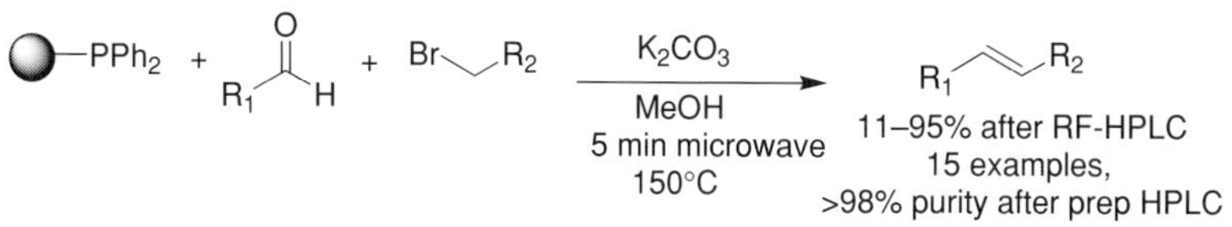

Scheme 6.7

6.3.7. Acylation reactions

The acylation of various functional groups is one of the most fundamental chemical manipulations in organic chemistry. This is particularly important when the formation of amide bonds are considered, as exemplified by the prevalence of the peptide bond within nature. Accordingly, there are numerous synthetic methods and reagents dedicated to the activation and coupling of these units. Unfortunately, many of the reagents and additives used in these reactions can be difficult to remove or display high toxicity. To simplify the isolation procedure and increase the safety profile, a number of these reagents have been immobilised on solid supports. It has been demonstrated that at least two of these reagent systems display shortened reaction times and improved yields when integrated with a microwave heating procedure[136,137].

Botta *et al.* have applied microwave heating procedures to both the preparation of their immobilised reagents and also to facilitate its subsequent reactions (Scheme 6.8)[137]. The initial preparation of the acylating reagents was carried out using 4 equiv. of the acid chloride **4** in a mixture of pyridine and dichloromethane with irradiation for 3 min at 200 W. The immobilised 4-acyloxypyrimidine **6** was then progressed through a washing sequence, resulting in a resin with approximately 68% of its potential theoretical loading. This activated material could then be suspended in dichloromethane with the required amine and irradiated for 3 min, again at 200 W. A rapid filtration and evaporation of the solvent provided the amide products as crystalline solids in good to excellent yield. The spent resin **5** from these reactions could be easily recycled for several runs without any detectable loss of activity.

In a similar way, Nicewonger and co-workers used microwave-mediated conditions to prepare a solid-phase imide **7** as an amine acylating reagent (Scheme 6.9)[136]. As in the above example, the spent resin could be easily isolated, washed and reactivated for further acylation cycles by repeating the microwave activation sequence. Using this reagent, excellent yields were obtained with both primary and secondary derivatives, but no reaction was observed with anilines. It is surprising that despite the emphasis the authors put on the ability of microwave heating to drive normally sluggish reactions to

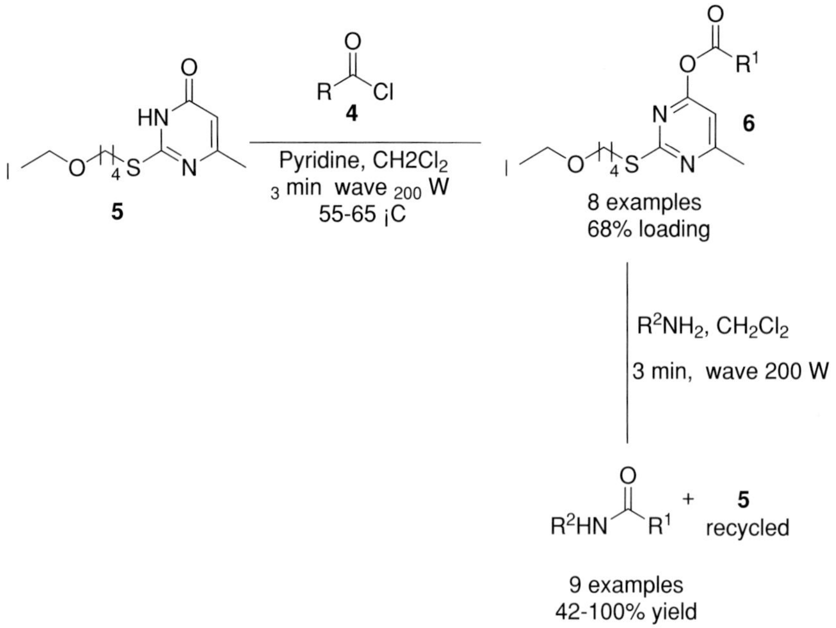

Scheme 6.8

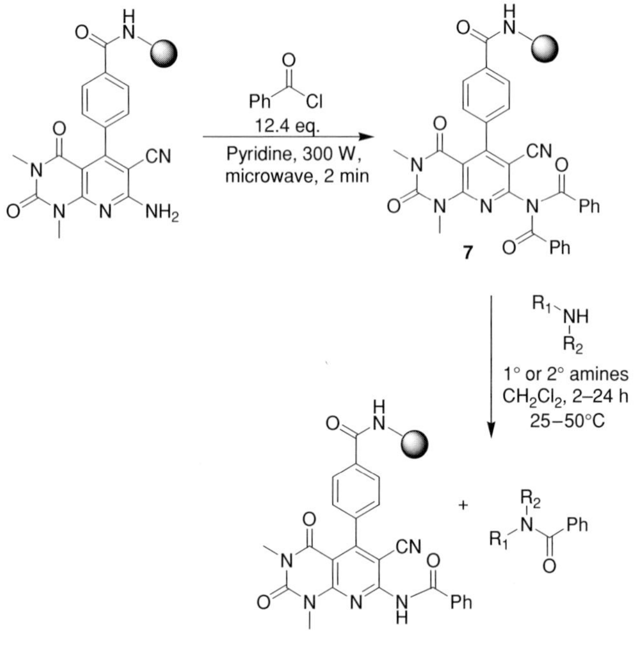

Scheme 6.9

completion, no reported attempt was made to apply this concept to the reaction of the anilines, with only temperatures of 25–50°C at atmospheric pressure being applied.

6.3.8. *Preparation of isocyanides*

The availability of novel monomer building blocks has become an important consideration in combinatorial chemistry because of the greater emphasis placed on diversity and novelty in modern library generation. Isocyanides have received particular attention because of their unique reactivity in multi-component coupling reactions, such as the Passerini[138] and Ugi[139,140] reactions. To augment the commercial supplies of these compounds, a number of groups have devised short and efficient protocols for their generation[141–145]. Unfortunately, the isolation of the isonitrile products can be problematic due to their high reactivity and in certain cases instability to traditional forms of extraction and purification. To overcome these difficulties, two independent approaches have been devised that combine the easy work-up protocols allowed by immobilised reagents and the rapid and efficient conversions attained by using flash microwave heating.

The first of these methods involves the *in situ* generation of isocyanides from the corresponding isothiocyanates using a supported [1.3.2] oxazaphospholidine species **8** (Scheme 6.10)[146]. This clean and efficient reagent was successfully used to generate, on demand and in parallel, a wide selection of isocyanide products. These materials were then submitted to an Ugi coupling reaction to prepare a focussed library of bicyclic amides (Scheme 6.11).

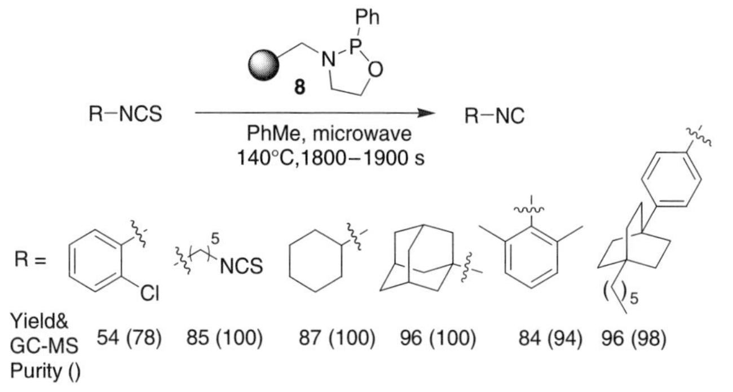

Scheme 6.10

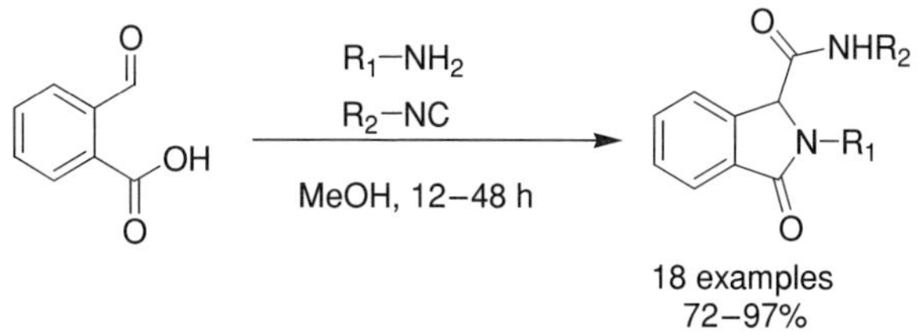

Scheme 6.11

Interestingly, the same reagent was also used in the conversion of isothiocyanates to isocyanides under conventional thermal conditions. Whilst isocyanides were successfully produced, the products were discovered to be highly susceptible to a competing rearrangement process, generating the equivalent nitriles after prolonged heating[147].

The second preparative procedure was used to transform formamides to isocyanides with an immobilised sulphonyl chloride (Scheme 6.12)[148]. The reactions were performed rapidly (ca. 10 min) under microwave heating (100°C), in the presence of excess pyridine to give the corresponding isocyanide in high overall purity, although in only modest yield. The only yield quoted in the paper is 70%, although independent attempts on other substrates have given similar results. The procedure described does, however, avoid the tedious requirement for chromatographic purification that was necessary in the original solution-phase procedure, although it still requires an aqueous extraction sequence to afford clean products. An important point to note is that the spent resin could be quantitatively regenerated by treatment with 5 equiv. of phosphorous pentachloride (PCl_5) in dichloromethane at ambient temperature. Although this is a very effective method, it is not always a simple process to remove the phosphorous by-products from the resin, which can result in contamination in subsequent reactions.

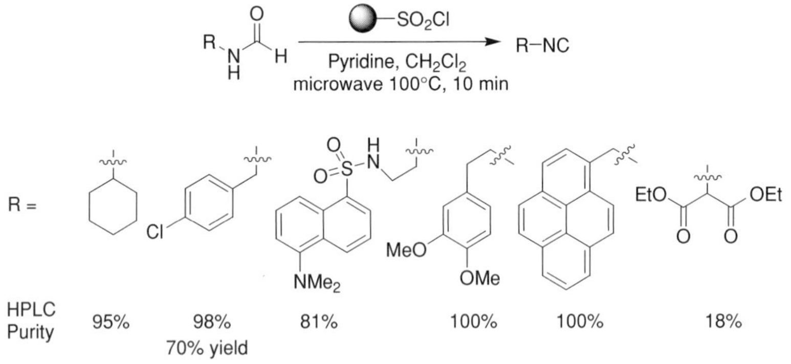

Scheme 6.12

6.3.9. Synthesis of thioamides

Thioamides are synthetically versatile molecules being key components in the preparation of many useful heterocycles. The traditional synthetic approaches to the generation of thioamides usually require rather harsh conditions, long reaction times and often result in difficult product isolation procedures. Moreover, the reagents and by-products associated with their synthesis can possess strong malodour and extremely high toxicity, making scale up or high throughput synthetic application technically very challenging.

Recognition of these drawbacks prompted the Ley group to design a new polymer-supported reagent **9** for direct amide thionation[149]. Preparation of this immobilised reagent was very simple and has been conducted on a relatively large scale (150 g of resin) (Scheme 6.13).

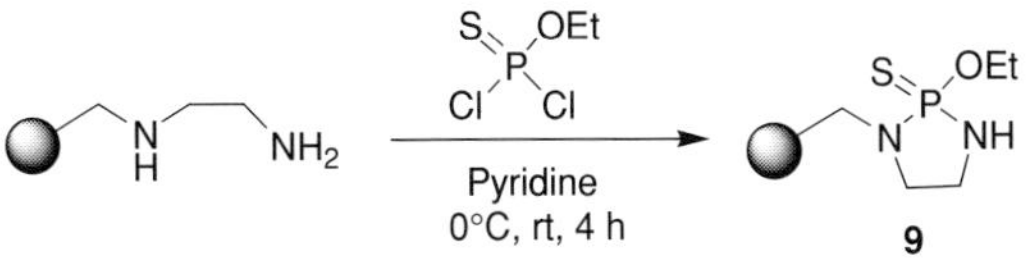

Scheme 6.13

This novel reagent cleanly converted both 2° and 3° amides into their corresponding thioamides under conventional thermal conditions (90°C, 30 h). However, the microwave-induced transformation was significantly faster. Heating the reactions in a sealed reaction vessel at 200°C for 15 min furnished the desired thioamides in almost quantitative yields with no purification step necessary apart from filtration of the resin (Scheme 6.14). These optimised conditions were obtained by conducting the reactions in a mixture of toluene and the ionic liquid 1-ethyl-3-methyl-1H-imidazolium hexafluorophosphate (20:1). This was found to be an excellent compromise between the requirements for the non-polar solvating properties of toluene for effective swelling of the resin and the necessity for a strong dipole to aid dielectric heating. To our knowledge, this represents the first reported application of an ionic liquid for increasing the thermal profile of a solution under microwave irradiation. For the same reason, acetonitrile, a strong dielectric heater, was also investigated as an alternative dopant, but proved to be less efficient than the ionic liquid. Under these conditions, treatment of 1° amides with the immobilised reagent led only to the corresponding nitrile – an observation that was consistent with results published for the solution-phase reaction of 1° amides with Lawesson's reagent[150].

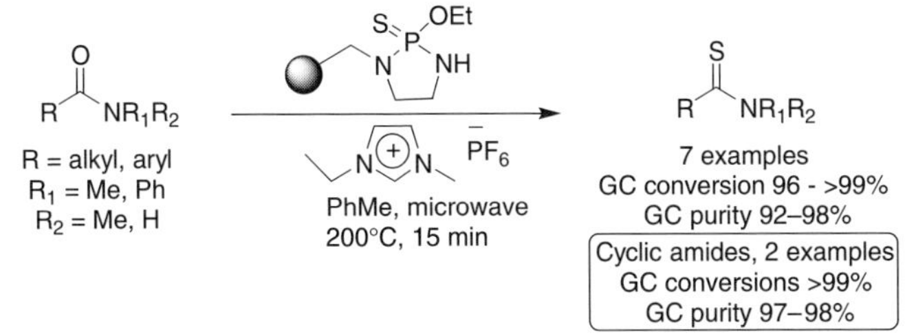

Scheme 6.14

Further studies have shown that the same reagent species can be effectively constructed on a silica backbone giving appreciable increased reactivity[151] (S.V. Ley, unpublished results). A silica-based support provides advantages in terms of more effective heating profiles because of the superior absorbency of the silica and increased compatibility with a wider range of solvents. By using these supplementary attributes it has been possible to process a wider variety of substrates, thus increasing the synthetic utility of this reagent.

6.3.10. *Esterification of alcohols using heterogeneous acid catalyst*

Kabza and co-workers have investigated the kinetics of the simple Fischer-type esterification of isopentyl alcohol and acetic acid using a heterogeneous polymer-supported reagent[152,153]. Amberlyst 15 sulphonic acid resin was used as a strong protic acid catalyst under microwave heating conditions to search for evidence of athermal effects, derived from the selective superheating of the immobilised catalytic sites during the transformation. A continuous-flow reactor as depicted in Fig. 6.5 was constructed to facilitate rapid and continuous monitoring of the reaction, while various experimental parameters were altered. From their findings, they concluded that the reaction investigated behaves analogously under both microwave-mediated and classically heated conditions. The reaction kinetics demonstrated no 'energy-type' preference or specific rate enhancements when microwave heating was applied. The authors report an additional interesting discovery, in that increasing the proportion of water present in the resin significantly retarded the reaction rate. In fact, the rate constant for the reaction was similar to that reported for the equivalent reaction carried out under fully aqueous conditions[153]. The experimental data obtained in the study suggested this effect was not purely due to an alteration of the equilibrium constant as a result of the additional water content, although this was making the assumption that the system behaves like an

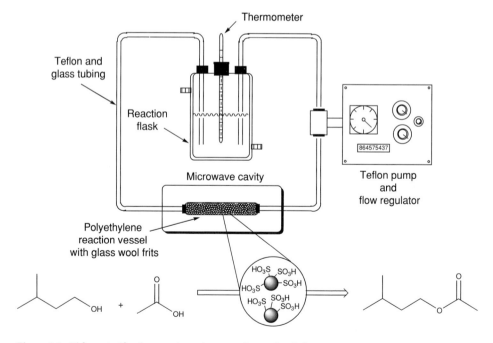

Figure 6.5 Fisher esterification reaction using a continuous batch flow reactor.

ideal homogeneous reaction. Kabza proposed that this observed aqueous impediment was actually as a consequence of trapped water within the porous cavities of the resin reducing accessibility to the catalytic sites[154]. This work represents only a preliminary study of this type of system and as the authors are quick to point out, a more detailed and better understanding of heterogeneous environments is still required.

6.3.11. Chemoselective bromomethoxylation

The selective bromomethoxylation of a variety of alkenes under microwave irradiation conditions have been demonstrated using two polymer-supported macro-porous bromine resins. The reactions have been easily achieved in moderate to good yields and high regio-and chemoselectivity (Scheme 6.15)[155]. All the experiments were carried out in a modified domestic microwave oven fitted with a standard reflux condenser. Reactions were complete within a range of 30 s to 5 min when methanol was used as the solvent. Ten different compounds were reported, including a number of sensitive natural product derivatives such as methyl angolensate derivative **10**, azadirachtin-A derivative **11** and carvone derivative **12**. It is quite remarkable that molecules containing such sensitive functionality as azadirachtin-A can survive these conditions, although we too have noticed the benefits of using microwave heating in connection with this molecule[28]. Unfortunately, no comparison was made with a conventionally heated reaction at the equivalent temperature, or any indication of the length of time such reactions required under ambient conditions.

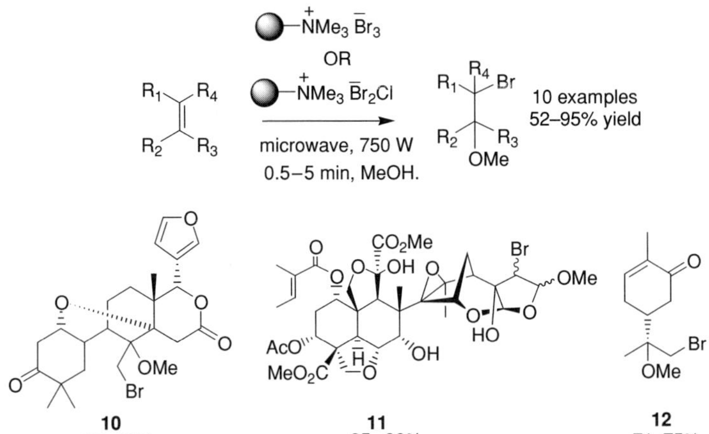

Scheme 6.15

6.3.12. Beckmann rearrangement

Another transformation that has received considerable interest is the Beckmann rearrangement. Traditionally, this process requires the presence of a strong acid with heating at high temperatures over extended periods of times. It therefore represents an ideal candidate for the application of microwave flash heating. Recently, a number of publications have shown that the reaction can be accelerated using microwave-assisted heating, when conducted under solvent-free conditions using mineral clays as acidic supports[18,156]. Although a very valuable addition to the repertoire of available transformations, this approach is rather limited in its versatility. For example, this approach precludes the ability to intercept the transient intermediates, which would otherwise allow propagation of alternative reaction pathways. The reaction can, however, be carried out in solution employing a sulphonyl chloride resin as the oxime activator. Brown and colleagues have demonstrated that a standard Beckmann rearrangement can be promoted using this immobilised reagent, and further more, the selectivity of the migration can be affected by the specific conditions employed (Scheme 6.16) (R.M. Hughes and R.C.D. Brown, unpublished results; I.R. Baxendale and S.V. Ley, unpublished results).

rt 20 h	30%	0%
microwave 10 min	42%	0%
microwave H$_2$O (3 equiv.) 20 min	40%	8%

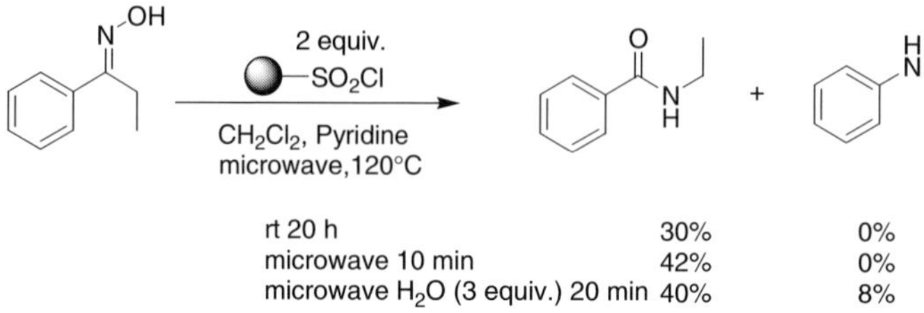

Scheme 6.16

The Ley group were simultaneously involved in a more diverse investigation of this reaction[151] (S.V. Ley, unpublished results). They have shown that it is possible to directly transform the oximes to the corresponding amides, by simply mixing with sulphonyl chloride resin and irradiating the resulting suspension (Scheme 6.17). They also demonstrated that the reactive intermediates can be trapped with a number of nucleophilic components to gain access to valuable heterocyclic products and other useful intermediates (Scheme 6.18). Further research to fully exploit this approach, including the expansion of the range of nucleophiles, is currently underway.

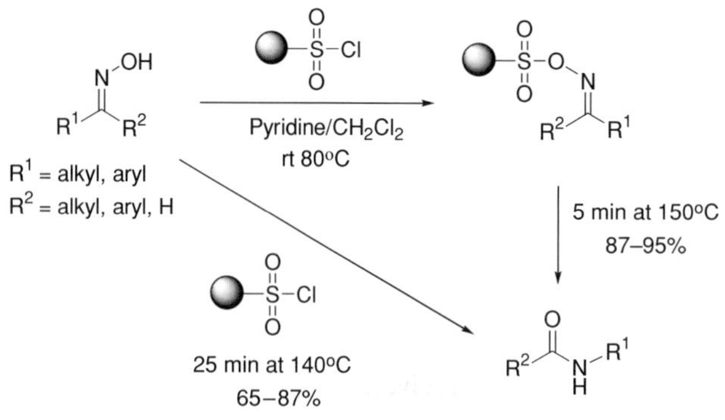

Scheme 6.17

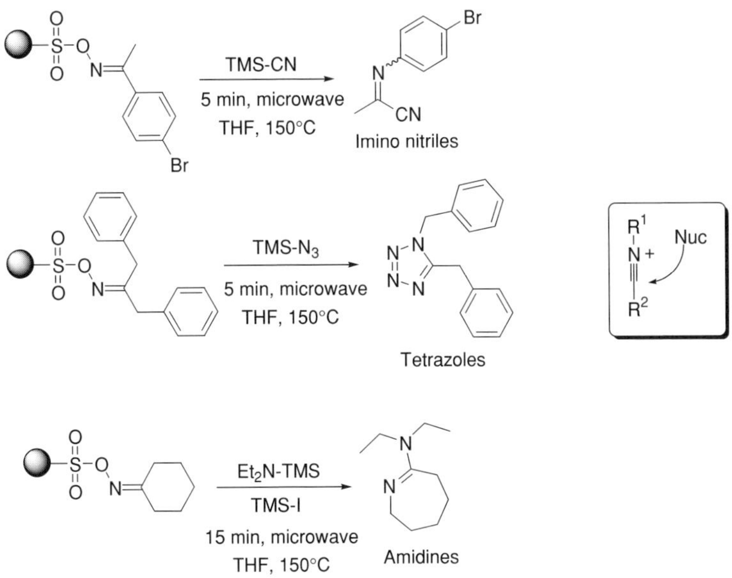

Scheme 6.18

6.3.13. Hydrogenation of electron-deficient alkenes

Catalytic transfer hydrogenation is a crucial reaction in organic synthesis. Most applications of this reaction employ ammonium formate as a hydrogen source in the presence of a catalyst, such as palladium charcoal (Pd/C) or Wilkinson's catalyst RhCl(PPh$_3$)$_3$. However, there are a number of difficulties associated with the use of ammonium formate, such as its ability to sublime, which leads to blocking of the reaction apparatus. In addition, the reaction can generate significant quantities of ammonia, which is problematic when working on a large scale. In an attempt to overcome these problems, Desai and Danks (see Chapter 4) have reported on the use of a polymer-supported formate (on Amberlite IRA 938) for the reduction of electron-deficient alkenes in the presence of Wilkinson's catalyst (Scheme 6.19)[157]. Once again, in comparison to non-microwave conditions, dielectric heating offered a tremendous rate enhancement and increased product purity. It is worth making the general observation that many reactions involving metal catalysts show enhanced substrate reactivity and higher turnover numbers, when employed in conjunction with microwave irradiation. The involvement

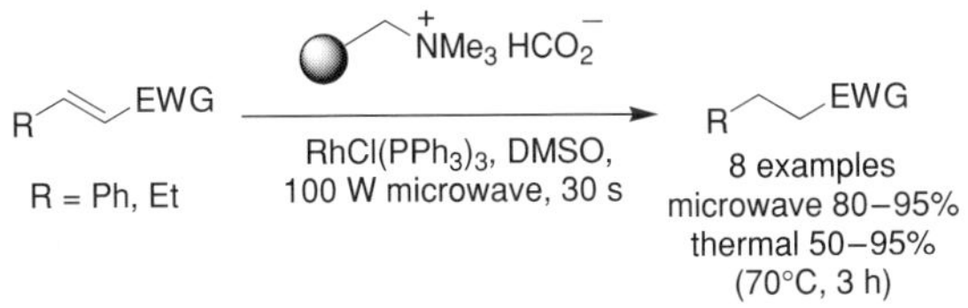

Scheme 6.19

of selective superheating of the metal particles combined with temperature-mediated self-cleaning mechanisms (decoking) certainly contributes to these observed benefits.

6.3.14. *Heck reactions*

The development of novel supports and presentation formats has become an important consideration in solid-phase chemistry, particularly in the area of immobilised catalysts. The ability to bind transition metals such as palladium to the surface of an extractable solid support facilitates rapid and convenient recycling. It has also become apparent that certain immobilised metal species can gain significantly enhanced reactivity from selective heating mechanisms produced by microwave energy transfer. Buchmeiser and co-workers have prepared two alternative supports based on ring-opening metathesis polymerisation (ROMP) grafting techniques. These macromolecular structures contain ligating sites, which have been used to immobilise palladium dichloride to create catalysts for use in Heck type reactions (Fig. 6.6)[158].

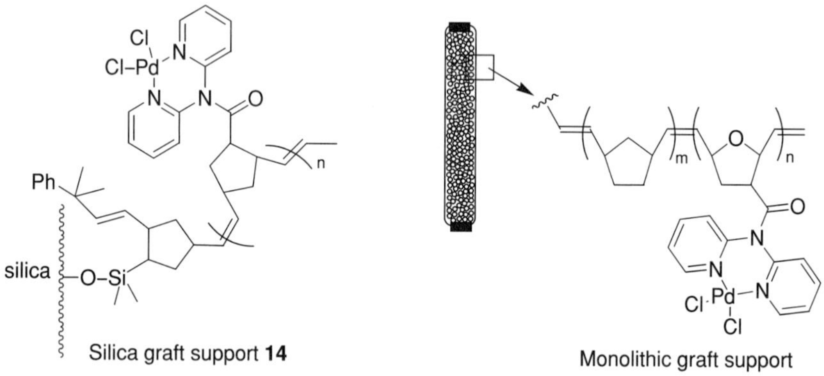

Silica graft support **14**   Monolithic graft support

Figure 6.6   (a) A silica immobilised palladium catalyst for Heck reactions. (b) The same catalyst prepared in a Monolithic graft support.

Slurries of the silica-based material **13** were used for Heck coupling of aryl halides and styrene-based alkenes under standard as well as microwave heating conditions (Scheme 6.20). Microwave irradiation leads to drastic reductions in the reaction times, proving, on average, to be six times faster. The supports also displayed very low levels of leaching, typically <2.5%, although it was not clear from the paper if this was affected by the method of heating. Interestingly, the silica support **14** and the monolithic

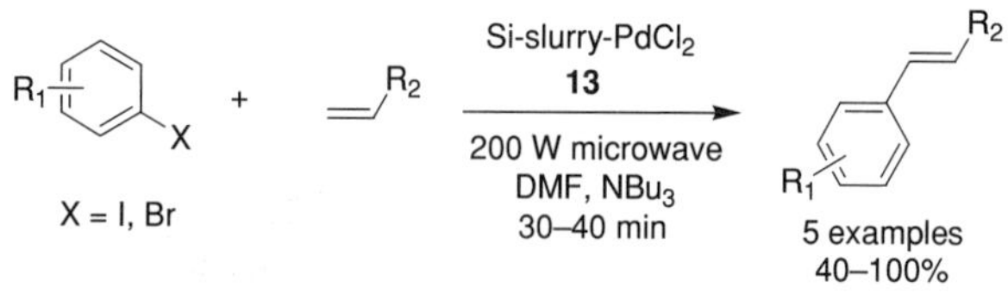

Scheme 6.20

material displayed in Fig. 6.6 could be packed into column cartridges and used in a continuous flow process. In these systems, only traditional heating methods were attempted, although the design of reactors to facilitate microwave irradiation of such systems has already been proven (see Chapter 9), so additional work on this area will not be long in coming.

6.3.15. Ketone–ketone rearrangements using polymer-supported AlCl₃

6.3.15. *Ketone–ketone rearrangements using polymer-supported AlCl$_3$*

A recent publication from an Indian group has described the synthesis of a small selection of 3-(4-alkoxyphenyl)-3-methylbutan-2-ones, which are useful starting materials in the synthesis of a wide range of natural products. A polymer-supported aluminium chloride species[159] was used to catalyse the rearrangement of ketone 15, followed by O-alkylation of the trapped intermediate with simple alkyl halides (bromo and chloro) as well as dimethyl and diethyl sulphate (Scheme 6.21)[160]. Both transformations were found to occur with improved efficiency, when the heating was conducted using microwave irradiation. The authors claim this procedure is the simplest and most economical methodology used to synthesise these intermediates to date.

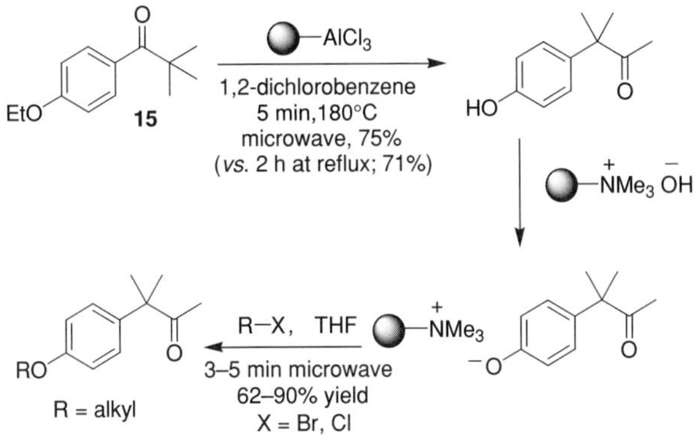

Scheme 6.21

6.3.16. Synthesis of 1,3,4-oxadiazoles using polymer-supported Burgess reagent

6.3.16. *Synthesis of 1,3,4-oxadiazoles using polymer-supported Burgess reagent*

On the basis of a wealth of information concerning the dehydrating properties of the solution-phase Burgess reagent, Brain and co-workers at Novartis have demonstrated that the analogous polymer-supported equivalent, first synthesised by Wipf[161], could be ameliorated when used in conjunction with microwave heating protocols (see Chapter 3)[162, 163]. In this way, they were able to facilitate the swift cyclo-dehydration of 1,2-diacylhydrazines to the corresponding oxadiazoles (Scheme 6.22). These products were easily isolated in high overall yield and purity by a simple filtration through a frit of silica gel to remove the PEG-supported materials.

Although this protocol was very mild and efficient, the methodology was subsequently revised by the generation of a new heterogeneous polystyrene immobilised

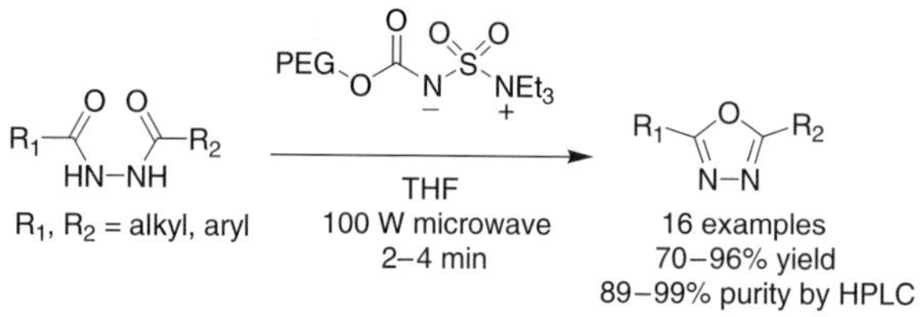

Scheme 6.22

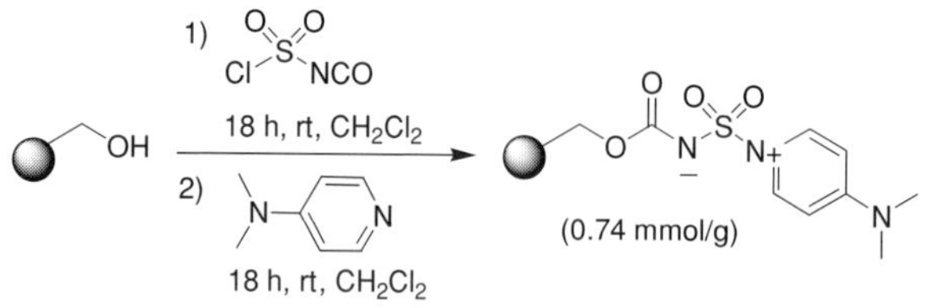

Scheme 6.23

variant of the Burgess reagent (Scheme 6.23)[162]. The synthesis of this modified material alleviated the need for a silica gel purification step, and in addition, aided the recovery and recycling of the spent resin. It also demonstrated good stability and could be stored for over a month when refrigerated under a nitrogen atmosphere, although it could be conveniently handled in air during dispensing. Initial experiments employing this species have been successful, affording relatively high yields of the heterocyclic product. However, a further improvement was achieved with the introduction of the strongly basic guanidine additive DBU. Using the optimised reaction conditions, a rapid and near quantitative formation of 14 structurally diverse oxadiazoles was achieved. Reaction clean up was achieved using an immobilised sulphonic acid (Amberlyst 15), which was added directly to the crude reaction mixture to expediate the removal of the excess guanidine base. A single filtration step followed by evaporation of the solvent yielded the heterocyclic products (Scheme 6.24). Throughout this work traditional thermal heating was conducted in parallel with the microwave experiments, although in the majority of cases this resulted in substantially lower yields and purities, along with the requirement for prolonged reaction times.

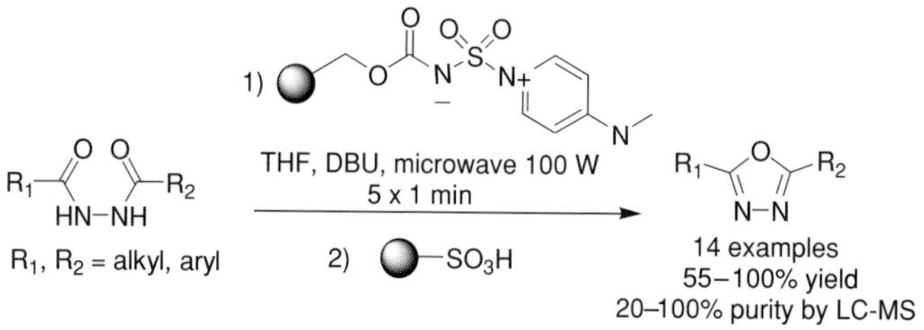

Scheme 6.24

In the same publication, it was reported that this cyclo-dehydration could also be affected by using tosyl chloride and the polymer-supported phosphazene base PS-BEMP, and again, microwave heating was found to be advantageous (Scheme 6.25). In utilising this protocol, no scavenger purification strategy was deemed necessary and the authors note that this is the most efficient 1,3,4-oxadiazole synthesis of the three polymer-supported methods described.

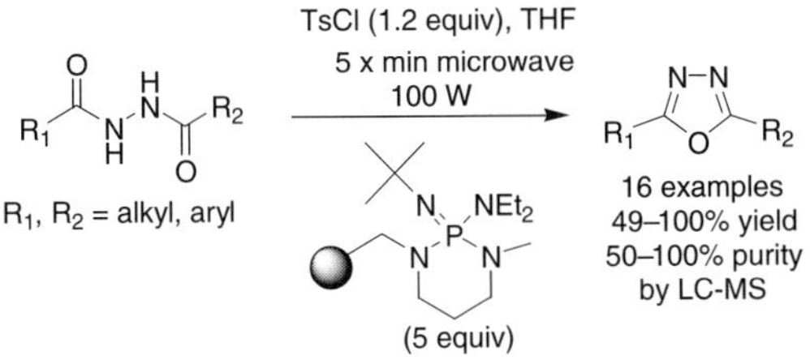

Scheme 6.25

6.3.17. *Preparation of a substituted 2-amino-1,3,4-oxadiazole library*

A related series of 5-substituted-2-amino-oxadiazole compounds have also been prepared in a one-pot procedure using a microwave-assisted cyclisation procedure (Scheme 6.26)[164]. Rapid preparation of the pre-requisite ureas from the mono acyl hydrazines and various isocyanates (or the isothiocyanate) was easily achieved by simple mixing. The resulting products were then cyclodehydrated by one of the two procedures: either by the addition of polymer-supported DMAP and tosyl chloride or alternatively with an immobilised carbodiimide and catalytic sulphonic acid. Purity in most cases was excellent after only filtration through a small plug of silica but an SCX-2 cartridge (sulphonic acid functionalised – catch and release) could be used in the cases where reactions required additional purification.

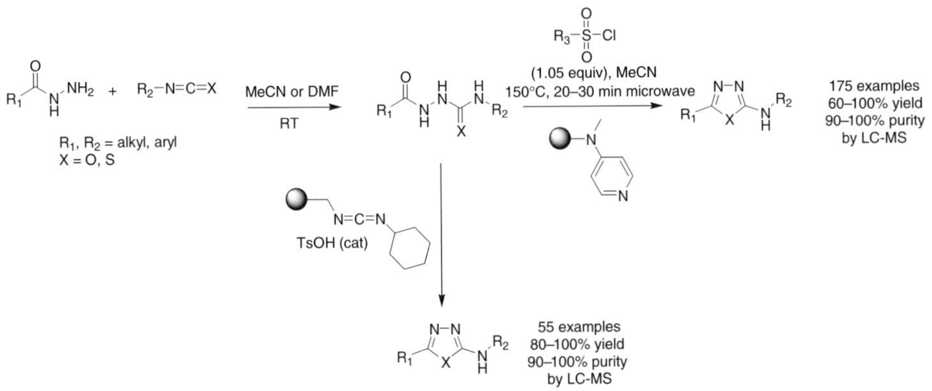

Scheme 6.26

As an extension of this work Ley *et al.* demonstrated that it is possible to prepare the corresponding *N*-protected sulphonamides in an analogues single-pot procedure (Scheme 6.27). The addition of a 2.1 equiv. excess of a sulphonyl chloride to the urea in the presence of the stronger immobilised base PS-BEMP facilitates both the cyclisation and subsequent sulphonylation steps. The isolated products following filtration were obtained in both high yields and excellent purities although certain reagent combinations resulted in the formation of the alternative regioisomer as a major product. It was noted that the solvent played an important role in the generation of the regioisomeric structures with changes in mixture compositions of 35:1 to 1:1 being observed when swopping from MeCN to THF. Additional investigations are under way to further understand and exploit these findings.

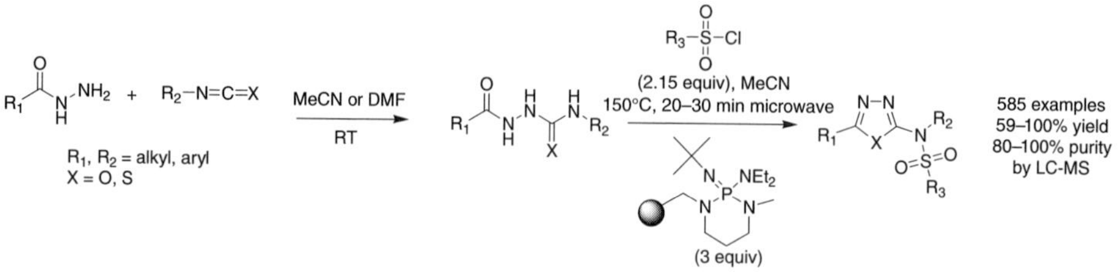

Scheme 6.27

6.3.18. Synthesis of thiohydantoins

In another heterocycle forming procedure, Westman *et al.* (see Chapter 5) employed microwave dielectric heating to a sequential one-pot, three-step reaction sequence leading to an array of structurally diverse thiohydantoins (Scheme 6.28)[165]. Although the final sequence is a purely solution-phase protocol, involving no immobilised reagents,

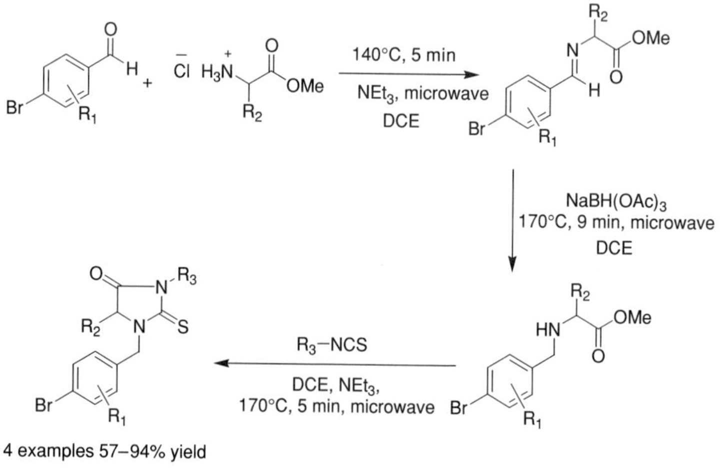

4 examples 57–94% yield

Scheme 6.28

the authors do describe their initial investigations, which involved the direct coupling and cyclisation of amino acids with isothiocyanates, catalysed by a polymer-supported DMAP reagent (Scheme 6.29). They state that this method provided a much cleaner reaction mixture when compared with the final solution-phase step, which uses triethylamine; however, this supported reagent step was unfortunately less compatible with the other steps of the synthesis.

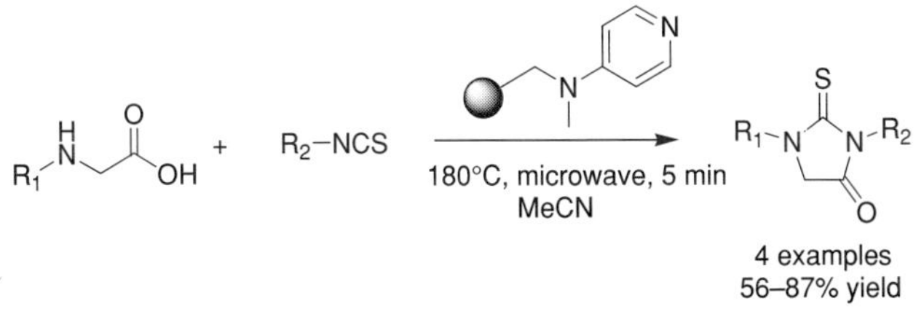

4 examples
56–87% yield

Scheme 6.29

6.3.19. *Hydrolysis of sucrose to fructose*

The hydrolysis of sucrose to form fructose and glucose, catalysed by the strongly acidic cation-exchange resins Doulite C25D[166] and Amberlite 200C (protic forms)[167–169], has also been studied. The principle work by Wang and co-workers[166] used a modified domestic microwave oven containing a Teflon coil as a readily adaptable continuous flow reactor. The coil situated within the microwave cavity (ca.10 ml in volume) was packed with the immobilised Doulite particles and a solution of sucrose was slowly pumped through the system. The reaction required only a 10 min residence time at 72 W to achieve a 95% conversion. Of significant interest is the observation that the use of the strongly acidic Doulite C25D resin gave better results than comparable experiments with a mineral acid. More recently, Plazl[167–169] has investigated the same reaction under both conventional and microwave heating with a view to analyse the response of the reaction to changes in the heating mechanism. Initial reactions were conducted in a stirred tank reactor, which was housed in the chamber of a domestic microwave oven. Later a more sophisticated fixed bed reactor was constructed. In this set up the immobilised catalyst (Amberlite 200C) was located in a Pyrex glass tube sited in the microwave cavity (Fig. 6.7). Automated on-line monitoring of the various fluid inputs and outputs allowed a detailed analysis and the development of a mathematical model to predict the heating requirements and thermal profile of the reactor. In this way, it was possible for the researchers to optimise the power settings, flow rates and temperature requirements in order to produce a given conversion of the sucrose or successfully predict a given level of transformation from a set of experimental parameters.

6.3.20. *Microwave-promoted enzymatic reactions*

The use of enzymes as reagents in general organic synthesis is becoming more widespread as the scope of the transformations for which they can be applied is more thoroughly investigated and understood. The thermal acceleration of enzymatic transformations

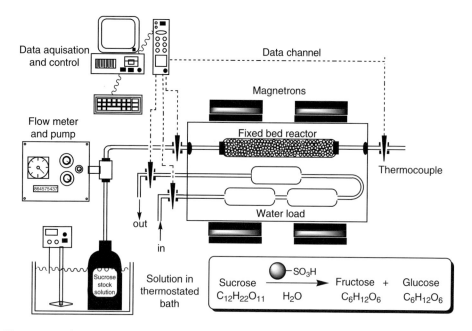

Figure 6.7   A diagrammatic representation of a fixed bed reactor used for the degradation of sucrose.

is generally perceived to reduce the enantiospecificity of the process. However, a growing population of literature indicates that many biological catalysts are not as strictly
temperature dependant as previously assumed[170–173]. In addition, the further use of
designer thermostable enzymes prepared by recombinant DNA technology now per-
mits regular application of enzymes at relatively high temperatures[174]. Loupy *et al.* have
identified the potential gains to be leveraged by using microwave heating in combination with enzymatic processes[175,176]. Their approach relies on the ability to induce
equilibrium shifts in the composition of the biological mixture, by evaporation of low-
molecular-weight polar molecules. This is achieved by selective excitation of the small
molecule fractions as a result of their strong coupling interaction with the microwave
irradiation. In their original work, they investigated the lipase-catalysed resolution of
1-phenylethanol under 'dry'/solvent-free conditions, with the enzymes immobilised
on a solid support (various supports were tested including Accurel a polypropylene
resin) (Scheme 6.30). The working temperatures for the investigated esterification and
transesterfication reactions were in the range of 70–100°C; this was achieved using a
single-mode microwave reactor with a controllable power setting for maintaining the
temperature. The reactions showed very pronounced increases in their general rate of
acyl transfer under these conditions, which could be directly related to the exclusion
of the volatile by-products from the equilibrium. In this way, it was also possible to
demonstrate that normally low-yielding reactions (not fully resolved) could be driven
to completion, obviating the need for further enantiomeric separation. In most cases,
when compared to classical heating methods, the microwave-promoted reactions gave
higher stereoselectivity and improved kinetics, although with essentially identical levels
of conversion for a given temperature. The supported enzymes were also preserved
under the microwave conditions and could be recycled without loss of activity.

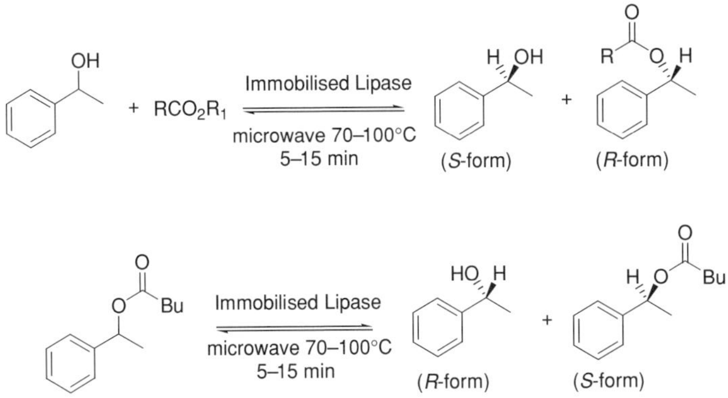

Scheme 6.30

As an extension of their initial work, Loupy and co-workers reported on the regios-elective lipase-catalysed esterification of a selection of α-D-glucopyranosides with var-ious fatty acids, under similar microwave-promoted reaction conditions[177]. The lipase *Candida antartica*, immobilised on Novozym 435, was impregnated with the coupling component (fatty acid) and the dried solid material was irradiated with the pyrano-sides. The comparison with conventional heating conditions at equivalent recorded temperatures and for the same duration of heating showed substantially higher levels of conversion in the cases of microwave heating. For example, the reaction shown in Scheme 6.30 gave >95% conversion after 5 h of microwave heating, whereas the same reaction heated in a thermostatically controlled oil bath was only 55% complete after the same period. Various other sugars were also successfully esterified using this approach, and in all cases the catalysed transformation showed high levels of regioselectivity under microwave conditions (e.g. Scheme 6.31). A probable explanation for this is that the rapid kinetics furnish a high yield of the desired product, which could then be iso-lated, before any additional equilibria leading to alternatively substituted products were sufficiently established.

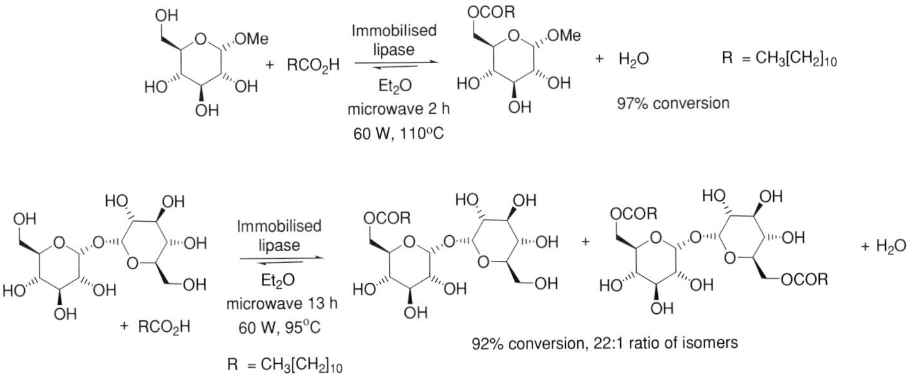

Scheme 6.31

In a third paper, the same group continued their studies by evaluating enzymatic glycosidation reactions (Scheme 6.32)[176]. Their general findings were that all the reactions investigated demonstrated enhanced kinetics, when compared to the literature reported analogous transformation. Furthermore, the authors were able to optimise the conditions for transglycosidation reactions resulting in a complete conversion within 3 h while minimising the level of competing hydrolysis – a significant problem that is normally encountered under forcing conditions. It was also found that the proportion of glycoside acceptor could also be significantly reduced to only 2 equiv. Overall, this method of glycosidation represents a very significant improvement over the previous literature examples.

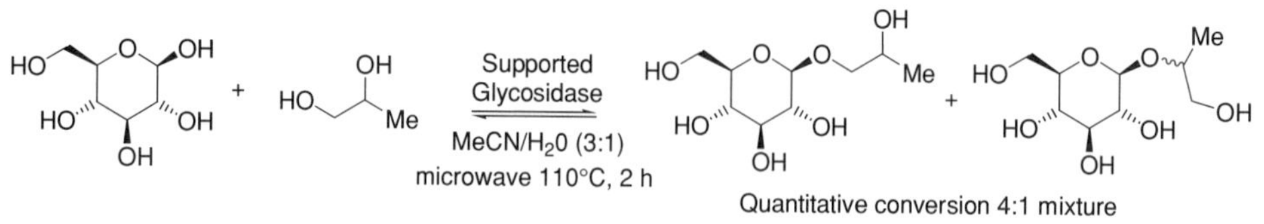

Scheme 6.32

6.3.21. *Spectroscopic estimation of polymer-supported functional groups*

Microwave-assisted heating has also been employed in the determination of polymer-supported functional groups[178]. The reagent (4,4'-dimethoxytrityl)-3-mercaptopropanoic acid (DMPA) in conjunction with a redox coupling mixture of triphenylphosphine and bromotrichloromethane (TTP/BTCM) can be used to assess the total number of assessable hydroxyl, amino or mercapto sites in a given resin. The process is greatly facilitated by the microwave-enhanced coupling of the DMPA to the resin-bound functionality (<4 min). The extent of the coupling can then be estimated in reverse by spectrophotometrically monitoring the levels of dimethoxytrityl cations, which possess a high-molar absorption ($\varepsilon_{498} = 700\,001$ mol^{-1} cm^{-1}), liberated from a known quantity of the resin upon acid treatment (Scheme 6.33). This method has proved advantageous not only for identifying functional group loading on resins for use in solid-phase synthesis, but also for preparing and analysing affinity chromatography matrices.

6.3.22. *The synthesis of (+)-plicamine*

The improved productivity derived through microwave heating in combination with polymer-supported reagents has been used extensively in the total synthesis of the natural product (+)-plicamine as well as a number of related spirocyclic templates[179,180]. Microwave dielectric heating was used as the primary means of accelerating a number of slow reactions, in order to maximise the quantities of intermediates that could be progressed through the synthetic sequence. The rapid optimisation and screening permitted by the adoption of automated microwave reaction handling was crucial to the successful completion of this total synthesis.

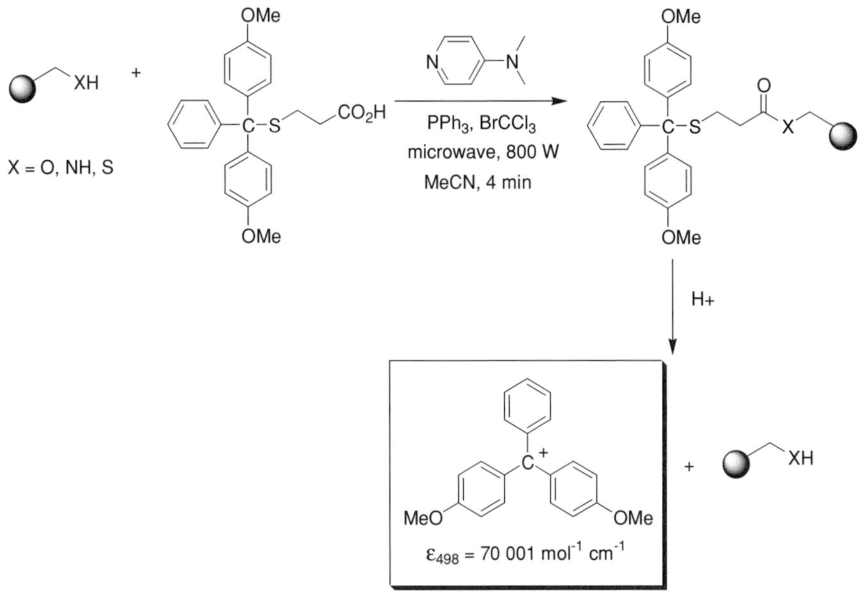

Scheme 6.33

Preparation of the key methoxy-substituted intermediate **16** was achieved by one of the two alternative pathways (Scheme 6.34). The initial route involved the reduction of the dimethyl acetal species **17**, which was prepared by microwave heating a mixture of the ketone **18**, trimethyl orthoformate (or dimethoxy acetone) and Nafion SAC-13 in methanol at 100°C for 45 min. This transformation was extremely efficient, giving quantitatively and cleanly the acetal adduct **17** as a crystalline material following only filtration and solvent evaporation. The subsequent reduction step followed a modified procedure from the pioneering work of Olah and co-workers for reductively cleaving acetals using triethylsilane[181]. By conducting the same reaction under microwave heating conditions, the corresponding methyl ether **16** was obtained in an excellent yields, 95% conversion (only yields of 43–56% were obtained following the literature procedure using classical thermal heating). Alternatively, the same material **16** could be synthesised by direct alkylation of the reduced alcohol **19** (*via* reduction with an immobilised borohydride resin), by an activated alkyl donor. The most effective combination of reagents and conditions discovered for this transformation was a blend of methyl trifluorosulphonate and poly(4-vinyl-2,6-*tert*-butylpyridine) under microwave irradiation. Applying normal reflux conditions gave rather disappointing yields as a result of the formation of a diene side-product resulting from elimination of the hydroxyl or methoxy group as a consequence of the prolonged thermal heating (>4 h). This contamination was almost completely avoided with the drastic reduction in heating times permitted when using the microwave system (35 min instead of 8 h).

The cleavage of various trifluoroacetate functionalised amines (e.g. see structures **16** and **19**) has also been accomplished using base-catalysed hydrolysis with Ambersep 900 (OH⁻ form) in methanol. Again, it was found expedient to heat these reactions under

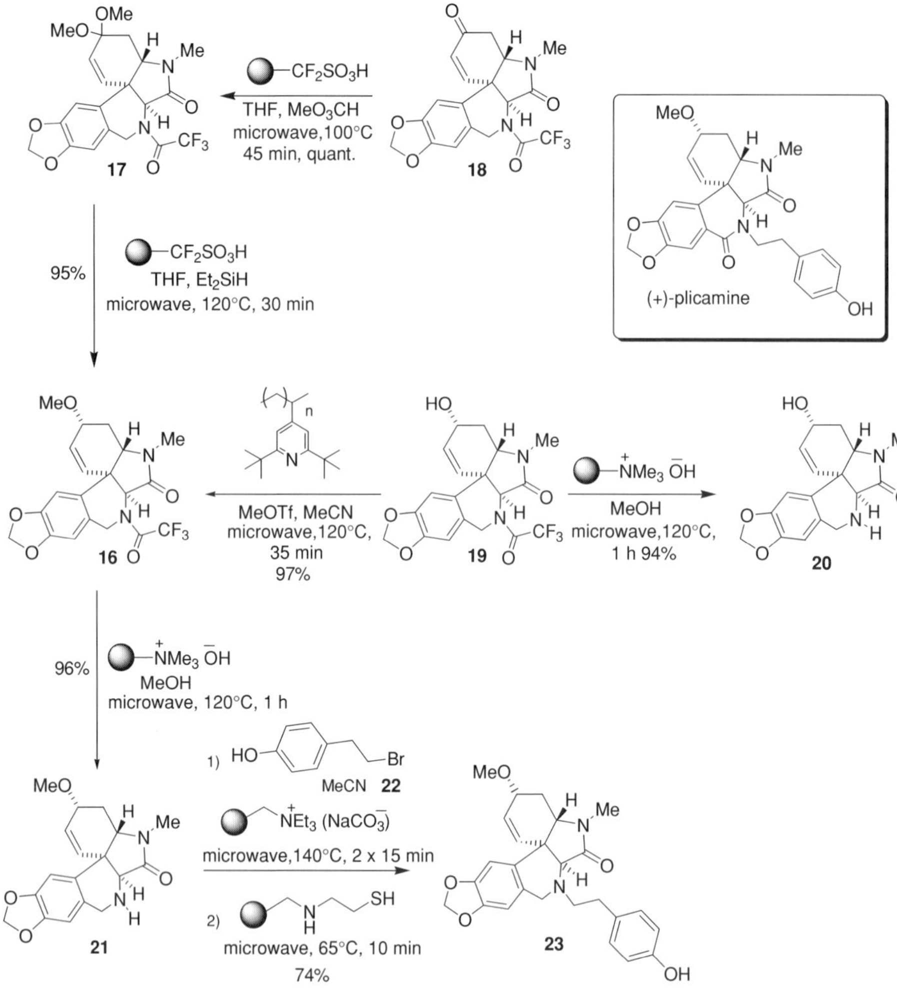

Scheme 6.34

microwave irradiation rather than conventional thermal conditions reducing reaction times from 6 h to less than 1 h, whilst maintaining an almost quantitative yield.

The advanced intermediate **23** also proved to be an ideal candidate for microwave-assisted synthesis. In the penultimate step of the synthesis, the secondary amine **21** was required to undergo alkylation with the phenolic halide **22**; however, this proved to be a very slow and troublesome reaction. Traditional thermal heating of the reaction with various immobilised bases produced complex mixtures, which were difficult to separate. Alternatively, a procedure using a supported carbonate reagent was found to be ideal for this conversion when used in acetonitrile under microwave irradiation conditions. Unusually, a heating sequence employing 2 × 15 min pulses maintained at 140°C with cooling to ambient temperature between cycles gave a higher yield than a single heating event. In an extended study, microwave-accelerated purification was also attempted to enhance the rate of scavenging of excess alkyl halide **22** by employing thio-amine resin. At ambient temperature and pressure the process required 2 h, but employing a short microwave heating sequence (optimised to 10 min at 65°C) facilitated a more rapid purification, but was accompanied by a lower isolated yield (74% compared to 90%).

6.3.23. *Microwave-assisted scavenging reactions*

The general use of microwave heating to promote higher rates of scavenging in the purification of mixtures has received almost no coverage in the chemical literature. This is despite the extensive investigations that have been pioneered by many industrial medicinal chemistry and combinatorial groups. One of the most amenable reactions to this form of assistance is the removal of residual aldehydes, ketones and amines, by imine (enamine) formation with a reciprocally functionalised resin. Normally these reactions can be quite slow, especially when bulky or electronically retarded substrates are involved, and in these cases flash microwave heating has been identified as being highly beneficial. As previously mentioned, microwave-promoted reactions involving imine formation and the subsequent reduction by immobilised reducing agents have been used to prepare diverse libraries of functionalised amines. In many of these cases, it was found to be more efficient to use an excess of one component resulting in impure products. It is then possible to add a secondary purification step, involving the dispensing of an appropriately substituted resin (to form the imine/enamine) followed by a repeated heating cycle. The solvents of choice for these transformations have usually been THF or toluene, although a few examples have utilised alcoholic solvents for reasons of enhanced solubility and increased reactivity. Again many of these reactions are carried out at temperatures above the normal boiling points of the solvents in pressurised reaction vessels, which can assist in establishing advantageous equilibrium positions.

One especially interesting experiment that the authors have encountered involved the equilibrium transfer of an amine, which was immobilised on a resin as the enamine species (Scheme 6.35) (I.R. Baxendale, and S.V. Ley, unpublished results). This resin was added in stoichiometric excess to a 'wet' THF solution of a ketone and an immobilised cyanoborohydride resin, followed by a relatively short microwave heating sequence. This resulted in an essentially pure solution of the required amine product, which was both cleaner and was also obtained in a higher yield than any other method attempted. It is worth pointing out that this particular case represented a very specific set of examples and no additional effort was made to determine its generic applicability.

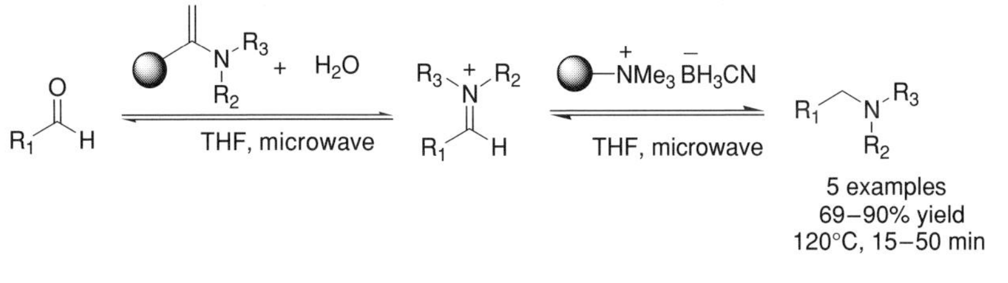

Scheme 6.35

Additional sequestering protocols have been conducted using both polymer-supported and silica-supported isocyanates, isothiocyanates, anhydrides and acid chloride species. The latter two have proved especially effective for the accelerated removal of normally difficult heterocyclic amines and anilines. Again, flash heating techniques have been applied resulting in high internal temperatures and pressures. This form of microwave heating is also conducive to the removal of electrophilic components from reaction mixtures. Anhydrides, acid halides, isocyanates and activated α-halo compounds have all been successfully scavenged from solution in shorter periods of time, when mediated by microwave dielectric heating. Kappe and co-workers[182] have recently published the synthesis of acylated dihydropyrimidines, where in part of their synthetic scheme they employed a microwave-assisted scavenging procedure to remove excess anhydride from the mixture (Scheme 6.36). This approach drastically reduced the synthesis time by shortening the scavenging periods from between 4 and 6 h at ambient temperature to less than 10 min under the directed microwave heating.

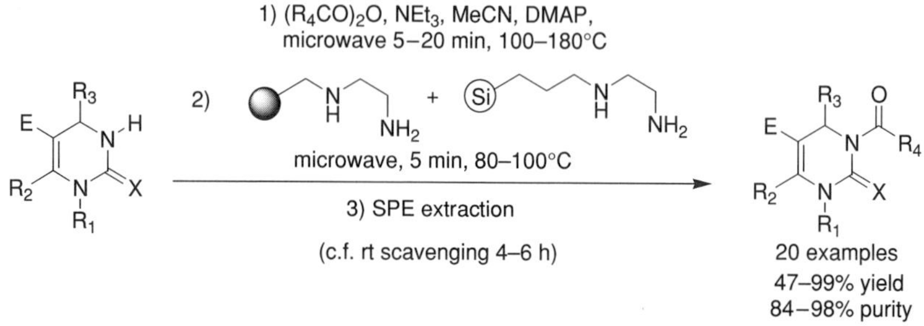

Scheme 6.36

With the multiple applications of microwave-assisted scavenging being readily applied to chemical synthesis, it is initially slightly surprising that no mention so far has been made of its use in a 'catch and release' procedure. On further consideration, the rate enhancements pertaining to the majority of the above chemistry would certainly not be as pronounced. Much of the standard catch and release methodology is based on ionic exchange interactions, which inherently display relatively high operating kinetics. However, a few examples of catch and release procedures exist, which involve the formation and cleavage of covalent bonds, where the potential arises for enhancing the reactions. This has indeed been realised with a number of processes, although in general the cases described are from industrial applications where the scavenged material was either a very valuable component or an advanced intermediate (Scheme 6.37) (I.R. Baxendale and S.V. Ley, unpublished results). The first two examples illustrate the removal of small quantities of unreacted starting materials that could be released and recycled in subsequent conversions. The third and fourth sequences represented a solution to a slightly different problem. In these two cases, an impurity in the reactions could not be separated from the product by any conventional methods attempted, and the following reactions were incompatible with the contaminating species. Here a catch and release

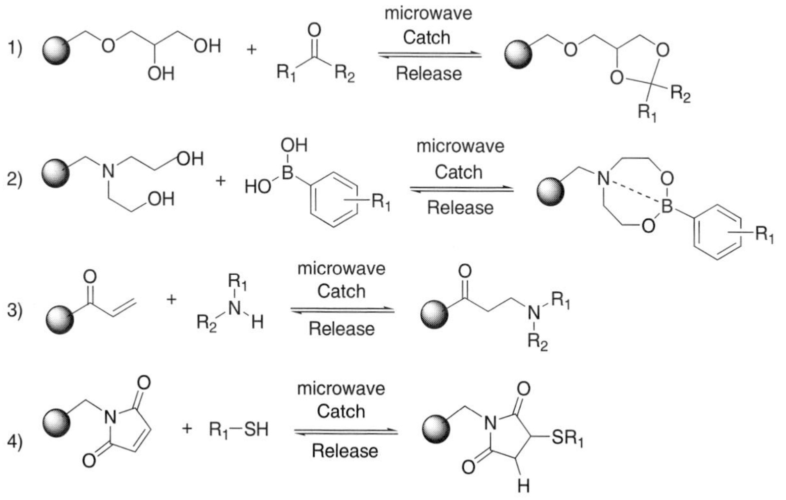

Scheme 6.37

sequence was performed using a stoichiometric amount of the immobilised reagent to sequester and purify the desired product. In principle, all the examples above could be used in a reverse sense to scavenge the reciprocally functionalised species. For example, an immobilised boronic acid or aldehyde could easily be applied for the separation of diols; such reactions have already been proven to work at room temperature although they require relatively long reaction times.

6.4. Conclusion

Microwave flash heating can provide a set of reaction parameters that permit the stimulation of a reaction, without the requirement for a prolonged or ecliptic heating profile. In this way, many reactions that would normally suffer significant thermal degradation can be performed as a result of the shortened reaction times. Gross and local pressure changes created during the reaction heating can also be important in enhancing the rate of reaction. At present, this is an area that is recognised but is undervalued and under utilised. Although extremely advantageous, it should be noted that any approach reliant on a purely thermal enhancement, independent of the heating mechanism, depends significantly on the stability of the product and reagents used.

Many of the examples mentioned have been directly concerned with, or have a consequential importance, to the area of parallel high throughput chemical processing. Any methodology that offers the ability to accelerate the synthesis of large chemical libraries or to access new chemical entities in a clean fashion needs careful consideration. It is certainly true that we have only just begun to comprehend and contemplate the possibilities that the combination of these two powerful technologies could offer. It is therefore hoped that this short chapter will in some way stimulate you, the reader, to have a new outlook on these synthetic methods.

6.5. References

1. Gedye, R., Smith, F., Westaway, K., Ali, H., Baldisera, L. Laberge, and, L., Rousell, J., The use of microwave ovens for rapid organic synthesis, *Tetrahedron Lett.,* 1986, **27**, 279.
2. Giguere, R.J., Bray, T.L., Duncan, S.M. and Majetich, G., Application of commercial microwave ovens to organic synthesis, *Tetrahedron Lett.,* 1986, **27**, 4945.
3. Maoz, R., Cohen, H. and Sagiv, J., Specific nonthermal chemical structural transformation induced by microwaves in a single amphiphilic bilayer self-assembled on silicon, *Langmuir,* 1998, **14**, 5988.
4. Stuerga, D.A.C. and Gaillard, P., Microwave athermal effects in chemistry: a myth's autopsy 2. Orienting effects and thermodynamic consequences of electric field, *J. Microw. Power Electromagn. Energy,* 1996, **31**, 101.
5. Saillard, R., Poux, M., Berlan, J. and Audhuypeaudecerf, M., Microwave-heating of organic-solvents – thermal effects and field modeling, *Tetrahedron,* 1995, **51**, 4033.
6. Pagnotta, M., Pooley, C.L.F., Gurland, B. and Choi, M., Microwave activation of the mutarotation of alpha-D-glucose – an example of an intrinsic microwave effect, *J. Phys. Org. Chem.,* 1993, **6**, 407.
7. Stuerga, D., Gonon, K. and Lallemant, M., Microwave heating as a new way to induce selectivity between competitive reactions – application to isomeric ratio control in sulfonation of naphthalene, *Tetrahedron,* 1993, **49**, 6229.
8. Molina, A., Vaquero, J.J., Garcianavio, J.L. and Alvarezbuilla, J., One-pot Graebe–Ullmann synthesis of gamma-carbolines under microwave irradiation, *Tetrahedron Lett.,* 1993, **34**, 2673.
9. Raner, K.D., Strauss, C.R., Vyskoc, F. and Mokbel, L., A comparison of reaction-kinetics observed under microwave irradiation and conventional heating, *J. Org. Chem.,* 1993, **58**, 950.
10. Raner, K.D. and Strauss, C.R., Influence of microwaves on the rate of esterification of 2,4,6-trimethylbenzoic acid with 2-propanol, *J. Org. Chem.,* 1992, **57**, 6231.
11. Adamek, F. and Hajek, M., Microwave-assisted catalytic addition of halocompounds to alkenes, *Tetrahedron Lett.,* 1992, **33**, 2039.
12. Abramovitch, R.A., Abramovitch, D.A., Iyanar, K. and Tamareselvy, K., Application of microwave-energy to organic-synthesis – improved technology, *Tetrahedron Lett.,* 1991, **32**, 5251.
13. Wei, J.H. and DeLong, J.D., Comparison of microwave and thermal cure of epoxy-resins, *Polym. Eng. Sci.,* 1993, **33**, 1132.
14. Wei, J., DeMuse, M. and Hawley, M.C., Kinetics modelling and time-temperature-transformation diagram of microwave and thermal cure of epoxy-resins, *Polym. Eng. Sci.,* 1995, **35**, 461.
15. Fang, X., Hutcheon, R. and Scola, D.A., Microwave syntheses of poly(epsilon-caprolactam-co-epsilon-caprolactone), *J. Polym. Sci., Part A: Polym. Chem.,* 2000, **38**, 1379.
16. Westaway, K. and Gedye, R.N., The question of specific activation of organic reactions by microwaves, *J. Microw. Power Electromagn. Energy.,* 1995, **30**, 219.
17. Haswell, S.J. and Howarth, N., Perturbation of a solid phase separation process by a non-thermal microwave effect, *Anal. Chim. Acta,* 1999, **387**, 113.
18. Caddick, S., Microwave-assisted organic-reactions, *Tetrahedron,* 1995, **51**, 10403.
19. Strauss, C.R. and Trainer, R.W., Invited review – developments in microwave-assisted organic-chemistry, *Aust. J. Chem.,* 1995, **48**, 1665.
20. Galema, S.A., Microwave chemistry, *Chem. Soc. Rev.,* 1997, **26**, 233.
21. Ranu, B.C., Guchhait, S.K., Gosh, K. and Patre, A., Construction of bicyclo[2.2.2]octanone systems by microwave-assisted solid phase Michael addition followed by Al_2O_3-mediated intramolecular aldolisation. An eco-friendly approach, *Green Chem.,* 2000, **2**, 5.
22. Loupy, A. and Perreux, L., A tentative rationalization of microwave effects in organic synthesis according to the reaction medium, and mechanistic considerations, *Tetrahedron,* 2001, **57**, 9199.
23. Lidström, P., Tierney, J., Wathey, B. and Westman, J., Microwave assisted organic synthesis – a review, *Tetrahedron,* 2001, **57**, 9225.
24. Loupy, A., *Microwaves in Organic Synthesis,* Wiley-VCH, Berlin, 2002.
25. Baxendale, I.R., Lee, A.-L. and Ley, S.V., A concise synthesis of carpanone using solid-supported reagents and scavengers, *J. Chem. Soc., Perkin Trans. 1,* 2002, 1850.
26. Baxendale, I.R. and Ley, S.V., Polymer-supported reagents for multi-step organic synthesis: application to the synthesis of sildenafil, *Bioorg. Med. Chem. Lett.,* 2000, **10**, 1983.
27. Baxendale, I.R., Storer, R.I. and Ley, S.V., Supported reagents and scavengers in multi-step organic synthesis, In: Buchmeiser, M. (Ed.) *Supported Reagents and Scavengers in Multi-Step Organic Synthesis,* Wiley-VCH, Berlin, 2003.
28. Durand-Reville, T., Gobbi, L.B., Gray, B.L., Ley, S.V. and Scott, J.S., Highly selective entry to the azadirachtin skeleton *via* a Claisen rearrangement/radical cyclization sequence, *Org. Lett.,* 2002, **4**, 3847.
29. Habermann, J., Ley, S.V. and Scott, J.S., Synthesis of the potent analgesic compound (±)-epibatidine using

an ochestrated multi-step sequence of polymer-supported reagents, *J. Chem. Soc., Perkin Trans. 1,* 1999, 1253.

30. Kiyota, H., Dixon, D.J., Luscombe, C.K., Hettstedt, S. and Ley, S.V., Synthesis, structure revision, and absolute configuration of (+)-didemniserinolipid B, a serinol marine natural product from a *Tunicate Didemnum sp.*, *Org. Lett.*, 2002, **4**, 3223.

31. Ley, S.V. and Mynett, D.M., Microwave promoted hydrolysis of esters absorbed on alumina: a new deprotection method for pivaloyl groups, *Synlett,* 1993, 793.

32. Siu, J., Baxendale, I.R. and Ley, S.V., Microwave assisted Leimgruber–Batcho reaction for the preparation of indoles, azaindoles and pyrroylquinolines, *Bioorg. Med. Chem.*, 2004, 2,(2) 160.

33. Ley, S.V., Baxendale, I.R., Bream, R.N., Jackson, P.S., Leach, A.G. *et al.*, Multi-step organic synthesis using solid-supported reagents and scavengers: a new paradigm in chemical library generation, *J. Chem. Soc., Perkin Trans. 1,* 2000, 3815.

34. Ley, S.V., Baxendale, I.R., Brusotti, G., Caldarelli, M., Massi, A. *et al.*, Solid-supported reagents for multi-step organic synthesis: preparation and application, *Farmaco,* 2002, **57**, 321.

35. Ley, S.V. and Baxendale, I.R., Organic synthesis in a changing world, *Chem. Rev.,* 2002, **2**, 377.

36. Ley, S.V., Baxendale, I.R. and Sherrington, D.S. (Ed.), *Supported Catalysts Applications*, Royal Society of Chemistry, Cambridge, 2000.

37. Bose, A.K., Manhas, M.S., Banik, B.K. and Robb, E.W., Microwave-induced organic-reaction enhancement (MORE) chemistry – techniques for rapid, safe and inexpensive synthesis, *Res. Chem. Intermed.,* 1994, **20**, 1.

38. Banik, B.K., Manhas, M.S., Kaluza, Z., Barakat, K.J. and Bose, A.K., Microwave-induced organic-reaction enhancement chemistry 4. Convenient synthesis of enantiopure α-hydroxy-β-lactams, *Tetrahedron Lett.,* 1992, **33**, 3603.

39. Bose, A.K., Manhas, M.S., Ghosh, M., Shah, M., Raju, V.S. *et al.*, Microwave-induced organic-reaction enhancement chemistry 2. Simplified techniques, *J. Org. Chem.,* 1991, **56**, 6968.

40. Banik, B.K., Manhas, M.S., Newaz, S.N. and Bose, A.K., Facile preparation of carbapenem synthons *via* microwave-induced rapid reaction, *Bioorg. Med. Chem. Lett.,* 1993, **3**, 2363.

41. Manhas, M.S., Banik, B.K., Mathur, A., Vincent, J.E. and Bose, A.K., Vinyl-β-lactams as efficient synthons. Eco-friendly approaches *via* microwave assisted reactions, *Tetrahedron,* 2000, **56**, 5587.

42. Bose, A.K., Banik, B.K., Mathur, C., Wagle, D.R. and Manhas, M.S., Polyhydroxy amino acid derivatives *via* beta-lactams using enantiospecific approaches and microwave techniques, *Tetrahedron,* 2000, **56**, 5603.

43. Baghurst, D.R. and Mingos, D.M.P., Superheating effects associated with microwave dielectric heating, *J. Chem. Soc., Chem. Commun.,* 1992, 674.

44. Baghurst, D.R. and Mingos, D.M.P., A new reaction vessel for accelerated syntheses using microwave dielectric super-heating effects, *J. Chem. Soc., Dalton Trans.,* 1992, 1151–1155.

45. Marand, E., Baker, K.R. and Graybeal, J.D., Comparison of reaction-mechanisms of epoxy-resins undergoing thermal and microwave cure from in situ measurements of microwave dielectric-properties and infrared-spectroscopy, *Macromolecules,* 1992, **25**, 2243.

46. Baker, K.R., Marand, E. and Graybeal, J.D., *Abstr. Pap. Am. Chem. Soc.,* 1992, **66**, 422.

47. Wei, J.H., DeLong, J.D. and Hawley, M.C., Comparison of microwave and thermal cure of epoxy-resins, *Polym. Eng. Sci.,* 1993, **33**, 1132.

48. Wei, J., DeLong, J.D. and Hawley, M.C., Kinetics modelling and time-temperature-transformation diagram of microwave and thermal cure of epoxy-resins, *Polym. Eng. Sci.,* 1995, **35**, 461.

49. Lewis, D.A., Summers, J.D., Ward, T.C. and McGrath, J.E., *Polym. Prepr.,* 1987, **28**, 330.

50. Lewis, D.A., Summers, J.D., Ward, T.C. and McGrath, J.E., Accelerated imidization reactions using microwave-radiation, *J. Polym. Sci., Pol. Chem.,* 1992, **30**, 1647.

51. Matsumoto, M. and Miyata, Y. EMC '98 ROMA, 1998, pp 523.

52. Kato, Y., Sugimoto, S., Shinohara, K., Tezuka, N., Kagotani, T. *et al.*, Magnetic properties and microwave absorption properties of polymer-protected cobalt nanoparticles, *Mater. Trans., JIM,* 2002, **43**, 406.

53. Maeda, T., Sugimoto, S., Kagotani, T., Book, D., Homma, M. *et al.*, Electromagnetic microwave absorption of α-Fe microstructure produced by disproportionation reaction of Sm_2Fe_{17} compound, *Mater. Trans., JIM,* 2000, **41**, 1172.

54. Yoshida, S., Sato, M., Sugawara, E. and Shimada, Y., Permeability and electromagnetic-interference characteristics of Fe-Si-Al alloy flakes polymer composite, *J. Appl. Phys.,* 1999, **85**, 4636.

55. Roussy, G., Jassm, S. and Thiebaut, J.M., Modeling of a fluidized-bed irradiated by a single or a multi-mode electric microwave field distribution, *J. Microw. Power Electromagn. Energy,* 1995, **30**, 178.

56. Thomas, J.R. and Faucher, F., Thermal modeling of microwave heated packed and fluidized bed catalytic reactors, *J. Microw. Power Electromagn. Energy,* 2000, **35**, 165.

57. Metaxas, A.C. and Meredith, R.J., *Industrial Microwave Heating,* Peter Peregrinus Ltd., London, 1983.

58. For an excellent review on the area the authors recommend Lanz, J.E., *A Numerical Model of Thermal Effects in a Microwave Irradiated Catalyst Bed* (PhD thesis), Virginia Polytechnic Institute and State University, Blacksburg, Virginia, 1998.

59. Barringer S.A., Ayappa, G., Davis, E.A., Davis, H.T. and Gordon, J., *Int. J. Food Sci.*, 1995, **60**, 5, 1132.
60. Sasaki, K., Shimada, A., Hatae, K. and Shimada, A., *Agric. Biol. Chem.*, 1988, **52**, 2273.
61. Stratton, J.A., *Electromagnetic Theory*, McGraw-Hill, New York, 1941, p. 1.
62. Thomas, C., Perl, J.P. and Wasan, D.T.J., *Colloid Interface Sci.*, 1990, **139**, 1.
63. Ayappa, K.G. and SenGupta, T.J., Microwave heating in multiphase systems: evaluation of series solutions, *Eng. Math.*, 2002, **44**, 155.
64. Daniels, V., *Dielectric Relaxation*, Academic Press, London, 1967.
65. Fröhlich, H., *Theory of Dielectrics*, Oxford University Press, London, 1958.
66. Debye, P., *Polar Molecules*, Chemical Catalog, New York, 1929.
67. Whittaker, A.G. and Mingos, D.M.P., The application of microwave-heating to chemical synthesis, *J. Microwave Power Electromagn. Energy*, 1994, **29**, 195.
68. Maxwell, J.C., *A Treatise on Electricity and Magnetism, Vol. 1*, 3rd. ed., Clarendon Press, Oxford, 1891, Ch. IX.
69. Wagner, K.W., Erklärung der dielektrischen Nachwirkungs vorgänge auf Grund Maxwellscher Vorstellungen, *Arch. Elektrotech.*, 1914, **2**, S371.
70. Sillars, R.W.J., Dielectric properties, *Proc. Inst. Elect. Eng.*, 1937, **100**, 199.
71. Gibson, C., Matthews, I. and Samuel, A., Microwave enhanced diffusion in polymeric materials, *J. Microwave Power Electromagn. Energy*, 1998, **23**, 17.
72. Hedrick, J.C., Lewis, D.A., Lyle, G.D., Wu, S.D., Ward, T.C. *et al.*, Microwave processing of functionalized poly(arylene ether ketones), *Abstr. Pap. Am. Chem. Soc.*, 1989, **197**, 92-PMSE.
73. Haggag, K., Tinctoria 1991, **88**, 66.
74. Kim, S.S., Leem, S.G., Ghim, H.D., Kim, J.H., and Lyoo, W.S., Microwave heat dyeing of polyester fabric, *Fibers Polymers*, 2003, **4**, 204.
75. Jeon, J.Y. and Kim, H.Y., Microwave irradiation effect on diffusion of organic molecules in polymers, *Eur. Polym. J.*, 2000, **36**, 895.
76. Lai, C.P., Tsai, M.H., Chen, M., Chang, H.S., and Tay, H.H., Morphology and properties of denture acrylic resins cured by microwave energy and conventional water bath, *Dental Mater.*, 2004, **20**, 133.
77. Zong, L.M., Zhou, S.J., Sun, R.S., Kempel, L.C. and Hawley, M.C., Dielectric analysis of a crosslinking epoxy resin at a high microwave frequency, *J. Polym. Sci. Part B: Polym. Phys.*, 2004, **42**, 2871.
78. Jacob, J., Chia, L.H.L. and Boey, F.Y.C., Thermal and nonthermal interaction of microwave-radiation with materials, *J. Mater. Sci.*, 1995, **30**, 5321.
79. Jacob, J., Chia, L.H.L. and Boey, F.Y.C., Comparative study of methyl-methacrylate cure by microwave-radiation versus thermal-energy, *Polym. Test*, 1995, **14**, 343.
80. Sengwa, R.J., Chaudhary, R. and Kaur, K., Microwave dielectric relaxation study of poly(propylene glycol) in dilute solution, *Polym. Int.*, 2000, **49**, 1308.
81. Sengwa, R.J. and Kaur, K., Microwave absorption in oligomers of ethylene glycol, *Indian J. Biochem. Biophys.*, 1999, **36**, 325.
82. Semenov, A.N., Dynamics of irregular copolymers, *Phys. Rev. E*, 1999, **60**, 3076.
83. Leon, C., Ngai, K.L. and Roland, C.M., Relationship between the primary and secondary dielectric relaxation processes in propylene glycol and its oligomers, *J. Chem. Phys.*, 1999, **110**, 11585.
84. Kakizaki, M. and Hideshima, T., Effect of distribution of free volume on concentration dependence of dielectric relaxation in water mixtures with poly(ethylene glycol) and glucose, *Jpn. J. Appl. Phys., Part 1*, 1998, **37**, 900.
85. Sengwa, R.J., Solvent effects on microwave dielectric relaxation in poly(ethylene glycols), *Polym. Int.*, 1998, **45**, 43.
86. Bergman, R., Borjesson, L., Torell, L.M. and Fontana, A., Dynamics around the liquid-glass transition in poly(propylene-glycol) investigated by wide-frequency-range light-scattering techniques, *Phys. Rev. B*, 1997, **56**, 11619.
87. Kakizaki, M., Anada, U. and Hideshima, T., Distribution of free-volume and dielectric alpha-relaxation in glass-forming liquid-mixtures, *Jpn. J. Appl. Phys., Part 1*, 1995, **34**, 3593.
88. Sengwa, R.J. and Kaur, K., Dielectric dispersion studies of poly(vinyl alcohol) in aqueous solutions, *Polym. Int.*, 2000, **49**, 1314.
89. Shinyashiki, N. and Yagihara, S., Comparison of dielectric relaxations of water mixtures of poly(vinylpyr-rolidone) and 1-vinyl-2-pyrrolidinone, *J. Phys. Chem. B*, 1999, **103**, 4481.
90. Shinyashiki, N., Yagihara, S., Arita, I. and Mashimo, S., Dynamics of water in a polymer matrix studied by a microwave dielectric measurement, *J. Phys. Chem. B*, 1998, **102**, 3249.
91. Bond, G., Moyes, R.B., Pollington, S.D. and Whan, D.A., The superheating of liquids by microwave-radiation, *Chem. Ind.*, 1991, 686.
92. Rault, S., Gillard, A.C., Foloppe, M.P. and Robba, M., A new rearrangement under microwave-heating conditions - synthesis of cyclopenta B 1,4 benzodiazepines and tetrahydrodibenzo B,F 1,4 diazepines, *Tetrahedron Lett.*, 1995, **36**, 6673.

93. Chemeat, F. and Esweld, E., *Chem. Eng. Technol.*, 2001, 7, 735.
94. Chemeat, F. and Esweld, E., ECOC-5, 2001, www.ecocexhibition2004.com.
95. Rao, K.J., Vaidhyanathan, B., Ganguli, M. and Ramakrishnan, P.A., Synthesis of inorganic solids using microwaves, *Chem. Mat.*, 1999, 11, 882.
96. Wan, J.K.S. and Koch, T.A., Application of microwave-radiation for the synthesis of hydrogen-cyanide, *Res. Chem. Intermed.*, 1994, 20, 29.
97. Wan, J.K.S., Microwaves and chemistry – the catalysis of an exciting marriage, *Res. Chem. Intermed.*, 1993, 19, 147.
98. Mingos, D.M.P. and Baghurst, D.R., Applications of microwave dielectric heating effects to synthetic problems in chemistry, *Chem. Soc. Rev.*, 1991, 20, 1.
99. Tse, M.Y., Depew, M.C. and Wan, J.K.S., Applications of high-power microwave catalysis in chemistry, *Res. Chem. Intermed.*, 1990, 13, 221.
100. Mingos, D.M.P., Microwaves in chemical synthesis, *Chem. Ind.*, 1994, 596.
101. Zhang, X.L., Hayward, D.O., and Mingos, D.M.P., Effects of microwave dielectric heating on heterogeneous catalysis, *Catalysis Lett.*, 2003, 88, 33.
102. Fan, X., Yuan, K., Hao, C., Li, N., Tan, G.Z., and Yu, X.D., Esterification by microwave irradiation on activated carbon, *Org. Prep. Proced. Int.*, 2000, 32, 287.
103. Hajek, M., Microwave activation of homogeneous and heterogeneous catalytic reactions, *Collect. Czech. Chem. Comm.*, 1997, 62, 347.
104. Tanner, D.D., Kandanarachi, P., Ding, D., Shao, H., Vizitiu, D., and Franz, J.A., The catalytic conversion of C1-Cn hydrocarbons; microwave assisted C-H' bond activation, *Energy Fuel*, 2000, 15, 197.
105. Abribat, B. and LeBigot, Y., Solvent-free selective etherification of hydroxyl. 2. catalytic role of polyethyleneg-lycols, *Tetrahedron*, 1997, 53, 2119.
106. Abribat, B., Lebigot, Y. and Gaset, A., Etherification of alcohols in the absence of solvent – catalytic function of polyethers in a solid–liquid medium, *Synth. Commun.*, 1994, 24, 2091.
107. Abribat, B., Lebigot, Y. and Gaset, A., A simple and inexpensive method for etherification of hydroxylated polyethers in the absence of solvent, *Synth. Commun.*, 1994, 24, 1773.
108. Berdague, P., Perez, F., Courtieu, J. and Bayle, J.P., Selective etherification of polyhydroxybenzenes using PEG-200 as solvent or cosolvent, *Bull. Soc. Chim. Fr.*, 1993, 130, 475.
109. Blettner, C.G., Konig, W.A., Stenzel, W. and Schotten, T., Microwave-assisted aqueous Suzuki cross-coupling reactions, *J. Org. Chem.*, 1999, 64, 3885.
110. Neumann, R. and Sasson, Y., Poly(ethylene glycols) as phase-transfer catalysts in the alkoxylation of halobenzenes of alkyl aryl ethers, *Tetrahedron*, 1983, 39, 3437.
111. Sauvagnat, B., Kulig, K., Lamaty, F., Lazaro, R. and Martinez, J., Soluble polymer supported synthesis of α-amino acid derivatives, *J. Comb. Chem.*, 2000, 2, 134.
112. Sauvagnat, B., Lamaty, F., Lazaro, R. and Martinez, J., Poly(ethylene glycol) as solvent and polymer support in the microwave assisted parallel synthesis of aminoacid derivatives, *Tetrahedron Lett.*, 2000, 41, 6371.
113. Sauvagnat, B., Lamaty, F., Lazaro, R. and Martinez, J., Polyethylene glycol (PEG) as polymeric support and phase-transfer catalyst in the soluble polymer liquid phase synthesis of α-amino esters, *Tetrahedron Lett.*, 1998, 39, 821.
114. Totten, G.E. and Clinton, N.A., Poly(ethylene glycol) derivatives as phase-transfer catalysts and solvents for organic-reactions, *J. Macromol. Sci., Rev. Macromol. Chem. Phys.*, 1988, C28, 293.
115. Totten, G.E., Clinton, N.A. and Matlock, P.L., Poly(ethylene glycol) and derivatives as phase transfer catalysts, *J. Macromol. Sci., Rev. Macromol. Chem. Phys.*, 1998, C38, 77.
116. Gokel, G.W., Goli, D.M. and Schultz, R.A., Binding profiles for oligoethylene glycols and oligoethylene glycol monomethyl ethers and an assessment of their abilities to catalyze phase-transfer reactions, *J. Org. Chem.*, 1983, 48, 2837.
117. Harris, J.M., Hundley, N.H., Shannon, T.G. and Struck, E.C., Polyethylene glycols as soluble, recoverable, phase-transfer catalysts, *J. Org. Chem.*, 1982, 47, 4789.
118. Harris, J.M. (Ed.), *Poly(ethylene glycol) Chemistry. Biotechnological and Biomedical Applications*, Plenum Press, New York, 1992, p. 3.
119. Takagi, T., *Appl. Polym. Sci.*, 1975, 19, 1774.
120. Blackwell, H.E., Out of the oil bath and into the oven – microwave-assisted combinatorial chemistry heats up, *Org. Biomol. Chem.*, 2003, 1, 1251.
121. Kappe, C.O., High-speed combinatorial synthesis utilizing microwave irradiation, *Curr. Opin. Chem. Biol.*, 2002, 6, 314.
122. Yang, G.C., Chen, Z.X. and Hu, C.L., Microwave-assisted polymer-supported organic reaction, *Chin. J. Org. Chem.*, 2002, 22, 289.
123. Varma, R.S. and Dahiya, R., Microwave-assisted facile synthesis of imines and enamines using envirocat EPZG as a catalyst, *Synlett*, 1997, 1245.

124. Perez, R., Perez, E.R., Suarez, M., Gonzalez, L., Loupy, A. *et al.*, Synthesis of aminotoluenesulfonamide derivatives using conventional heating or microwave-assisted methods, *Org. Prep. Proced. Int.*, 1997, **29**, 671.
125. Kidwai, M., Kumar, P., Goel, Y. and Kumar, K., Microwave assisted synthesis of 5-methyl-1,3,4-thiadiazol-2-ylthio/tetrazol-1-yl substituted pyrazoles, 2-azetidinones, 4-thiazolidinones, benzopyran-2-ones and 1,3,4-oxadiazoles, *Indian. J. Chem., Sect. B*, 1997, **36**, 281.
126. Tierney, J., In: *Advances in Productive Chemistry Development, Second Conference on Coherent Synthesis*, San Diego, USA, 2002.
127. Perreux, L., Loupy, A. and Delmotte, M., Microwave effects in solvent-free ester aminolysis, *Tetrahedron*, 2003, **59**, 2185.
128. Simoni, D., Invidiata, F.P., Manfredini, S., Ferroni, R., Lampronti, I. *et al.*, Facile synthesis of 2-nitroalkanols by tetramethylguanidine (TMG)-catalyzed addition of primary nitroalkanes to aldehydes and alicyclic ketones, *Tetrahedron Lett.*, 1997, **38**, 2749.
129. Simoni, D., Rondanin, R., Morini, M., Baruchello, R. and Invidiata, F.P., 1,5,7-triazabicyclo 4.4.0 dec-1-ene (TBD), 7-methyl-TBD (MTBD) and the polymer-supported TBD (P-TBD): three efficient catalysts for the nitroaldol (Henry) reaction and for the addition of dialkyl phosphites to unsaturated systems, *Tetrahedron Lett.*, 2000, **41**, 1607.
130. Merriman, G., In: *Advances in Productive Chemistry Development, Second Conference on Coherent Synthesis*, San Diego, USA, 2002.
131. Vasudevan, A., Gentles, R., Sowin, T. and Wodka, D., In: *Advances in Productive Chemistry Development, Second Conference on Coherent Synthesis*, San Diego, USA, 2002.
132. Vasudevan, A., *Microwave Synthesis: Chemistry at the Speed of Light*, Hayes, B.L., ed., CEM Corp Publishing, 2002.
133. Crosignani, S., White, P.D. and Linclau, B., Microwave-accelerated O-alkylation of carboxylic acids with O-alkylisoureas, *Org. Lett.*, 2002, **4**, 2961.
134. Crosignani, S., White, P.D. and Linclau, B., Polymer-supported O-methylisourea: a new reagent for the O-methylation of carboxylic acids, *Org. Lett.*, 2002, **4**, 1035.
135. Westman, J., An efficient combination of microwave dielectric heating and the use of solid-supported triphenylphosphine for wittig reactions, *Org. Lett.*, 2001, **3**, 3745.
136. Nicewonger, R.B., Ditto, L., Kerr, D. and Varady, L., Synthesis of a novel, recyclable, solid-phase acylating reagent, *Bioorg. Med. Chem. Lett.*, 2002, **12**, 1799.
137. Petricci, E., Botta, M., Corelli, F. and Mugnaini, C., An improved synthesis of solid-supported reagents (SSRs) for selective acylation of amines by microwave irradiation, *Tetrahedron Lett.*, 2002, **43**, 6507.
138. Passerini, M., Formation of α-hydroxycarboxamides on treatment of an isonitrile with a carboxylic acid and an aldehyde or ketone, *Gazz. Chim. Ital.*, 1921, **51**, 126.
139. Ugi, I., Betz, W. and Fetzer, U., 1961, **94**, 2814.
140. Ugi, I., Domling, A. and Werner, B., Since 1995 the new chemistry of multi-component reactions and their libraries, including their heterocyclic chemistry, *J. Heterocycl. Chem.*, 2000, **37**, 647.
141. Domling, A. and Ugi, I., The 7-component reaction, *Angew. Chem., Int. Ed. Engl.*, 1993, **32**, 563.
142. Bauer, S.M. and Armstrong, R.W., Total synthesis of Motuporin (Nodularin-V), *J. Am. Chem. Soc.*, 1999, **121**, 6355.
143. Weber, L., Illgen, K. and Ailmletter, M., Discovery of new multi component reactions with combinatorial methods, *Synlett*, 1999, 366.
144. Domling, A. and Ugi, I., Multi-component reactions with isocyanides, *Angew. Chem., Int. Ed. Engl.*, 2000, **39**, 3169.
145. Domling, A., The discovery of new isocyanide-based multi-component reactions, *Curr. Opin. Chem. Biol.*, 2000, **4**, 318.
146. Ley, S.V. and Taylor, S.J., A polymer-supported [1.3.2]oxazaphospholidine for the conversion of isothiocyanates to isocyanides and their subsequent use in an Ugi reaction, *Bioorg. Med. Chem. Lett.*, 2002, **12**, 1813.
147. Rüchardt, C., Meier, M., Haaf, K., Pakusch, J., Wolber, E.K.A. *et al.*, The isocyanide-cyanide rearrangement – mechanism and preparative applications, *Angew. Chem.-Int. Ed. Engl.*, 1991, **30**, 893.
148. Launay, D., Booth, S., Clemens, I., Merritt, A. and Bradley, M., Solid phase-mediated synthesis of isonitriles, *Tetrahedron Lett.*, 2002, **43**, 7201.
149. Ley, S.V., Leach, A.G. and Storer, R.I., A polymer-supported thionating reagent, *J. Chem. Soc., Perkin Trans. 1*, 2001, 358.
150. Perregaard, J. and Lawesson, S.-O., Studies on organophosphorus compounds – XVI. On O,O-Dialkyl dithiophosphoric acids and N,N,N′,N′-tetramethylthiophosphoric diamide as thiation agents: conversion of carboxamide into thiocarboxamides, alkyl dithiocarboxylates, or nitriles, *Bull. Soc. Chim. Belg.*, 1977, **86**, 321.

151. Pellletier, J.C., Khan, A., and Tang, Z., A tandem three-phase reaction for preparing secoundary amines with minimal side products, *Org. Lett.*, 2002, **4**, 26, 4611.
152. Kabza, K.G., Chapados, B.R., Gestwicki, J.E. and McGrath, J.L., Microwave-induced esterification using heterogeneous acid catalyst in a low dielectric constant medium, *J. Org. Chem.*, 2000, **65**, 1210.
153. Zu, Z.P. and Chuang, K.T., Kinetics of acetic acid esterification over ion exchange catalysts, *Can. J. Chem. Eng.*, 1996, **74**, 493.
154. Toteja, R.S.D., Jangida, B.L., Sundaresan, M. and Venkataramani, B., Water sorption isotherms and cation hydration in Dowex 50W and Amberlyst-15 ion exchange resins, *Langmuir*, 1997, **13**, 2980.
155. Gopalakrishnan, G., Kasinath, V., Singh, N.D.P., Krishnan, V.P.S., Solomon, K.A. *et al.*, Microwave assisted regioselective bromomethoxylation of alkenes using polymer supported bromine resins, *Molecules*, 2002, **7**, 412.
156. Varma, R.S., Solvent-free accelerated organic syntheses using microwaves, *Pure Appl. Chem.*, 2001, **73**, 193.
157. Desai, B. and Danks, T.N., Thermal- and microwave-assisted hydrogenation of electron-deficient alkenes using a polymer-supported hydrogen donor, *Tetrahedron Lett.*, 2001, **42**, 5963.
158. Buchmeiser, M.R., Lubbad, S., Mayr, M. and Wurst, K., Access to silica- and monolithic polymer supported C-C-coupling catalysts *via* ROMP: applications in high-throughput screening, reactor technology and biphasic catalysis, *Inorg. Chim. Acta*, 2003, **345**, 145.
159. Neckers, D.C., Kooistra, D.A. and Green, G.W., Polymer-protected reagents. Polystyrene-aluminium chloride, *J. Am. Chem. Soc.*, 1972, **94**, 9284.
160. Gopalakrishnan, G., Kasinath, V. and Singh, N.D.P., Microwave-assisted ketone-ketone rearrangement: an improved synthesis of 3-(4-alkoxyphenyl)-3-methylbutan-2-ones, *Org. Lett.*, 2002, **4**, 781.
161. Wipf, P. and Venkatramen, S., An improved protocol for azole synthesis with PEG-supported Burgess reagent, *Tetrahedron Lett.*, 1996, **37**, 4659.
162. Brain, C.T. and Brunton, S.A., Synthesis of 1,3,4-oxadiazoles using polymer-supported reagents, *Synlett*, 2001, 382.
163. Brain, C.T., Paul, J.M., Loong, Y. and Oakley, P.J., Novel procedure for the synthesis of 1,3,4-oxadiazoles from 1,2-diacylhydrazines using polymer-supported Burgess reagent under microwave conditions, *Tetrahedron Lett.*, 1999, **40**, 3275.
164. Baxendale, I.R., Ley, S.V. and Martinelli, M., *J. Med. Chem.*, (in press).
165. Ohberg, L. and Westman, J., One-pot three-step solution phase syntheses of thiohydantoins using microwave heating, *Synlett*, 2001, 1893.
166. Chen, S.T., Chiou, S.H. and Wang, K.T., Preparative scale organic-synthesis using a kitchen microwave-oven, *J. Chem. Soc., Chem. Commun.*, 1990, 807.
167. Leskovsek, S., Plazl, I. and Koloini, T., Hydrolysis of sucrose with microwaves in a tubular flow reactor, *Chem. Biochem. Eng. Q.*, 1996, **10**, 21.
168. Plazl, I., Leskovsek, S. and Koloini, T., Hydrolysis of sucrose by conventional and microwave-heating in stirred-tank reactor, *Chem. Eng. J. Biochem. Eng. J.*, 1995, **59**, 253.
169. Plazl, I., Pipus, G. and Koloini, T., Microwave heating of the continuous flow catalytic reactor in a nonuniform electric field, *AIChE J.*, 1997, **43**, 754.
170. Hut, Y.M. and Ju, L.K., Lipase-mediated deacetylation and oligomerization of lactonic sophorolipids, *Biotechnol. Prog.*, 2003, **19**, 303.
171. Tardioli, P.W., Fernandez-Lafuente, R., Guisan, J.M. and Giordano, R.L.C., Design of new immobilized-stabilized carboxypeptidase a derivative for production of aromatic free hydrolysates of proteins, *Biotechnol. Prog.*, 2003, **19**, 565.
172. Mabrouk, P.A., The use of poly(ethylene glycol) enzymes in nonaqueous enzymology in poly(ethylene glycol), In: *Chemistry and Biological Applications* (ACS Symposium Series, No 680), J. Milton Harris, Samuel Zalipsky, eds., American Chemical Society Division of Polymer Chemistry, Calif. American Chemical Society Meeting 1997, San Francisco, pp 118.
173. Shock, E.L., Stability of peptides in high-temperature aqueous-solutions, *Geochim. Cosmochim. Acta*, 1992, **56**, 3481.
174. Adams, M.W.W. and Kelly, R.M., Enzymes from microorganisms from extreme environments, *Chem. Eng. News*, 1995, **73**, 32.
175. Carrillo-Munoz, J.-S., Bouvet, D., Eryka, G.-J., Loupy, A. and Petit, A., Microwave-promoted lipase-catalyzed reactions. Resolution of (+/−)-1-phenylethanol, *J. Org. Chem.*, 1996, **61**, 7746.
176. GeloPujic, M., GuibeJampel, E., Loupy, A. and Trincone, A., Enzymatic glycosidation in dry media under microwave irradiation, *J. Chem. Soc., Perkin Trans. 1*, 1997, 1001.
177. GeloPujic, M., GuibeJampel, E., Loupy, A., Galema, S.A. and Mathe, D., Lipase-catalysed esterification of some alpha-D-glucopyranosides in dry media under focused microwave irradiation, *J. Chem. Soc., Perkin Trans. 1*, 1996, 2777.

178. Rao, N.S., Agarwal, S.K., Chauhan, V.K., Bhatia, D., Sharma, A.K. *et al.*, Microwave-assisted spectrophotometric estimation of polymer-supported functional groups using a universal reagent, *Anal. Chim. Acta*, 2000, **405**, 247.
179. Baxendale, I.R., Ley, S.V. and Piutti, C., Total Synthesis of the Amaryllidaceae Alkaloid (+)-plicamine and its unnatural enantiomer by using solid-supported reagents and scavengers in a multistep sequence of reactions, *Angew. Chem., Int. Ed. Engl.*, 2002, **41**, 2194.
180. Baxendale, I.R., Ley, S.V., Nessi, M. and Piutti, C., Total synthesis of the amaryllidaceae alkaloid (+)-plicamine using solid-supported reagents, *Tetrahedron*, 2002, **58**, 6285.
181. Olah, G.A., Yamato, T., Iyer, P.S. and Prakash, G.K.S., Catalysis by solid superacids. 20. Nafion-H catalyzed reductive cleavage of acetals and ketals to ethers with triethylsilane, *J. Org. Chem.*, 1986, **51**, 2826.
182. Dallinger, D., Gorobets, N.Y. and Kappe, C.O., High-throughput synthesis of N3-acylated dihydropyrimidines combining microwave-assisted synthesis and scavenging techniques, *Org. Lett.*, 2003, **5**, 1205.

7 Microwave-assisted solid-phase synthesis

ALEXANDER STADLER AND C. OLIVER KAPPE

7.1. Combinatorial chemistry and solid-phase organic synthesis

Combinatorial chemistry[1], the art and science of rapidly synthesising and testing potential lead compounds for any desired property, has turned out to be one of the most
promising approaches in drug discovery. Because of the enormous progress made in ge-
nomic sciences, molecular biology and biochemistry, a large number of biologically important target proteins have now become available for screening purposes. This has led
to an ever-growing demand for large libraries of novel compounds that are being evaluated for their biological properties using appropriate screening protocols. The discovery
of novel target molecules was paralleled by the development of modern high throughput screening (HTS) technologies including miniaturised formats, allowing testing of
thousands of individual compounds per day. Traditional methods of organic synthesis
are orders of magnitude too slow to satisfy the increasing demand for these compounds.

The fundamental meaning of combinatorial synthesis is the ability to generate large
numbers of chemical compounds in a short time. Therefore, it is utilised to generate
libraries of potential lead compounds, which can immediately be screened on biological efficiency. Although there was certain resistance against this new technique it
has become a leading principle for chemical synthesis. Nevertheless, to establish this
topic for synthesis, several skills including high-speed purification and characterisation
strategies, respectively, had to be developed. While chemistry in the past has been characterised by slow, steady and painstaking work, combinatorial chemistry has changed
the characteristics of chemical research and permitted a level of productivity thought
impossible a few years ago.

One of the key technologies used in combinatorial chemistry is the solid-phase organic synthesis (SPOS)[2], originally developed by Merrifield in 1963 for the synthesis of
peptides[3]. In SPOS, a molecule (scaffold) is attached to a solid support, for example, a
'polymer resin'. In general, resins are insoluble base polymers with a 'linker' molecule
attached to them. Often, spacers are included to reduce steric hindrance by the bulk of
the resin. Linkers on the other hand are functional moieties, which allow the attachment and cleavage of scaffolds under controlled conditions. Subsequent chemistry is
then carried out on the molecule attached to a support, until at the end of the often
multi-step synthesis the desired molecule is released from the support (Figure 7.1).

Solid-phase organic synthesis (SPOS) shows several advantages compared with classical protocols in solution. In order to accelerate reactions and to drive them to completion, a large excess of reagents can be used, as these can easily be removed by filtration
and subsequent washing of the solid support. In addition, SPOS can easily be automated using appropriate robotics. Furthermore, SPOS can be applied to the powerful
'split-and-mix' strategy, which turned out to be an important tool for combinatorial
chemistry. Alongside the advent of combinatorial chemistry, the development of feasible

RESIN

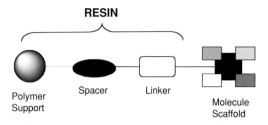

Figure 7.1   The concept of solid-phase organic synthesis.

characterisation strategies to support automated parallel synthesis was on demand, with high-throughput chromatography and high-performance liquid chromatography–mass spectrometry among them.

One of the pioneers of solid-phase synthesis is R. B. Merrifield, who developed his above-mentioned strategy for generating peptide molecules on polystyrene beads in 1963[3]. This technique was repeatedly applied to non-peptide syntheses starting in the early 1970s. During the 1980s, various protocols of solid-phase chemistry had been introduced, such as reactions on cellulose paper[4], on functionalised polypropylene pins[5] and the so-called 'tea-bag' strategy, where polypropylene mesh containers encapsulating polystyrene resin were utilised[6]. In the late 1980s, Furka and co-workers introduced the 'split synthesis' strategy[7], which deals with the 'one bead/one compound' concept, promising the delivery of million of compounds produced simultaneously on beads in a short period of time. From the 1990s onwards, an ever-growing number of publications dealing with all aspects of combinatorial synthesis were released, including reactions on soluble polymers or dendrimers carried out in homogeneous solution, or on a well-defined fluorous support. Nowadays a lot of common solution-phase reactions have been performed equally well on solid phase and a great variety of reagents have been attached to polymer supports (see Chapter 6). This allows one to study a large number of organic reactions in a combinatorial manner, leading to desired target molecules being synthesised with increased efficiency and productivity.

7.2. Microwave chemistry and solid-phase organic synthesis

Solid-phase organic synthesis (SPOS) exhibits several shortcomings because of the nature of the heterogeneous reaction conditions. Nonlinear kinetic behaviour, slow reactions, solvation problems and degradation of the polymer support as a result of the long reaction times are some of the problems typically experienced in SPOS[2]. Any technique that is able to address these issues and to speed up the process of solid-phase synthesis is of considerable interest, particularly for research laboratories involved in high-throughput synthesis.

In this context, microwave-enhanced organic synthesis has attracted a substantial amount of attention in recent years since its beginning in 1986[8,9]. As shown by the ever growing amount of publications and number of review articles available on this subject (for further information on microwave-assisted organic synthesis see: http://www.maos.net), high-speed microwave-assisted synthesis has been applied successfully in many fields of synthetic organic chemistry. In fact, it is becoming evident that microwave approaches can be developed for most chemical transformations requiring heat. The main benefits of performing reactions under microwave irradiation conditions are the significant rate-enhancements and the higher product yields that can frequently be observed. Not surprisingly, these features have recently also attracted interest from the combinatorial/medicinal chemistry community, where reaction speed is of great importance[10–14], and there is an ever growing number of publications reporting on rate-enhancements in SPOS utilising microwaves. This attractive linking between combinatorial processing and microwave heating is a logical consequence of the increased speed and effectiveness offered by the microwave approach.

7.2.1. *Microwave dielectric heating*

Using microwave dielectric heating[15], the microwave energy is introduced into the chemical reactor remotely and direct access by the energy source to the reaction vessel is obtained. Microwave irradiation penetrates the walls of the vessel and heats only the reactants and solvent, not the reaction vessel. The energy transfer is not produced by conduction or convection, but by dielectric loss. Thus, the propensity of a sample to undergo microwave heating depends on the dielectric properties and is represented by the so-called loss tangent (tan δ). Materials dissipate microwave energy by two main mechanisms: *dipole rotation* and *ionic conduction*. When molecules with a permanent dipole are submitted to an electric field, they become aligned. If this field oscillates, the orientation changes with each alternation. The strong agitation, provided by the reorientation of molecules, in phase with the electrical field excitation, causes an intense internal heating. During ionic conduction, as the dissolved charged particles in a sample (usually ions) oscillate back and forth under the influence of the microwave field, they collide with their neighbouring molecules or atoms. These collisions cause agitation or motion, generating heat. Further details on the rather complex theory of microwave dielectric heating are provided in Chapter 1.

7.2.2. *Solvents*

The choice of solvent in microwave-assisted SPOS is absolutely critical. Ideally, the solvent should have (i) good swelling properties for the resin involved, (ii) a high boiling point if reactions are to be carried out at atmospheric pressure, (iii) a high loss tangent (tan δ) for good interactions with microwaves and (iv) high chemical stability and inertness to minimise side reactions. Clearly, solvents such as dichloromethane (tan δ = 0.047), which are commonly used in SPOS under conventional conditions, may not be very useful in a microwave-assisted protocol. In those cases, a co-solvent with a high loss tangent or an entirely different solvent system needs to be used. In general, solvents with a loss tangent >0.1 are considered suitable for microwave dielectric heating. A summary of solvents useful for microwave-assisted solid-phase synthesis is given in Table 7.1.

For example, many microwave-assisted solid-phase coupling reactions utilise 1-methyl-2-pyrrolidone (NMP) or 1,2-dichlorobenzene (DCB) as the reaction solvent.

Table 7.1   Swelling behaviour, loss tangents (tan δ) and boiling points for solvents used in microwave-assisted solid-phase synthesis

Solvent	PS swelling[a]	Tentagel® swelling[a]	tan δ[b]	bp (°C)
Dimethylformamide (DMF)	5.2	4.4	0.161	153
Dimethylacetamide (DMA)	5.8	4.0	n.a.	166
1-Methyl-2-pyrrolidone (NMP)	6.4	4.4	0.275	202
Dimethylsulphoxide (DMSO)	4.2	3.8	0.825	189
Tetrahydrofuran (THF)	6.0	4.0	0.047	65
1,4-Dioxane	5.6	4.2	n.a.	106
1,2-Dichloroethane	4.4	5.4	0.127	83
Chlorobenzene	n.a.	n.a.	0.101	132
1,2-Dichlorobenzene (DCB)	4.8	5.2	0.280	180
Nitrobenzene	4.3	4.8	0.589	211
Methanol	1.6	3.6	0.659	65
Water	1.6	3.6	0.123	100

[a]Data from Ref. 2.
[b]Data from Ref. 8.

The main reason lies in the high-thermal stability of both solvents and their relatively high boiling points. The high boiling points of these solvents makes it unnecessary to carry out reactions in specialised sealed vessels under elevated pressure. Furthermore, polystyrene resins typically show excellent swelling characteristics in both NMP and DCB (Table 7.1). Because of the polar nature of these solvents a sufficiently strong absorption of microwave energy occurs. Indeed extremely rapid heating profiles can be obtained with NMP or DCB inside a microwave cavity. In contrast to a solution-phase reaction, a high-boiling point of a solvent is not considered a problem in the work-up/purification stage, since the desired target compound remains on the resin until the cleavage step, where a lower boiling solvent can be used. Since most dedicated microwave reactors nowadays offer the convenience of performing reactions under sealed vessel 'autoclave-like' conditions, lower boiling solvents can still be used at high reaction temperatures.

7.2.3. Thermal and mechanical stability of polymer supports

As far as polymer supports for microwave-assisted SPOS are concerned, the use of cross-linked macroporous or microporous polystyrene resins has been most prevalent. In contrast to the common belief that states that the use of polystyrene resins limits reaction conditions to temperatures below 130°C[16], it has been shown that these resins can withstand microwave irradiation for short periods of time, even at 200°C for 20–30 min in solvents such as NMP or DCB[17]. Figure 7.2 indicates the thermal stability of standard polystyrene Merrifield resin up to 220°C without any degradation of the macromolecular structure of the polymer backbone, which allows reactions even at significantly elevated temperatures.

Due to their mode of preparation, polymeric resin beads consist of a macroporous internal structure and of highly cross linked areas (>5%), respectively. The latter renders rigidity to the resin, whereas the porous areas provide a large internal surface for functionalisation, even in the dry state. These macroporous polystyrene-based resins are subsequently modified in various manners, which renders the accessibility to numerous

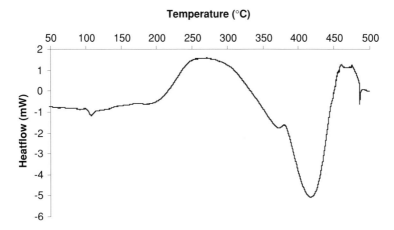

Figure 7.2 Thermal stability (differential scanning calorimetry) of polystyrene-based solid support (Merrifield resin) (A. Stadler and C.O. Kappe, unpublished results).

organic solvents. Furthermore they show high resistance towards osmotic shock, but can be brittle when not manipulated carefully.

Microporous resins, like the well-known Merrifield resin (Fig. 7.3), appearing only slightly cross linked (1–2%), show a more homogeneous network, as made evident by a glassy and transparent surface. In general, they are mechanically weak and can be easily subjected to damage. Furthermore, strange osmotic shock can occur, when pre-swollen resin beads are introduced into a poor swelling ('bad') solvent. The beads shrink rapidly under expulsion of the good swelling solvent and are subjected to stress, which leads to mechanical damage or considerable structural modifications. In addition, vigorous stirring over a longer period can cause the break down of the polymeric surface. However, these microporous resins tolerate a broad range of reaction conditions and show chemical stability over a wide range of reagents, such as strong bases and acids and even weak oxidants, while strong oxidants and electrophilic reagents should in general be avoided.

Tentagel® resins and cellulose membranes have also been used in microwave-assisted synthesis (see later), although some degradation has been observed during irradiation. It should be mentioned that most of the typically used polymer supports in SPOS (except, for example, polyacrylamides; see Fig. 7.3) do not carry a large number of polar functionalities and are therefore more or less transparent to microwave energy. In general, because of the rather short reaction times of microwave solid-phase reactions, even magnetic stirring can be performed with these resins without mechanical degradation effects. In contrast, solid-phase reactions utilising conventional thermal heating typically require longer reaction times and therefore the mechanical degradation of the solid support caused by extended stirring is significant.

One has to note, however, that other polymer composite materials also popular in solid-phase synthesis, for example, polyethylene or polypropylene tea bags such as IRORI™ kans, lanterns, crowns, or plugs are generally less suitable for high-temperature reactions (>160°C). Therefore, microwave irradiation is not typically a very suitable tool to speed up reactions, which utilise these materials as either a solid support or as containment for the solid support.

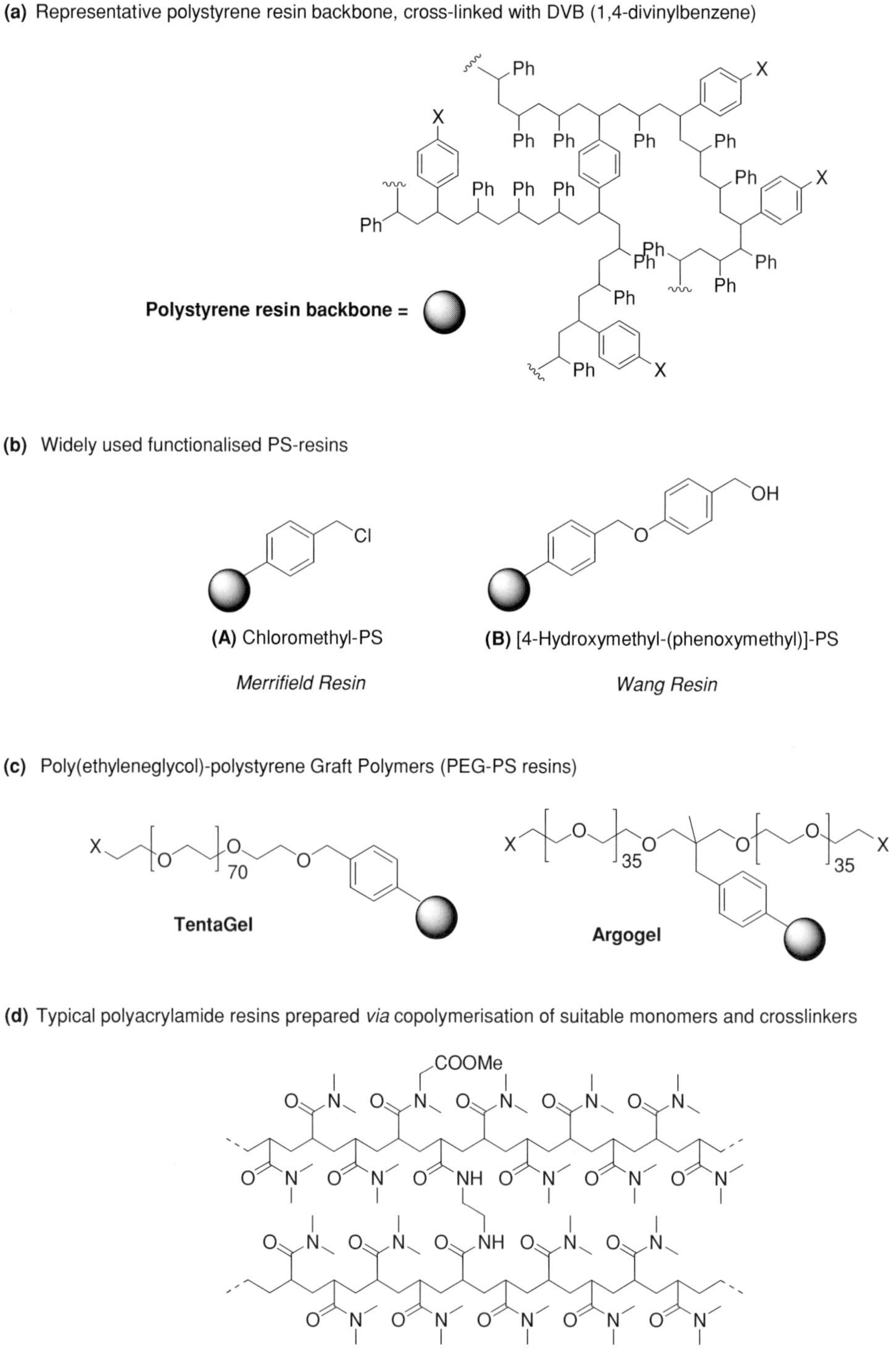

(a) Representative polystyrene resin backbone, cross-linked with DVB (1,4-divinylbenzene)

Polystyrene resin backbone =

(b) Widely used functionalised PS-resins

(A) Chloromethyl-PS

Merrifield Resin

(B) [4-Hydroxymethyl-(phenoxymethyl)]-PS

Wang Resin

(c) Poly(ethyleneglycol)-polystyrene Graft Polymers (PEG-PS resins)

TentaGel

Argogel

(d) Typical polyacrylamide resins prepared *via* copolymerisation of suitable monomers and crosslinkers

Figure 7.3 Some common supports used in solid-phase synthesis.

7.2.4. *Equipment*

Although the examples of microwave-assisted solid-phase reactions presented in this and other chapters demonstrate that rapid synthetic transformations can, in many cases, be achieved using microwave irradiation, the possibility of high-speed synthesis does not necessarily mean that these processes can also be adapted to a truly high-throughput format. In the past few years, all commercial suppliers of microwave instrumentation for organic synthesis have moved towards combinatorial/high-throughput platforms for conventional solution-phase synthesis (http://www.cemsynthesis.com; http://www.milestonesci.com; http://www.personalchemistry.com). Due to space limitations within this chapter, we shall not attempt to discuss these systems or basic multi-mode or single-mode microwave reactor design and technology here.

As of 2003, there were no commercially available dedicated reaction vessels for carrying out microwave-assisted solid-phase synthesis, that is, vessels that take advantage of bottom-filtration techniques. However, several articles in the area of microwave-assisted parallel synthesis have described irradiation of 96-well filter-bottom polypropylene plates in conventional household microwave ovens for high-throughput synthesis[18−21]. While some authors did not report any difficulties associated with the use of such equipment[21], others have experienced problems in connection with the thermal instability of the polypropylene material itself[19], and with respect to temperature gradients developing between individual wells upon microwave heating[19,20] (see Chapter 8). While Teflon (or similar materials like PFA) can eliminate the problem of thermal stability, the issue of bottom filtration reaction vessels has not been adequately addressed at present.

A recent article describes the construction and use of a parallel polypropylene reactor comprising of cylindrical, expandable reaction vessels with porous frits at the bottom (Fig. 7.4)[21]. This is the first description of reaction vessels for microwave-assisted synthesis that may be useful for carrying out solid-phase synthesis using bottom filtration techniques in conjunction with microwave heating. However, at the time when this chapter was being written no application of this system for solid-phase synthesis had been reported.

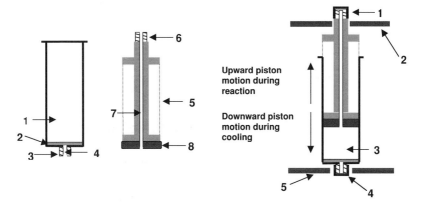

Figure 7.4 Microwave vessel components (left): (1) reaction chamber; (2) Frit; (3) Luer lock; (4) product outlet; (5) piston; (6) Luer lock; (7) hollow bore; (8) seal. Single microwave vessel during reaction (right): (1) central bore closed; (2) cover plate; (3) reaction components; (4) product outlet closed; (5) supporting plate. (Reproduced with permission from Ref. 21).

7.3. Literature survey

7.3.1. *Peptide synthesis and related examples*

One of the first dedicated applications of microwaves towards solid-phase chemistry was the synthesis of small peptide molecules, presented by Yu and co-workers[22]. As a preliminary test the authors coupled Fmoc-Ile and Fmoc-Val, respectively, with Gly-preloaded Wang resin using the corresponding symmetric anhydrides (Scheme 7.1).

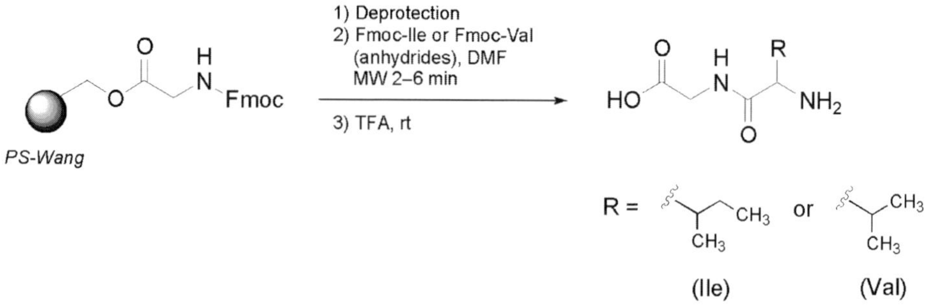

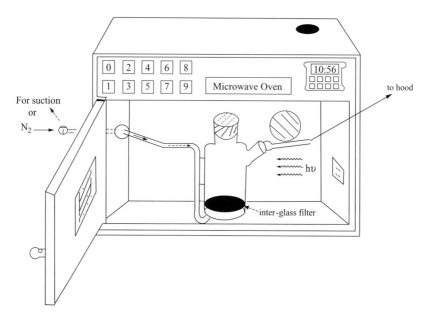

Scheme 7.1   Microwave-assisted peptide coupling.

The reactions were carried out within 2–6 min, using a modified domestic microwave oven (Fig. 7.5), employing a dedicated custom-made solid-phase reaction vessel under atmospheric pressure conditions.

Figure 7.5   Schematic description of reaction apparatus (reproduced with permission from Ref. 22).

The vessel was placed in the middle of the cavity, and a Teflon tube from the side arm was connected to a nitrogen source. During microwave irradiation, a stream of nitrogen gas was blown into the vessel with the gas bubbles serving as an agitator. After irradiation the reaction solution was filtered off *via* the side arm by suction.

The microwave protocol increased the reaction rate at least two- to threefold, as conversion was only 60–80% within 6 min under conventional heating. This improved coupling efficiency was duplicated with numerous amino acid derivatives and a further two peptide fragments were coupled with the Gly-Wang resin. These couplings were completed within 2 min as determined by quantitative ninhydrin assay.

For a more representative test, a fragment of the acyl carrier protein ([65-74]ACP) was synthesised using preformed active esters in *N,N*-dimethylformamide. Each coupling step included 4 min of irradiation and at the end of the elongation the authors determined an average coupling yield of 99.65% (Scheme 7.2).

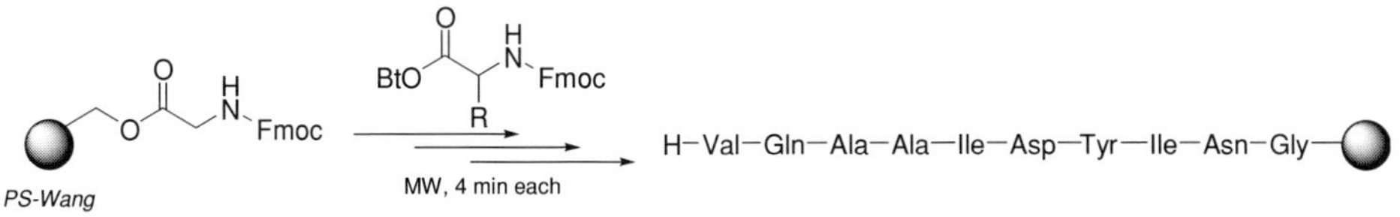

Scheme 7.2   Synthesis of [65-74]ACP employing stepwise coupling of amino acid esters.

Compared to the conventional protocol, where peptide bond formation proceeded within 30 min for each step, there was again a significant increase in the coupling efficiency. Importantly, it was demonstrated that no significant racemisation occurred in the peptide formation. Furthermore, the complete coupling of difficult sequence peptides could be accomplished within a few minutes and it was determined that under microwave irradiation conditions, peptide fragments had higher reactivity than single amino acid derivatives. However, in common with so many of the early publications in microwave-assisted chemistry, the exact reaction temperature during the irradiation period was not determined, presumably due to the lack of suitable instrumentation. As a result of the lack of temperature control, these reactions are unfortunately rarely reproducible, and the reasons for the observed rate-enhancements, as well as the possible involvement of so-called non-thermal microwave effects remains unclear.

In a more recent study, microwave irradiation has been applied to the coupling of sterically hindered amino acids, leading to di- and tripeptides (Scheme 7.3)[23]. Using a dedicated single-mode microwave instrument with temperature monitoring, Erdelyi and Gogoll investigated a variety of common coupling reagents, that is, PyBOP, HATU, TBTU and Mukaiyama's reagent for peptide synthesis.

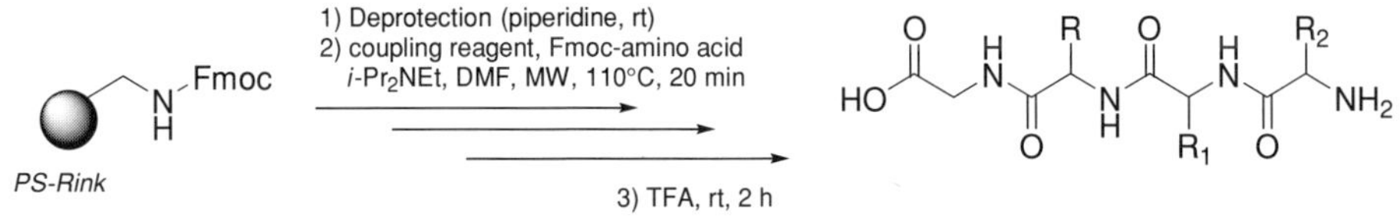

Scheme 7.3   Microwave-mediated tripeptide synthesis.

Under microwave conditions, coupling of the amino acids *via* the corresponding anhydrides or *N*-HOBt activated esters were completed in a few minutes (1.5–20 min, depending on the used reagent) without the need for double or triple coupling steps as in conventional protocols. The azobenzotriazole derivatives showed increased coupling efficiency up to 110°C. Above this temperature, decomposition of the reagents was indicated by the colour change of the reaction mixtures. However, no degradation of the solid support was observed. Furthermore, both LC–MS and ^{1}H NMR confirmed no racemisation, even after high temperature treatment in the presence of base.

A dramatic increase in the overall speed of peptoid synthesis was reported by Olivos and co-workers in a recent publication[24]. In the article, the authors present a multi-step protocol for the generation of various peptoids employing a domestic microwave oven (Scheme 7.4). Reaction times were drastically reduced, requiring less than 1 min for the coupling of each residue.

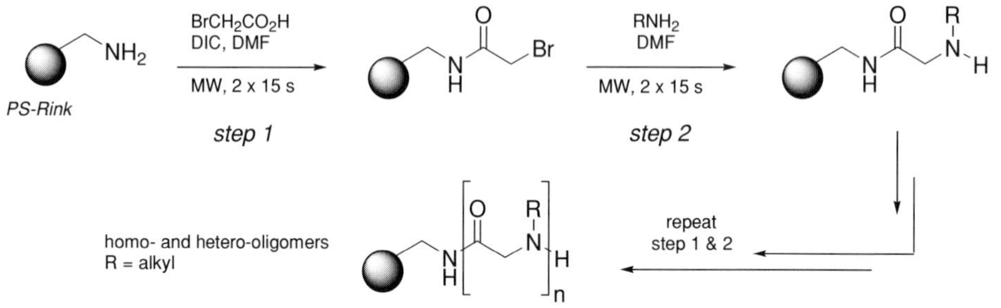

Scheme 7.4 Construction of peptoid sequences under microwave conditions. Cleavage was carried out using trifluoroacetic acid (TFA) at room temperature.

With this protocol nine different primary amines (Fig. 7.6) were used to generate different 9-residue homo-oligomers, one 20-residue homo-oligomer and one 9-residue hetero-oligomer.

Since an unmodified domestic microwave oven does not allow stirring of the reaction mixture, these transformations were performed in two 15 s runs with manual stirring between the irradiation steps. In this case, the temperature of the solutions did not exceed 35°C as determined by inserting a thermometer after the second 15 s run. For a comparative purpose, these couplings were also carried out by conventional thermal

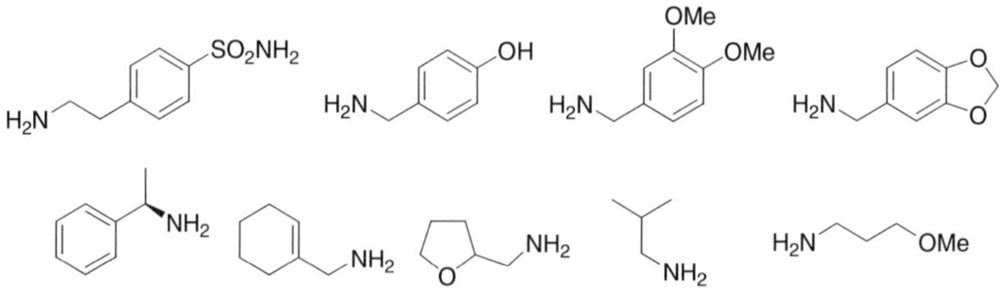

Figure 7.6 Amines used for peptoid synthesis.

heating at 37°C, leading to similar results. Both methods, however, provided better yields and purities than did conventional room temperature couplings. In general, this protocol allows a convenient method for high-throughput peptoid synthesis, as microwave acceleration reduces the overall production time. For example, a 9-residue peptoid can be synthesised in 3 h, compared to 20–32 h employing the standard protocol.

In another application, Finaru *et al.* reported the solid-supported synthesis of the indole core of melatonin analogues under microwave irradiation (Scheme 7.5)[25]. The reactions were carried out in a dedicated single-mode microwave reactor for organic synthesis. Coupling of the benzoic acid derivative to the Rink amide resin was achieved by using the convenient peptide coupling reagents 3-hydroxybenzotriazole (HOBt) and O-(benzotriazol-1-yl)-N,N,N′,N′-tetramethyluronium tetrafluoroborate (TBTU), respectively. Subsequent indole ring formation was carried out by palladium-catalysed coupling directly on the solid phase. Synthesis of the corresponding iodo-compound was accomplished by treatment of the polymer-bound indole derivative with N-iodosuccinimide (NIS). Finally, the desired compounds were released from the resin by using conventional trifluoroacetic acid/dichlormethane (TFA/DCM) cleavage conditions at room temperature.

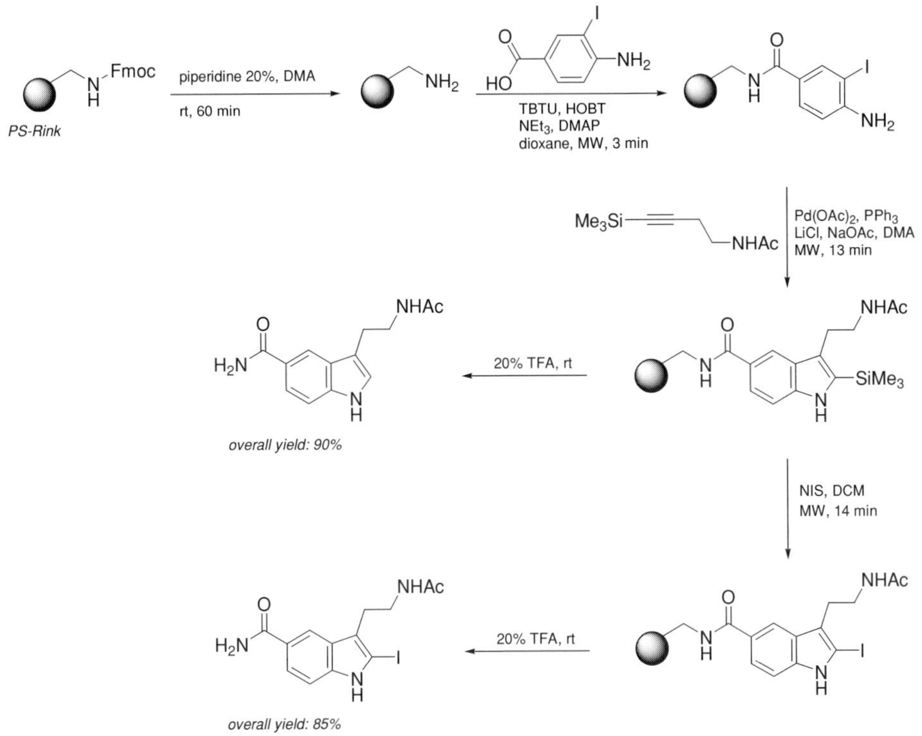

Scheme 7.5 Microwave-mediated solid-phase synthesis of 5-carboxamido-N-acetyltryptamines.

Carrying out these reactions under microwave conditions leads to a substantial increase in yields and a significant reduction of the reaction times, compared to conventional thermal heating, where each individual step takes 24–48 h. Due to the limited thermal stability of the solid support, the temperature during the reactions was kept below 140°C, which was achieved by strict power control of the microwaves.

The rapid microwave-assisted deprotection of N-benzyl carbamate (Cbz) and N-benzyl (Bn) derivatives in solution as well as on solid support was reported by Daga et al.[26] Within this report, amino groups protected as benzyl carbamates or with simple benzyl groups could be deprotected in a few minutes by microwave-assisted catalytic transfer hydrogenation with palladium charcoal in isopropanol, employing ammonium formate as the hydrogen donor (Scheme 7.6). Both MeO-PEG and PS Wang-resin were used as soluble and solid supports, respectively, in these reactions.

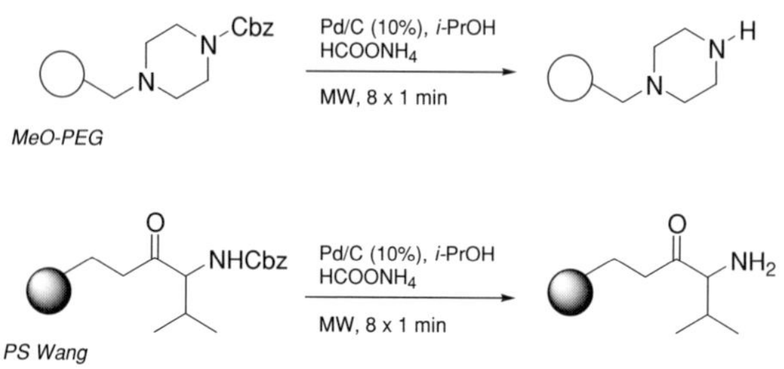

Scheme 7.6   Polymer-supported deprotection under microwave conditions.

The reactions were carried out in an Erlenmeyer flask, employing a conventional domestic microwave oven. To limit the presence of solvent vapours inside the cavity, the reaction mixtures were irradiated for 1 min and the deprotection reaction was subsequently monitored by thin layer chromatography (TLC). If starting material was present, additional cycles of 1 min were repeated until no traces of starting material could be detected by TLC.

This is a very simple and short method for the deprotection of N-Cbz and N-Bn groups, which is also applicable for N-Cbz protected amino acids and is compatible with Fmoc protecting groups, which remain unaffected under these conditions. Furthermore, the microwave protocol is fully compatible with enantiomerically pure amino acids and peptides, as no racemisation was observed in the resulting free amines.

7.3.2. Resin functionalisation

The functionalisation of commercially available standard solid supports is of common interest for combinatorial applications to enable a broad range of reactions to be studied. Since these transformations usually require long reaction times under conventional thermal conditions, it was obvious to combine microwave chemistry with the art of resin functionalisation.

As a suitable model reaction, the coupling of various substituted carboxylic acids to polymer supports has been investigated (Scheme 7.7)[27]. The resulting polymer-bound esters served as useful building blocks in a variety of further solid-phase transformations. In a preliminary experiment, benzoic acid was attached to Merrifield resin under

microwave conditions for 5 min. In addition, this functionalisation was used to determine the effect of microwave irradiation on the cleavage of substrates from polymer
supports (see Section 7.3.8).

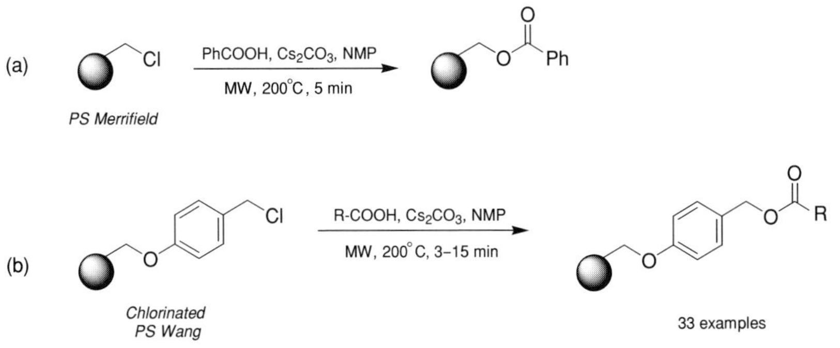

Scheme 7.7   Acid attachment employing microwave flash heating using (a) Merrifield resin and (b) commercially available chlorinated Wang resin as the polymer support.

The benzoic acid was quantitatively coupled within 5 min *via* its cesium salt by using
a dedicated multi-mode batch reactor, carried out in standard glassware under atmospheric reflux conditions. In a more extended study, various substituted carboxylic
acids (Fig. 7.7) were coupled to chlorinated Wang resin, employing an identical reaction protocol. In a majority of cases, the microwave-mediated conversion reached at
least 85% after 3–15 min. These microwave conditions represented a significant rate
enhancement, in contrast to the conventional protocol, which took 24–48 h. The microwave protocol has additional benefits in comparison to the conventional method, as
the amounts of acid and base equivalents can be reduced and potassium iodide as an
additive can be eliminated from the reaction mixture[27].

While no attempt was made to optimise all examples shown in Fig. 7.7, a number
of substituted benzoic acids were selected to compare their coupling behaviour under microwave conditions with the thermally heated protocol[27]. High loadings of the
resin-bound esters could be obtained very rapidly, even sterically demanding acids were
coupled successfully. Most importantly, in all the examples given[27], the loadings accomplished after 15 min of microwave irradiation were actually higher than that achieved
using the thermally heated protocols.

In general, the reasons for rate-enhancements in microwave-assisted transformations
in comparison to conventional heating are not always fully understood. Some authors
have postulated a specific 'non-thermal microwave effect' for those effects that could
not be rationalised as a simple consequence of superheated solvents and higher reaction
temperatures. Stadler and Kappe therefore carried out a kinetic comparison of the
thermal coupling of benzoic acid to chloro-Wang resin at 80°C, with the microwave-
assisted coupling *at the identical temperature of* 80°C and otherwise identical reaction
parameters. However, the reaction rates for the two runs were quite similar and the
small observed differences could not be attributed to non-thermal effects. In order
to confirm this hypothesis, the authors also carried out coupling experiments with

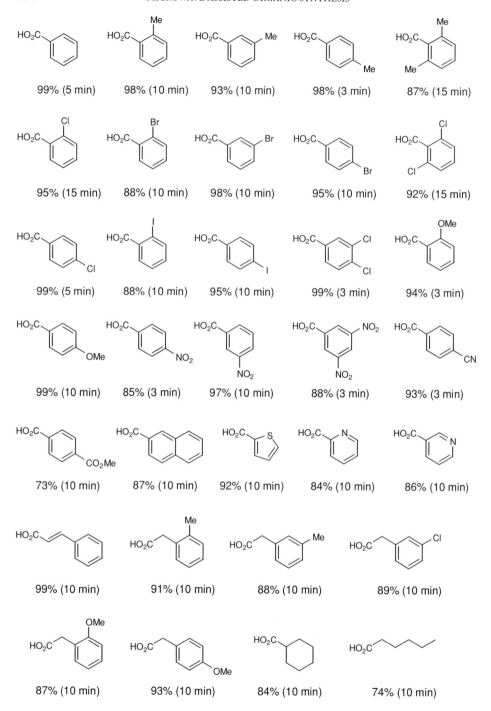

Figure 7.7 Loadings and reaction times (in parentheses) for the microwave-assisted coupling of carboxylic acids to chlorinated Wang resin.

benzoic acid at 200°C under standard thermally heated conditions. The loadings in these high-temperature thermally heated runs were comparable, but somewhat lower to the microwave-promoted protocols (e.g., 86% at 3 min and 95% at 10 min).

In a related study by the same authors, the effect of microwave irradiation on carbodiimide-mediated esterifications on solid support has been investigated employing benzoic acid[28]. Activation of the carboxylic acid was carried out using diisopropylcarbodiimide (DIC) *via* the O-acyl isourea or the symmetrical anhydride protocol, respectively (Scheme 7.8).

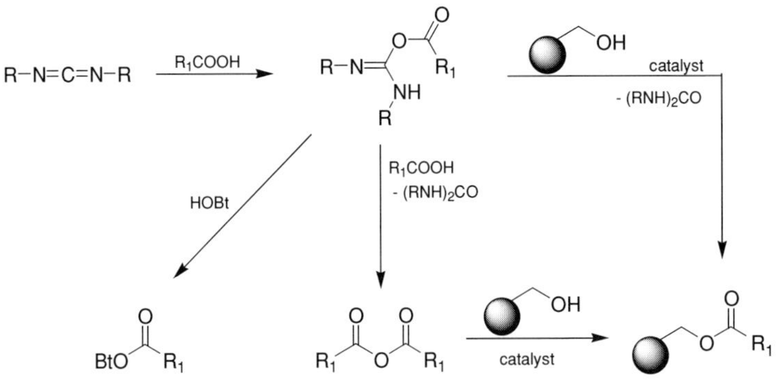

Scheme 7.8   Carbodiimide-mediated pathways for esterification reactions.

The isourea protocol was carried out in a 9:1 dichloromethane/N,N-dimethylformamide (DCM/DMF) solvent mixture in sealed vessels, whereas the anhydride reactions were carried out in 1-methyl-2-pyrrolidinone (NMP) under atmospheric pressure. In all experiments, the loading was estimated by on-bead Fourier transform infrared spectroscopy (FTIR) analysis and determined by cleavage from the PS Wang resin with 50% TFA in DCM.

Surprisingly, the isourea protocol showed some deficiencies, as complete conversion could not be obtained due to unexpected side reactions at higher temperatures. The anhydride protocol was superior to this method, as it could be carried out in simple glassware (open vessels in a dedicated multi-mode cavity) without the need for high pressure vessels. Quantitative coupling to the resin could be obtained with the anhydrides within 10 min, employing microwave heating at 200°C.

Yang and co-workers have reported the synthesis of a series of functionalised Merrifield resins[29]. The reactions were conducted in a multi-mode cavity refluxing system, a modified domestic oven. The reaction rates were dramatically enhanced over the conventional methods and high conversions were achieved in 7–25 min (Scheme 7.9). Since the choice of the solvent was pivotal in this procedure, the authors used a solvent mixture to balance between the aspects of good resin swelling properties and high microwave absorption efficiency (see Table 7.1). The conversions were calculated on the basis of the different chloride content of the polymer support before and after microwave irradiation. These microwave mediated pathways (Scheme 7.9) provide convenient methods for rapid and efficient solid-phase synthesis, using PS-Merrifield as either a support or a scavenger.

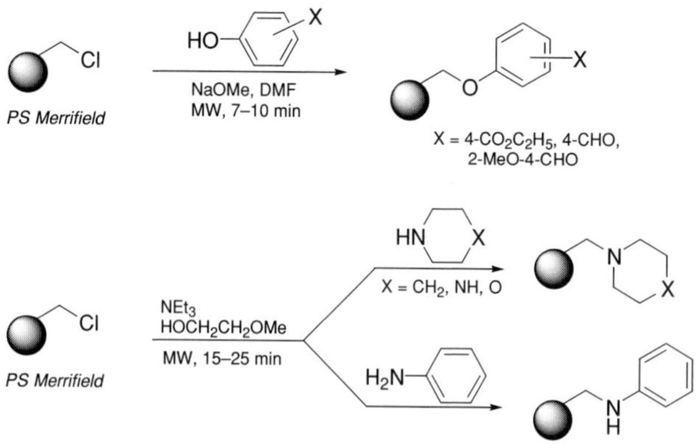

Scheme 7.9   Microwave-promoted preparation of functionalised resins.

In a more recent study, Westman and Lundin described the solid-phase synthesis of aminopropenones and aminopropenoates, respectively[30] as intermediates for hetero-cyclic synthesis. Two different three-step methods for the preparation of heterocycles have been developed. The first method involved formation of a polymer-bound es-ter from a *N*-protected glycine derivative and Merrifield resin (Scheme 7.10a), while the second method employed an interesting approach utilising simple aqueous methy-lamine solution for functionalisation of the solid support (Scheme 7.10b). In this latter approach, a variety of heterocycles were readily synthesised from the generated polymer-bound benzylamine using a two-step protocol (see Section 5.3.3).

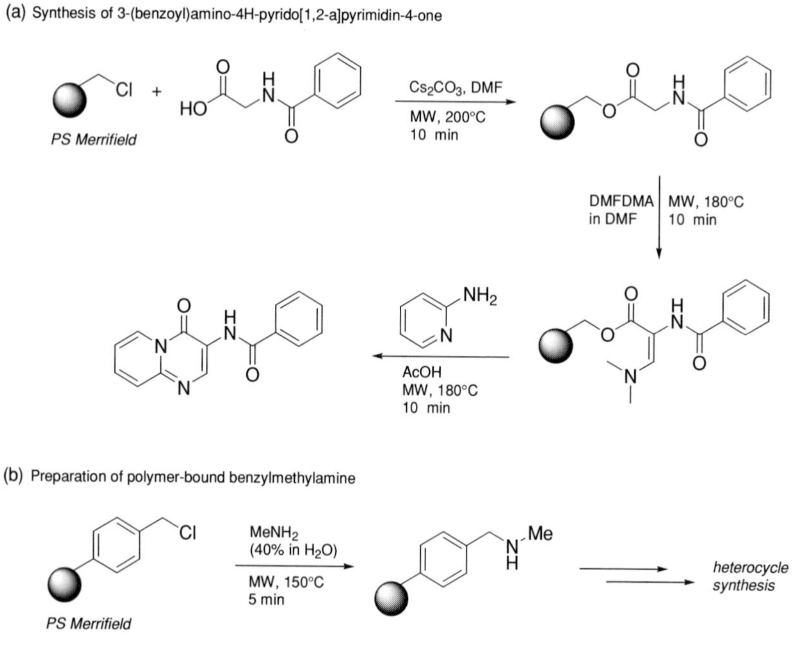

(a) Synthesis of 3-(benzoyl)amino-4H-pyrido[1,2-a]pyrimidin-4-one

(b) Preparation of polymer-bound benzylmethylamine

Scheme 7.10   Reaction strategies for the polymer-supportetd synthesis of dialkylaminopropenones.

The final step in the synthesis of the pyridopyrimidinones (Scheme 7.10a) involved the release of the products from the solid support by intramolecular cyclisation, whereupon the pure products were obtained in solution. All reaction steps were carried out in a dedicated single-mode microwave instrument under sealed vessel conditions.

In a dedicated combinatorial approach, Strohmeier and Kappe reported the rapid parallel synthesis of polymer-bound enones[31]. This approach involved a two-step protocol employing initial high-speed acetoacetylation of Wang resin with a selection of common β-ketoesters (Scheme 7.11) and subsequent microwave-mediated Knoevenagel condensations with a set of 13 different aldehydes (see Section 7.3.6).

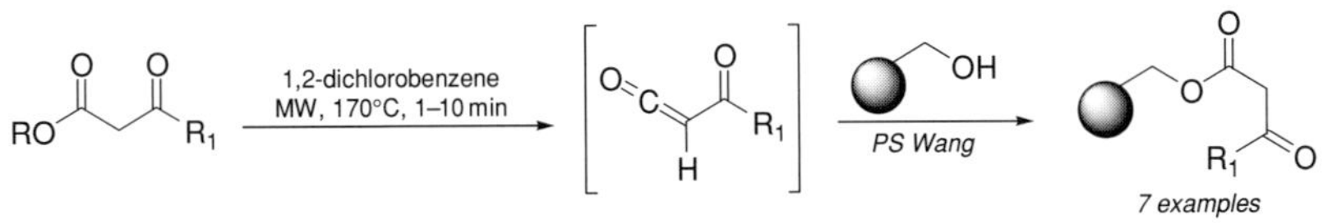

Scheme 7.11   Microwave-promoted acetoacetylation.

These transesterifications are believed to proceed by the initial formation of a highly reactive α-oxoketene intermediate with the elimination of the alcohol component of the acetoacetic ester being the limiting factor. Subsequent trapping of the ketene intermediate affords the transacetoacetylated products. For better handling of the polymer support, the reactions can be carried out under atmospheric pressure in open PFA vessels in a dedicated multi-mode batch reactor, using 1,2-dichlorobenzene (DCB) as the solvent. Acetoacetylations were performed successfully within 1–10 min under these microwave conditions. Furthermore, acetoacetylated products can be obtained in a parallel fashion in a single 10 min run employing a multi-vessel rotor system. Since these transesterifications usually require several hours, if conducted under conventional thermal heating, it has been demonstrated that microwave flash heating can be used as an effective tool to speed up this example of SPOS. It is worth highlighting that these transesterifications need to be carried out under open vessel conditions, so that the alcohol by-product can be removed from the reaction mixture.

7.3.3. Transition-metal catalysis

Palladium-catalysed cross-coupling reactions are one of the cornerstones in modern organic synthesis. It would therefore be interesting to apply both microwave irradiation and solid-phase synthesis to such chemical transformations. One of the first publications dealing with such reactions was presented by the group of Hallberg in 1996[32]. This group investigated the effect of microwave irradiations towards Suzuki- and Stille-type cross-coupling reactions on solid phase (Scheme 7.12).

The reactions were carried out in sealed Pyrex tubes, employing a single-mode microwave cavity. The reagents were added to the polymer-bound aryl halide under a nitrogen atmosphere and the reactions were irradiated for the time periods indicated. Very short reaction times (<4 min) provided almost quantitative conversion, as well as minimal decomposition of the solid support.

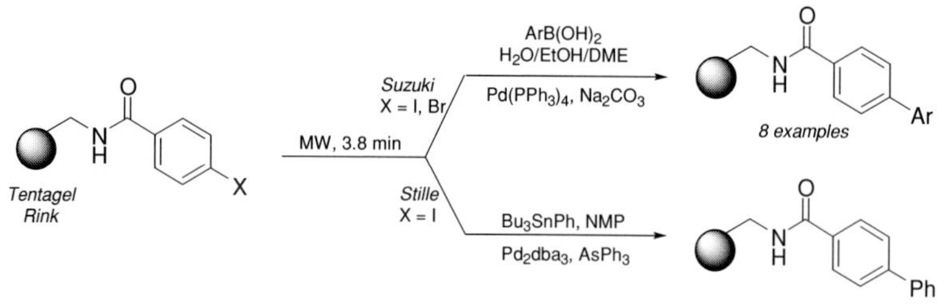

Scheme 7.12   Microwave-assisted cross-coupling reactions.

In a related study Hallberg *et al.* also investigated molybdenum-catalysed allylic alkylations in solution and on solid phase[33] demonstrating that microwave irradiation could also be applied to highly enantioselective reactions (Scheme 7.13).

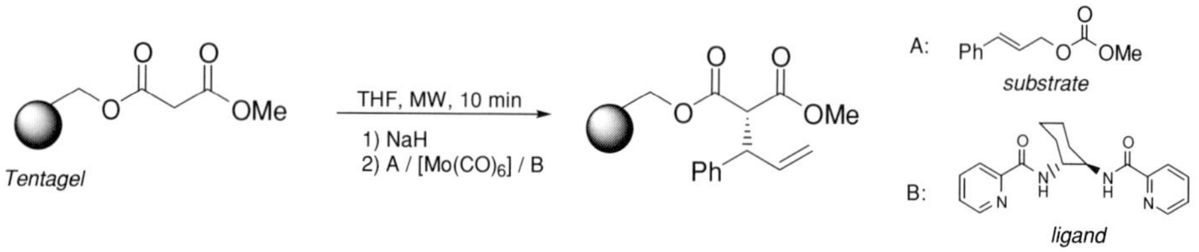

Scheme 7.13   Solid-phase molybdenum-catalysed allylic alkylation.

In these studies, commercially available and stable $[Mo(CO)_6]$ was used to generate the catalytic system *in situ*. The reactions in solution provided good yields. In contrast, the conversion rates for the solid phase examples were rather poor. However, the enantioselectivity was excellent (>99% ee) for both the solution- and solid-phase reactions.

In addition, further studies by Hallberg and co-workers reported the microwave-promoted preparation of tetrazoles employing organonitriles[34]. After establishing a solution-phase protocol, this protocol was also employed for solid-phase examples (Scheme 7.14).

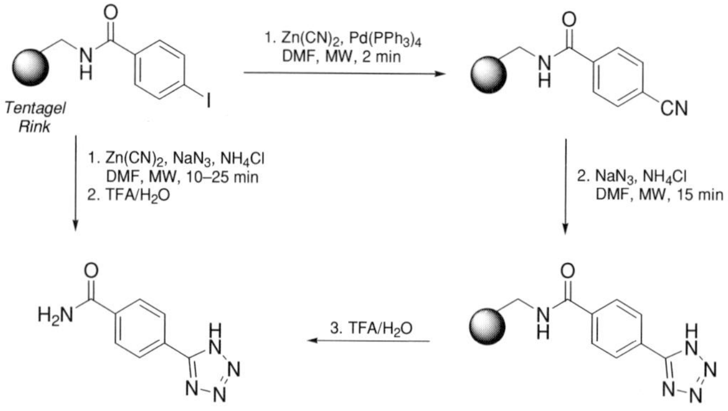

Scheme 7.14   Microwave-promoted cycloaddition reactions on solid-phase.

The reactions were performed in a commercially available single-mode microwave cavity in sealed Pyrex tubes. Full conversion to the corresponding nitriles was achieved after very short reaction times. During the reactions, temperatures reached 175°C as determined by measurement with a fibre optic probe. Subsequently, the nitriles were treated with sodium azide to form the desired tetrazoles. In this step, the reaction temperature reached up to 220°C after 15 min. Despite the rather high temperatures, only negligible decomposition of the solid support was observed. It is noteworthy that the formation of tetrazoles could be easily carried out as a 'one-pot' reaction in good yields, eliminating the need for bis(triphenylphosphine)palladium [Pd(PPh$_3$)$_4$] in the reaction mixture. Furthermore, reaction times were drastically reduced by using microwave heating, and comparable yields were achieved after 3–96 h using conventional thermal heating.

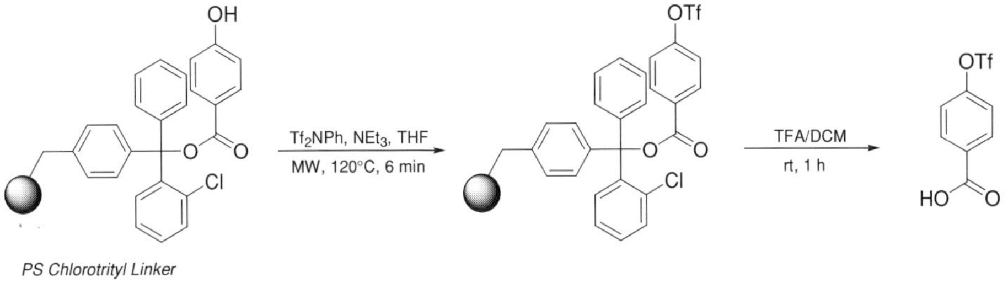

Scheme 7.15 Fast solid-phase triflate synthesis.

In a more recent study, microwave irradiation was employed in the accelerated solid-phase synthesis of aryl triflates[35] (Scheme 7.15). Aryl triflates are currently of major interest as they represent useful starting materials in several transition metal-catalysed reactions. The use of N-phenyltriflimide as a triflating agent in microwave-mediated protocols is an appropriate choice, since it is a stable, crystalline agent, which often results in improved selectivity. Reducing the reaction times from 3 to 8 h under conventional heating to only 6 min by employing the microwave protocol has made this procedure more amenable to high-throughput synthesis. Since many examples of the solution-phase synthesis of aryl triflates are known, this microwave-assisted method presented the first solid-phase approach for this chemistry. Use of the commercially available chlorotrityl linker as a solid support allows mild cleavage conditions to be employed to obtain the desired aryl triflates in good yields.

The first examples of microwave-mediated polymer supported C—N cross-coupling reactions were reported in 1999[22], using copper(II) as the catalytic agent (Scheme 7.16). The reactions were carried out in a domestic microwave oven at full power for 3 × 10 s. After five cycles of heating with the addition of fresh reagents, none of the remaining starting benzimidazole amide could be detected after cleavage from the solid support. This represented a reduction in the reaction time from 48 h under conventional heating at 80°C to less than 5 min by microwave heating. However, it should be noted that both possible N-arylated regioisomeric products were obtained with this microwave-heated procedure (Fig. 7.8).

To assess the versatility of this reaction on solid support, several heterocyclic carboxylic acids were coupled to the PS-PEG resin (PAL linker). Applying the

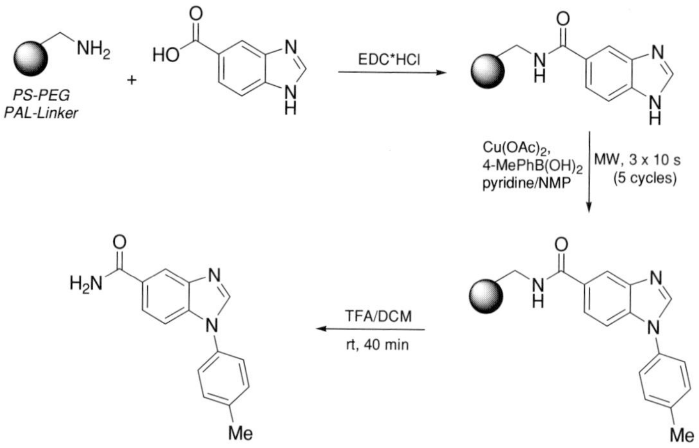

Scheme 7.16   Copper(II)-mediated N-arylation of polymer-bound benzimidazole.

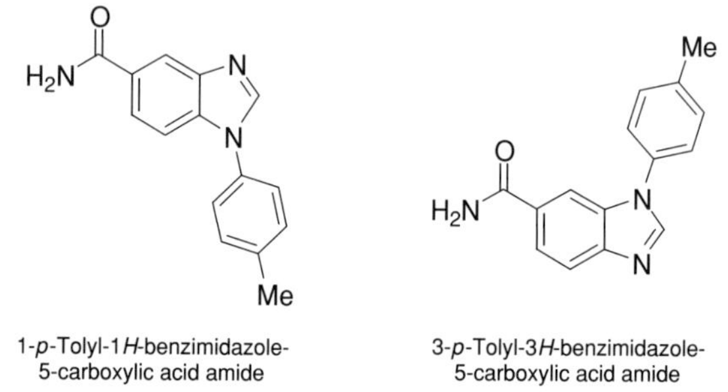

1-p-Tolyl-1H-benzimidazole-   3-p-Tolyl-3H-benzimidazole-
5-carboxylic acid amide   5-carboxylic acid amide

Figure 7.8   Observed regioisomeric isomers of the N-arylated benzimidazole product obtained from the microwave-heated protocol.

microwave conditions furnished the desired products in good yields and excellent purities (Scheme 7.17).

The ability to drive these solid-phase couplings to completion with multiple additions of excess reagents clearly demonstrates the utility of this method.

7.3.4. Substitution reactions

Another interesting field is that of microwave-mediated substitution reactions on solid phase. In this context, a very innovative study was presented by Scharn and co-workers[36]. They described the synthesis of trisamino- and amino-oxy-1,3,5-triazines on cellulose and polypropylene membranes, applying the SPOT-synthesis technique[36] (Scheme 7.18). This research demonstrated that further solid supports in addition to granulated polystyrene or PEG-resins could be used. The development of the SPOT-synthesis protocol allows rapid generation of highly diverse spatially addressed single compounds under mild conditions. This SPOT-synthesis protocol required the investigation of suitable planar polymeric supports bearing an orthogonal ester-free linker system, which were cleavable under dry conditions.

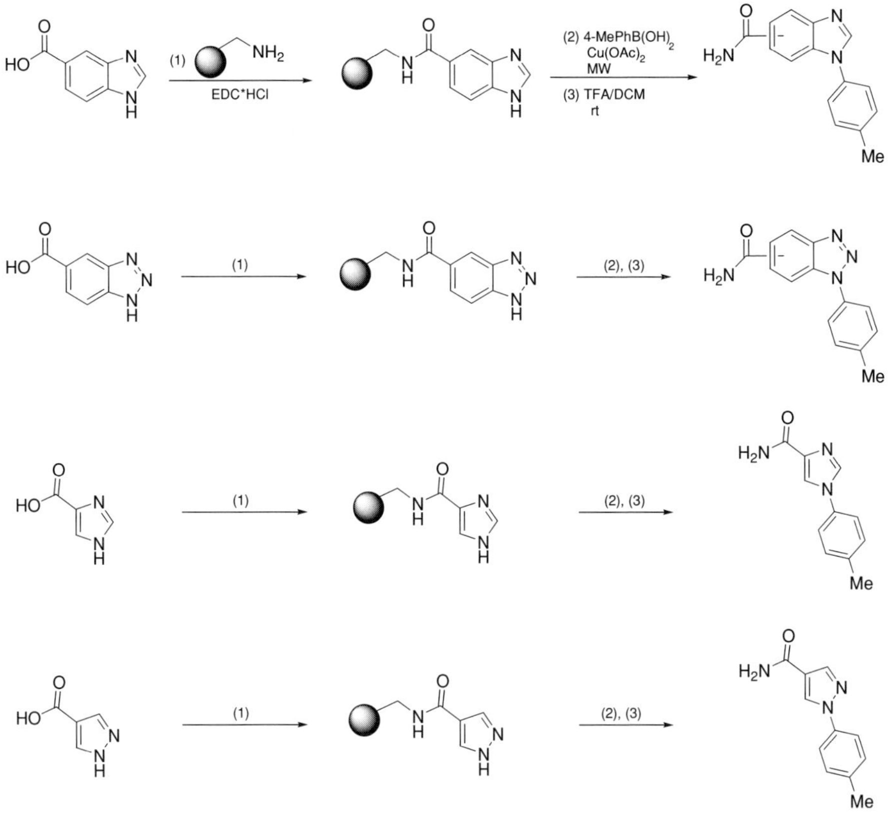

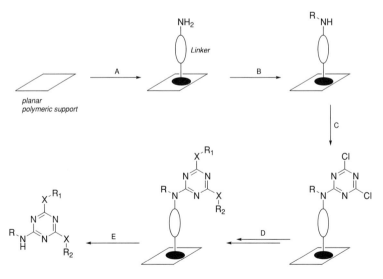

Scheme 7.17 Microwave assisted copper(II) mediated coupling of *p*-tolylboronic acid to various polymer bound heterocycles.

Scheme 7.18 General strategy of triazines on planar surfaces. (a) functionalisation/linker attachment; (b) introduction of the first building block; (c) attachment of cyanuric chloride; (d) stepwise chlorine-substitution; (e) cleavage.

Several functionalised membranes could be synthesised by conventional methods at room temperature. In contrast, microwave heating was employed for both the synthesis of a triazine membrane and the practical generation of an 8000-member library of triazines, bound to an amino-functionalised cellulose membrane (Scheme 7.19).

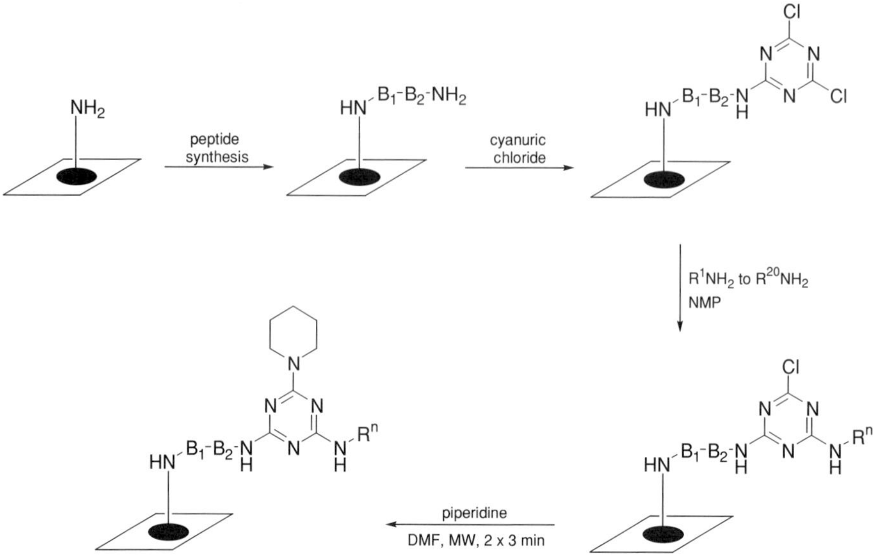

Scheme 7.19   Library generation on cellulose membrane employing the SPOT-technique.

For the preparation of the triazine membranes, the entire solid support (cellulose or polypropylene membrane) was treated with a 5 M solution of the corresponding amine in NMP and a 1 M solution of cesium phenolate in DMSO (2 μl each at one spot) and subsequently heated using microwave irradiation (domestic oven) for 3 min. After washing the support successively with DMF, methanol and DCM (three times each), the membrane was air dried.

All possible 400 dipeptides composed of the 20 proteinogenic L-amino acids (B^1 and B^2) were synthesised in 20 replica, as illustrated in Scheme 7.19. Subsequently, cyanuric chloride was attached to each dipeptide, followed by selective substitution of one of the two chlorine atoms by 20 different amines (Fig. 7.9). The second chlorine was ultimately replaced by piperidine under microwave irradiation conditions. Thereafter, the resulting library of functionalised dipeptides were tested directly on the cellulose sheet, for binding in the murine IgG mab Tab2 assay. Cleavage of the generated compounds could be achieved under mild conditions by treatment with TFA vapour, leaving the compounds adsorbed on the polymeric support. Since, the synthetic conditions described could be applied to the parallel assembly of 8000 cellulose-bound triazines; the method could be of potential interest for the parallel screening of small molecule compound libraries.

A similar approach has been described by the same authors for the synthesis of related cyclic peptidomimetics[37]. A set of ten nucleophiles were employed for the substitution

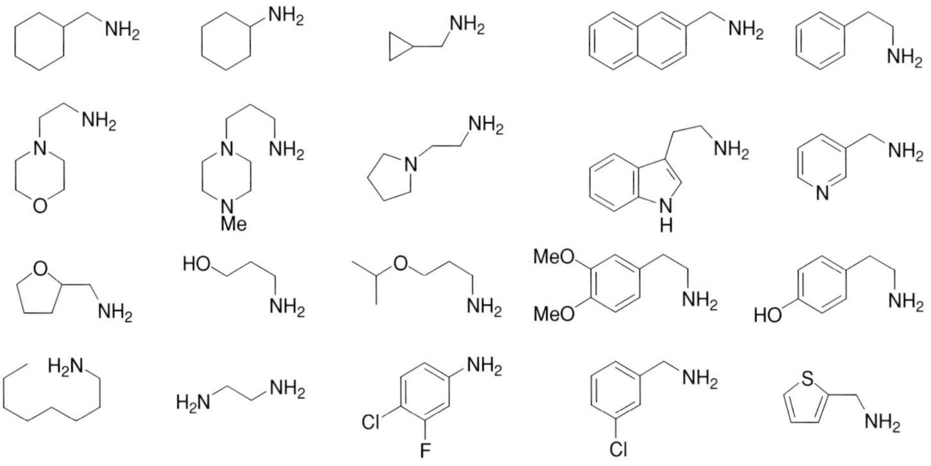

Figure 7.9 Structural representation of amines used in the triazines library generation.

of the chlorine atom of the cyclic triazinyl-peptide bound to the cellulose membrane. Due to the rate enhancement effects for nucleophilic substitution of the solid-supported monochloro triazines, these reactions were carried out rapidly by microwave heating. All products were obtained in high purities, enabling systematic modification of the molecular properties of the cyclic peptidomimetics.

This approach was tested for the cyclisation of peptides of various chain lengths. Following the SPOT-method described above[36], the N-terminal amino acids were attached to a photolinker-modified cellulose membrane (Scheme 7.20). 2,4,6-Trichloro-[1,3,5]-triazine was linked to the free N-terminus of the peptides, subsequently followed by deprotection of the Lys-side chain. Cyclisation was achieved by nucleophilic attack of the free amino group at the triazine moiety under basic conditions. Finally, the peptides were cleaved from the solid support by dry-state UV irradiation. For examples requiring microwave heating, the remaining chlorine functionality on the triazine was substituted by a set of ten different nucleophiles immediately before the cleavage step.

A more recent study has been reported involving the microwave-assisted solid-phase synthesis of purines[38]. The heterocyclic scaffold was first attached to the AMEBA-linked polystyrene (Fig. 7.10) via an aromatic nucleophilic substitution reaction by conventional heating in NMP in the presence of N,N-diisopropylethylamine. The key aromatic nucleophilic substitution of the iodine was conducted by microwave heating for 30 min at 200°C in NMP (Scheme 7.21). This microwave heating step was carried out in a dedicated multi-mode oven employing Teflon reaction vessels at 200°C for 30 min[38]. The resin was subsequently washed with tetrahydrofuran (THF) and methanol, and after drying the products were cleaved from the solid support using TFA/water at 60°C. After extraction, the products were purified by normal phase flash chromatography.

In conclusion, it has been shown that microwave heating is a powerful tool for nucleophilic substitutions on solid support, as conversion rates are significantly enhanced and reaction times can be drastically reduced compared to conventional heating.

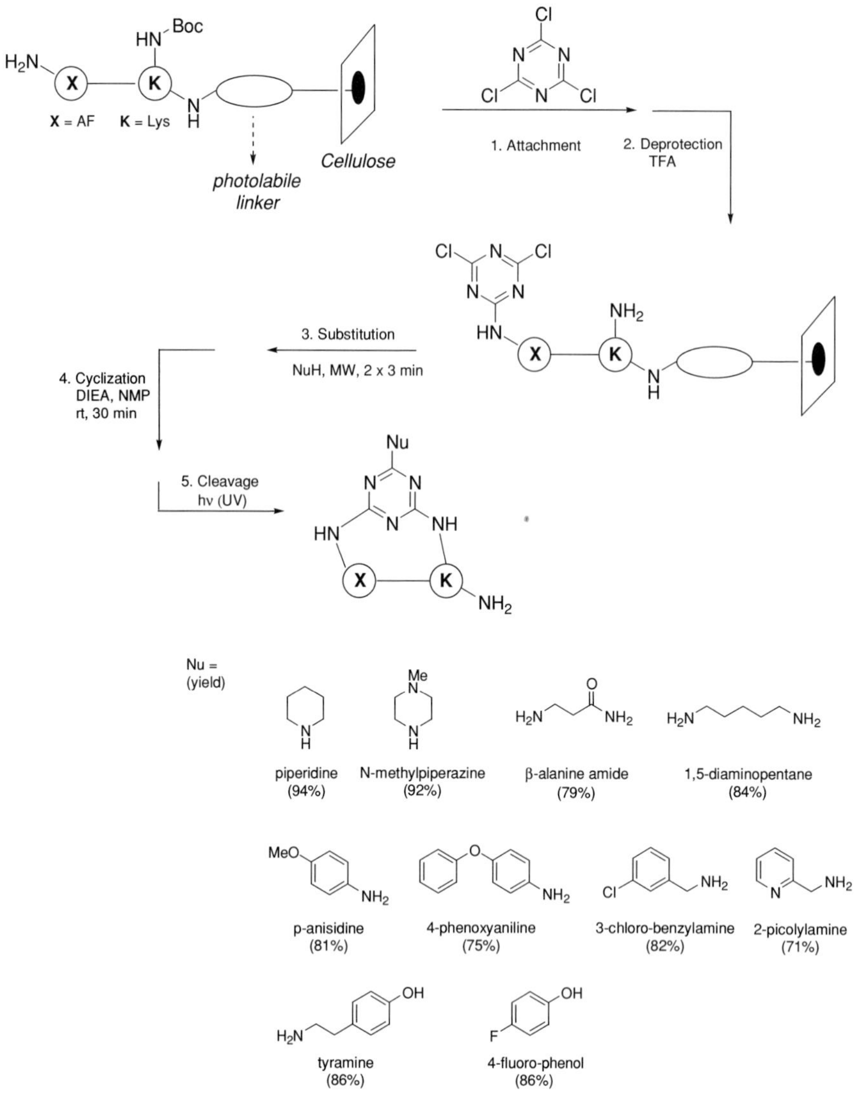

Scheme 7.20   Microwave-mediated synthesis and UV promoted cleavage of cyclic triazines on cellulose membrane (the numbers in brackets indicate the product purities obtained after cleavage).

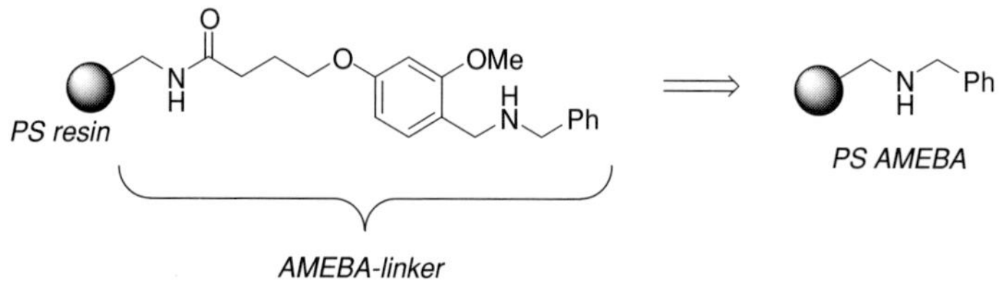

Figure 7.10   AMEBA-linked polystyrene resin.

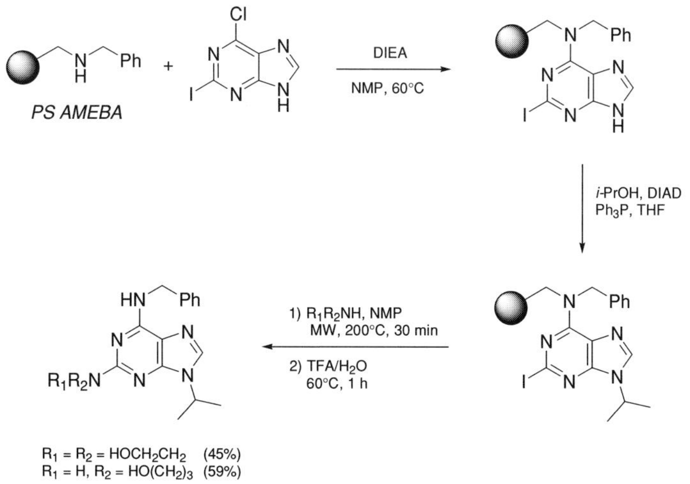

Scheme 7.21 Microwave-mediated purine synthesis on solid support (numbers in brackets denote yields after release from solid support).

7.3.5. Multi-component chemistry

Multi-component reactions where three or more components build a single product have received considerable interest for several years. Since most of these reactions tolerate a wide range of building block combinations, these types of reactions are frequently applied for combinatorial purposes. A solid-phase application towards the Ugi four component condensation (Ugi-4CC) generating a 18-member acylamino amide library appeared in 1999[39]. The acylamino amide library was synthesised using amino-functionalised PEG-polystyrene (TentaGel S RAM) as the solid support. (Scheme 7.22).

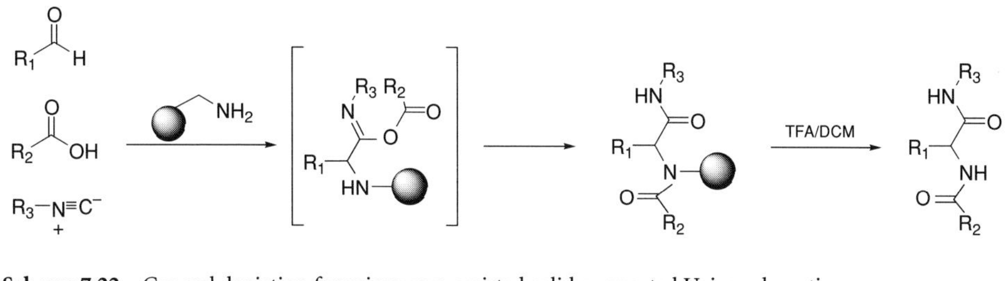

Scheme 7.22 General depiction for microwave-assisted solid-supported Ugi condensation.

A set of three aldehydes, three carboxylic acids and two isonitriles (Fig. 7.11) was used for the generation of the 18-member acylamino amide library.

In a typical procedure, the PS-TentaGel Fmoc-protected amino resin was deprotected using 20% piperidine in DMF (Scheme 7.23). After transferring the free amino resin into an appropriate microwave vial, the resin was swollen in a mixture of a 1 M solution of the corresponding aldehyde in DCM and a 1 M solution of the corresponding carboxylic acid in methanol. After 30 min, a 1 M solution of the corresponding isonitrile in DCM was

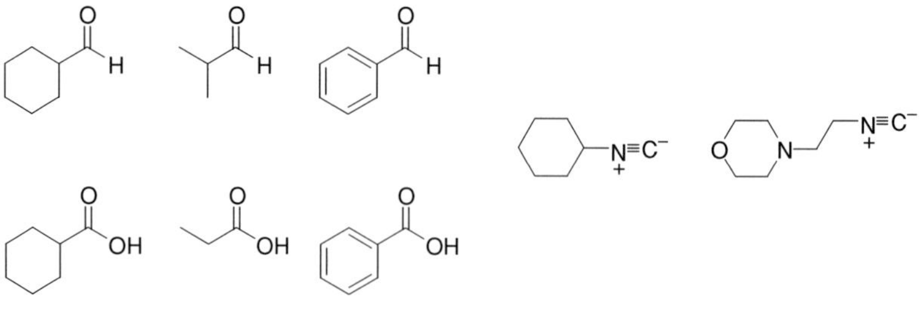

Figure 7.11 Building blocks for the solid-supported Ugi-4CC.

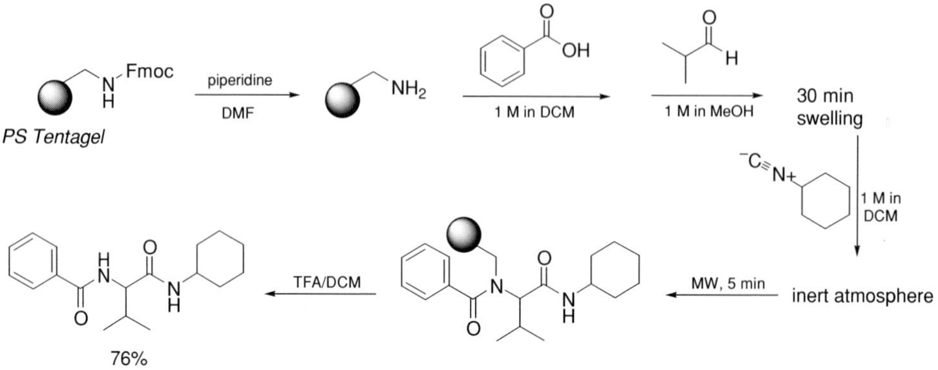

Scheme 7.23 Solid-supported Ugi-4CC under microwave conditions.

added to the pre-swollen resin mixture. The vial was flushed with an inert atmosphere of nitrogen and then sealed. The vial was then placed in the cavity of a single-mode microwave instrument and irradiated for 5 min (no temperature measurement given). The reaction mixture was cooled to room temperature and the resin was filtered off and washed, using a disposable syringe fitted with a polypropylene filter. Subsequent cleavage of the resins with TFA/DCM afforded the products in high purities but varying yields, after simple evaporation of the solvent.

A more recent study has dealt with the solid-phase mediated synthesis of isonitriles from formamides, utilising polystyrene-bound sulphonyl chloride as a suitable supported reagent[40]. Isonitriles are an important class of molecules because of their unusual bifunctional reactivity. However, the work-up and purification involved in reactions to generate isonitriles can be rather problematic, as isonitriles show a high reactivity over a wide range of conditions. Therefore, polymer supported sulphonyl chloride offers an efficient method of isonitrile generation under microwave irradiation, employing a simple filtration and acidic work-up (Scheme 7.24). Employing a dedicated single-mode instrument, six formamide derivatives could be converted into the corresponding isonitriles in good purities (as determined by HPLC). It is worthy to note that increasing substitution of the formamide has a decreasing effect on the purity of the desired products.

Interestingly the sulphonyl chloride resin could be quantitatively regenerated by treatment of the by-product sulphonic acid resin with phosphorous pentachloride (PCl_5) in

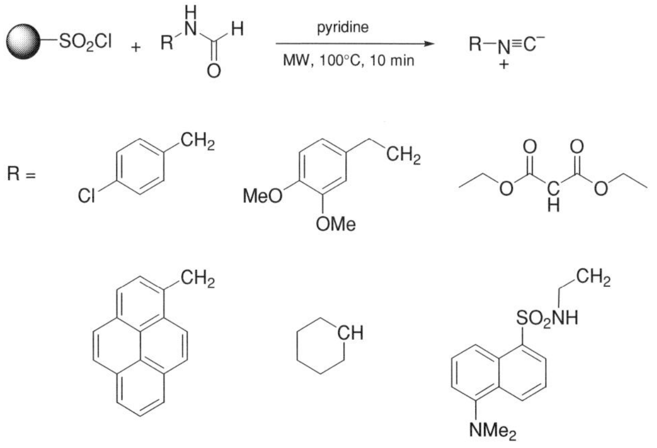

Scheme 7.24 Solid-phase mediated isonitrile synthesis under microwave irradiation.

DMF at room temperature. Solid-phase-mediated isonitrile synthesis under microwave irradiation led to much faster reactions and in many cases improved yields; therefore, allowing rapid access to this important class of compounds, amenable for a broad range of subsequent synthesis.

Another interesting multi-component reaction is the Gewald synthesis leading to 2-amino-3-acyl thiophenes (Scheme 7.25), which are of current interest since they are commercially used as dyes, conducting polymers and have shown extensive potential for pharmaceutical purposes.

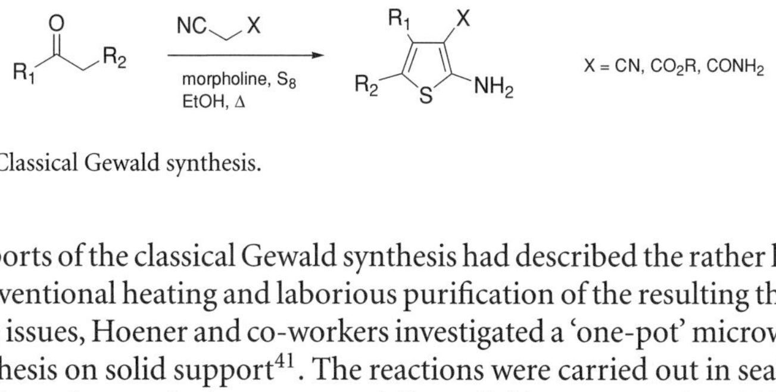

Scheme 7.25 Classical Gewald synthesis.

Earlier reports of the classical Gewald synthesis had described the rather long reaction times by conventional heating and laborious purification of the resulting thiophenes. In view of these issues, Hoener and co-workers investigated a 'one-pot' microwave-assisted Gewald synthesis on solid support[41]. The reactions were carried out in sealed vessels in a single-mode cavity employing commercially available cyanoacetic acid bound Wang resin as the solid support. Using the single-mode microwave, the overall two step reaction procedure including the acylation of the initially formed 2-aminothiophenes could be performed in less than 1 h. This solid phase 'one-pot' two step microwave-promoted process is an efficient route to 2-acylaminothiophenes (Scheme 7.26), which requires no filtration in between the two reaction steps.

As shown in Fig. 7.12, various aldehydes, ketones and acylating agents have been employed to generate the desired thiophene products in high yields (81–99%) and in generally good purities (70–99% as determined by HPLC). The resulting products could be purified by preparative HPLC if necessary to allow structural confirmation analysis.

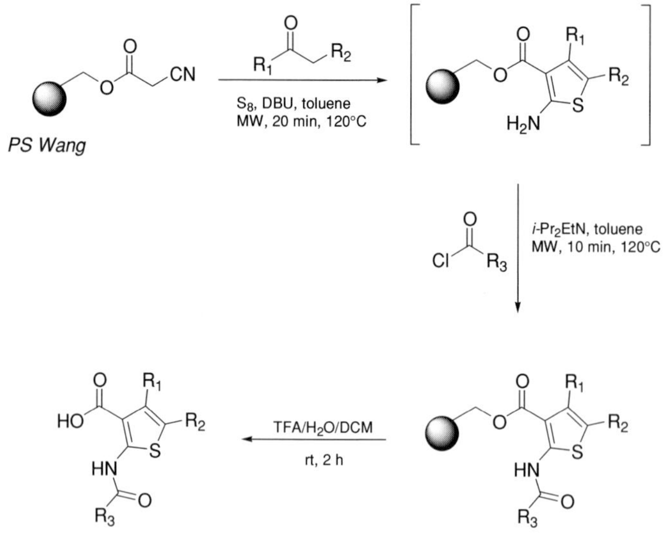

Scheme 7.26 'One-pot' microwave-assisted Gewald synthesis.

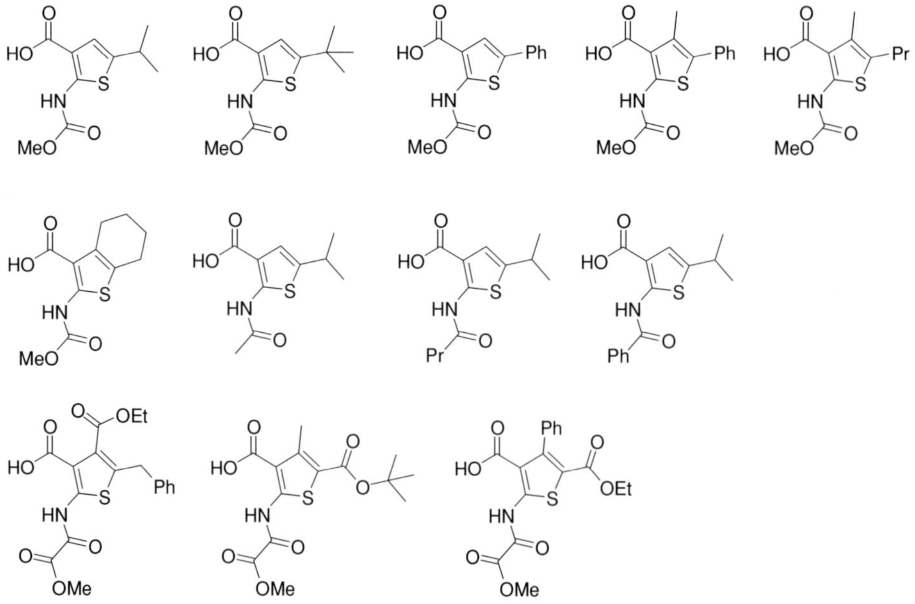

Figure 7.12 Structural representation of 2-acylaminothiophene products generated by microwave-assisted Gewald synthesis.

7.3.6. Condensation reactions

As already discussed in Section 7.3.2, polymer-bound acetoacetates can be used as precursors for the solid-phase synthesis of enones[31] (Scheme 7.27). For these Knoevenagel condensations, the crucial step is to initiate enolisation of the CH acidic component. In

general, enolisation can be initiated with a variety of catalysts (e.g., piperidine, piperi-
dinium acetate, ethylendiamine diacetate).

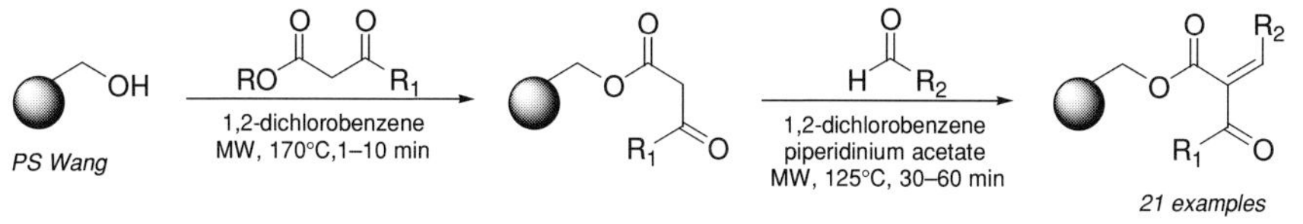

Scheme 7.27   Parallel synthesis of polymer-bound enones.

For these Knoevenagel condensations performed under microwave heating, piperi-
dinium acetate was found to be the catalyst of choice, provided temperatures were
kept below 130°C. At higher reaction temperatures, significant premature cleavage of
material would occur from the resin. Further studies found that at 125°C, virtually
quantitative conversion could be achieved within 30 min. To ensure complete conversion for all examples of a 21-member library, an irradiation time of 60 min was
used employing a multi-vessel rotor system for parallel microwave-assisted synthesis.
Despite the different properties of the polymer-bound acetoacetates and the diverse
nature of the aldehydes used, the Knoevenagel condensations went to near completion
for each example of the library. The results were confirmed by on-bead FTIR-analysis,
accurate weight gain measurements of washed and dried resins and post-cleavage analysis of the resulting enones. These examples of Knoevenagel condensations illustrated
that reaction times could be reduced from 1–2 days to 30–60 min by employing parallel microwave-promoted synthesis in open vessels, without affecting the purity of the
resin-bound products.

Microwave-assisted Knoevenagel reactions have been utilised in the preparation of
resin-bound nitroalkenes[42]. The generation of various resin-bound nitroalkenes was
described employing resin-bound nitroacetic acid, which was condensed with a variety
of aldehydes under microwave conditions (Scheme 7.28).

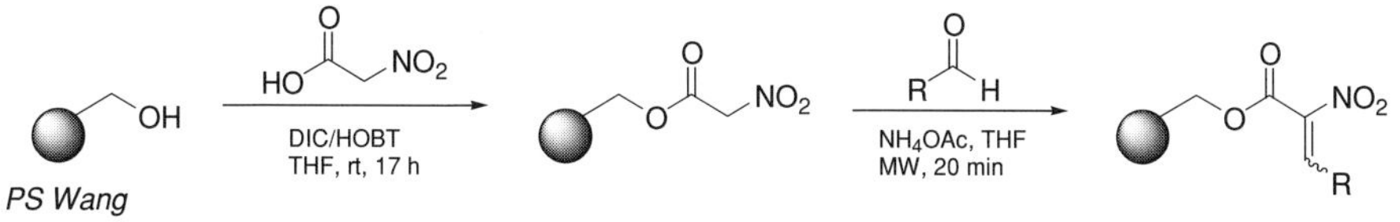

Scheme 7.28   Microwave-assisted Knoevenagel condensation.

In order to demonstrate the potential of these resin-bound products for combinatorial applications, the nitroalkenes were employed in a Diels–Alder reaction with
2,3-dimethylbutadiene (Scheme 7.29). In addition, the resin-bound nitroalkenes were
also used in a 'one pot' three-component tandem [4+2]/[3+2] reaction with ethyl vinyl
ether and styrene (see Scheme 7.30).

Subsequent cleavage of the resin-bound Diels–Alder adducts employing lithium aluminium hydride (LiAlH$_4$) *via* a traceless linker strategy afforded the cyclic phenylethy-
lamines. Alternatively, selective reduction of the nitro group using tin(II) chloride

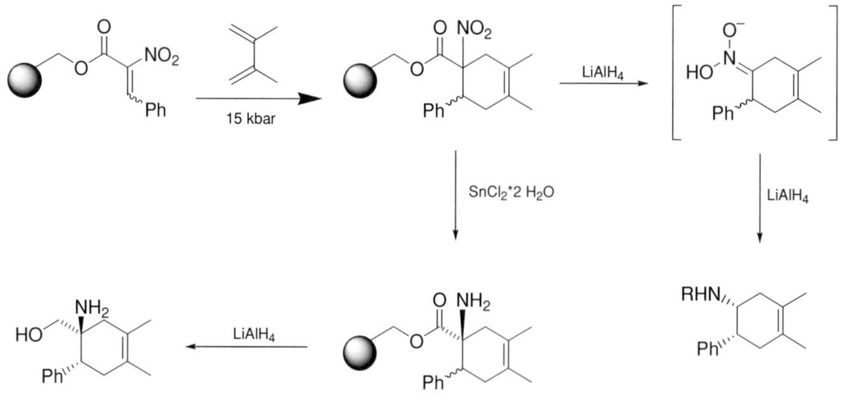

Scheme 7.29　High-pressure promoted Diels–Alder reaction on solid-phase.

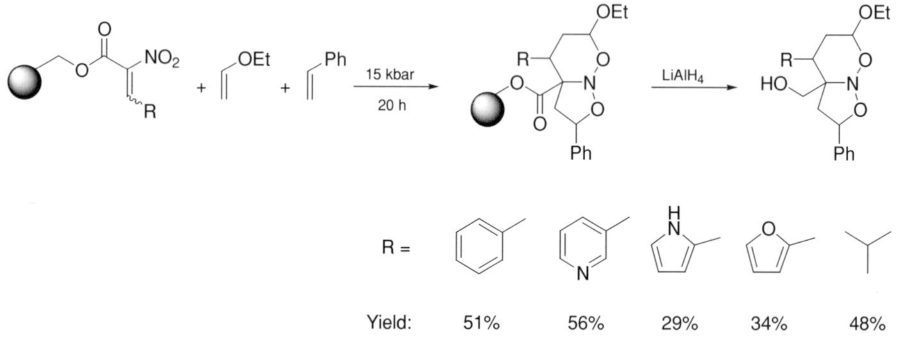

Scheme 7.30　High pressure promoted solid-supported synthesis of nitroso acetals (yields quoted are after reductive cleavage).

dihydrate (SnCl·2H$_2$O) before cleavage using lithium aluminium hydride furnished the corresponding cyclic phenylethylamino alcohols (Scheme 7.29).

Bicyclic nitroso acetals were able to be synthesised by employing ethyl vinyl ether (dienophile), styrene (dipolarophile) and the previously discussed resin-bound nitroalkenes in a one-pot tandem [4+2]/[3+2]. As illustrated in Scheme 7.30, several aromatic and aliphatic substituents could be introduced to the bicyclic scaffold. Reductive cleavage of the cycloadducts with lithium aluminium hydride (LiAlH$_4$) gave rise to the 3α-methyl alcohol substituted nitroso acetals in moderate overall yields. All these examples demonstrate that resin-bound nitroalkenes can be readily synthesised by microwave synthesis and thereafter can be used as starting materials, in a variety of high pressure-promoted cycloadditions.

7.3.7. *Rearrangements*

Microwave-assisted rearrangement reactions on solid-phase have not been discussed in the literature very often. At the time when this chapter was being written only one example dealing with Claisen rearrangements had been described[43]. Within this report, Merrifield resin bound *O*-allylsalicylic esters were rapidly rearranged to the corresponding ortho-allylsalicylic esters employing microwave heating (Scheme 7.31). Acid-mediated cleavage of these resin-bound ester products afforded the corresponding ortho-allylsalicylic acids.

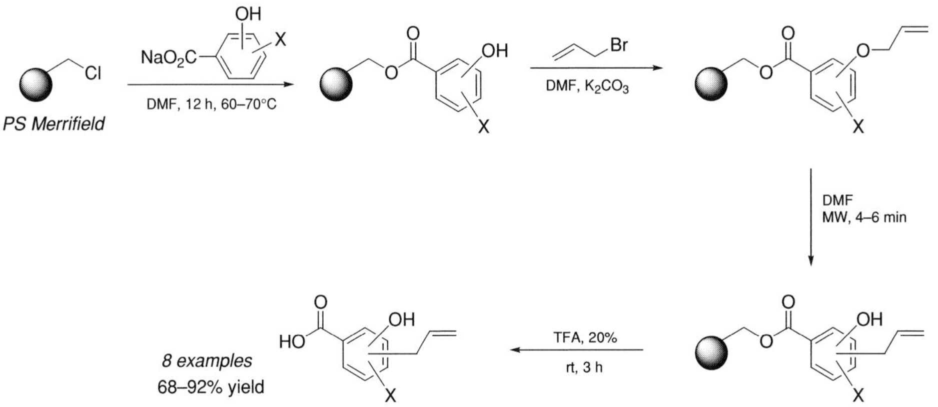

Scheme 7.31 Microwave-assisted Claisen rearrangement on solid-phase.

The resin-bound salicylic esters were suspended in DMF and placed in an Erlenmeyer flask within a domestic microwave cavity. After microwave irradiation for 4–6 min (1 min cycles), the reaction mixture was allowed to cool to ambient temperature and the resin was collected by filtration and washed with methanol and DCM. The resin-bound esters were subsequently cleaved with TFA/DCM. Removal of the solvent by evaporation gave the corresponding acid products (see Fig. 7.13) in high yields. Compared to conventional thermal heating (DMF, 140°C), reaction times could be drastically reduced

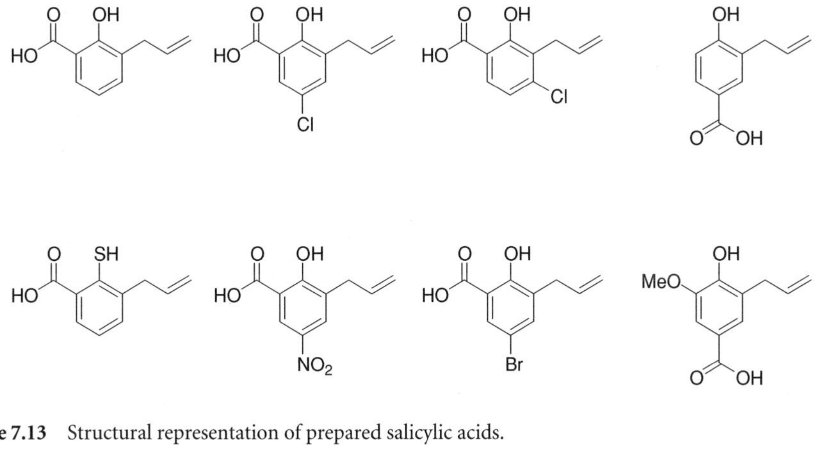

Figure 7.13 Structural representation of prepared salicylic acids.

from 10–16 h to a few minutes using microwave flash heating. In addition, higher yields of products were obtained by microwave heating.

As can be seen from Fig. 7.13, this method is also compatible with thiosalicylic acid; ortho-allylthiosalicylic acid was obtained high yield (89% after 4 min).

7.3.8. Cleavage reactions

One of the key steps in combinatorial solid-phase synthesis is clearly the cleavage of the desired product from the solid support. A variety of cleavage protocols have been investigated, depending on the nature of the employed linker. However to our knowledge, microwave-promoted cleavage reactions have rarely been reported so far. It would appear that a complete microwave-assisted protocol including attachment of the starting material to the solid support, scaffold preparation, scaffold decoration and cleavage of the resin-bound product would be desirable.

An interesting protocol for microwave-assisted acid mediated resin cleavage appeared in 2001[27]. Pre-coupled carboxylic acids (see Section 7.3.2) were cleaved from traditionally non-acid sensitive Merrifield resin employing TFA/DCM (1:1) under microwave heating (Scheme 7.32).

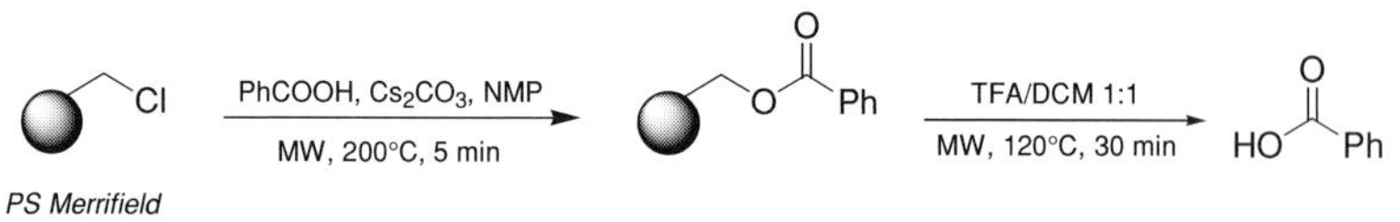

Scheme 7.32   Microwave-assisted acidic cleavage under elevated pressure.

In general, acidolysis of the Merrifield linker requires acids with a high ionising power, such as hydrogen fluoride, trifluoromethanesulphonic acid or hydrogen bromide/acetic acid. Therefore, under conventional conditions, cleavage does not take place with TFA. This feature can be used in a beneficial manner to enable the selective deprotection of Boc-protected amino groups with TFA without cleavage of the product from the resin. Employing microwave flash heating enables these cleavages to become possible, under elevated pressure/temperature using sealed vessels. Therefore, the resin-bound ester and TFA/DCM (10 ml per 250 mg resin) were placed inside a 100 ml PFA sealed reactor vessel and irradiated with slight stirring for 30 min with a pre-selected temperature of 120°C in a dedicated multi-mode batch reactor. After cooling the pressure container in an ice bath for 20 min, the system was vented and the solution was collected by filtration and the resin was washed with DCM. Evaporation of the filtrate and combined washings to dryness furnished the recovered benzoic acid in quantitative yield and excellent purity. Interestingly, no degradation of the polymer support could be detected, albeit the reaction conditions are often rather harsh for solid-phase chemistry. However, microwave heating allows a novel acid-mediated approach to be used for the cleavage of products from acid-stable linkers. It is worthy to note that only partial cleavage of the desired product occurred after treating the polymer-bound ester for 2 h in pure TFA under reflux by thermal heating.

A more innovative study was presented by Glass and Combs, elaborating the Kenner safety-catch principle for the generation of amide libraries[19,20]. In another application of microwave-assisted resin cleavage, N-benzoylated alanine attached to 4-sulphamylbutyryl resin was cleaved (after activation of the linker with bromoacetonitrile employing Kenner's safety catch principle) with a variety of amines (Scheme 7.33). Cleavage rates in dimethyl sulphoxide (DMSO) were investigated for diisopropylamine and aniline under different reaction conditions using both microwave (domestic oven) and conventional (oil bath) heating. The results showed that microwave heating did not accelerate reaction rates over conventional heating, when experiments were run at the same temperature (ca. 80°C). However, using microwave heating, even cleavage with normally unreactive aniline could be accomplished within 15 min at ca. 140°C.

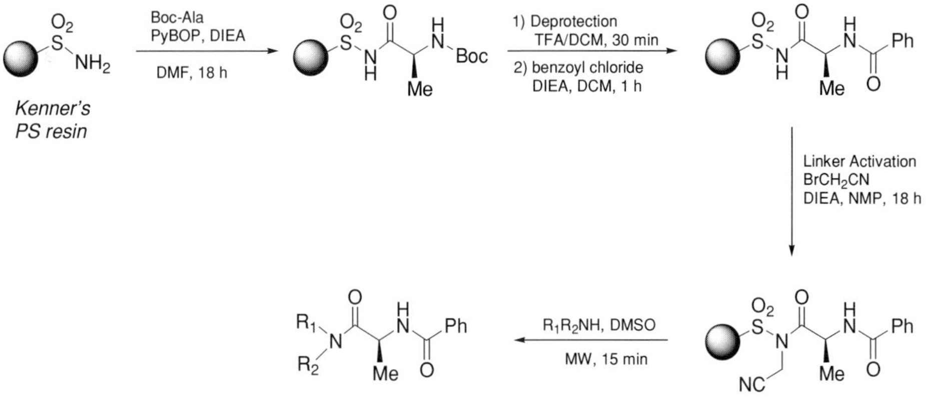

Scheme 7.33 Solid-supported amide synthesis employing the safety catch principle.

The microwave approach was used in the extension to the parallel synthesis of an 880-member library utilising 96 well plates, employing ten different amino acids coupled to 4-sulphambutyryl resin, each bearing a different acyl group, and using 88 diverse amines in the cleavage step. After linker activation, each resin was suspended in a solvent mixture of 3:2 DCM/DMF and split between 88 wells of an appropriate filter plate using a 96-channel pipettor. DMSO was then added to each well as a solvent, followed by the addition of a set of primary and secondary amines from a stock microtitre plate transferred again using a 96-channel pipettor. Sets of four plates were placed in a domestic microwave oven and were heated at a temperature of 80°C for 60 s. After the plates had cooled down, the solutions from the wells were drained into a collection microtitre plate and were combined with the DMSO resin washings from the respective wells to afford 10 mM solutions of the products in DMSO for biological screening. Existing temperature gradients between wells of the microtitre plates (see Fig. 7.14) did not represent a significant issue for this type of chemistry.

In closely related work, similar solid-phase chemistry was employed by the same research group to also prepare biaryl urea compound libraries via microwave-assisted Suzuki couplings, followed by cleavage from the resin with amines (Scheme 7.34).[20]

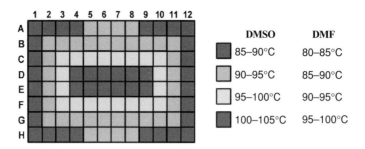

Figure 7.14　Temperature gradients within a microwave-heated microtitre plate (adapted from Ref. 20).

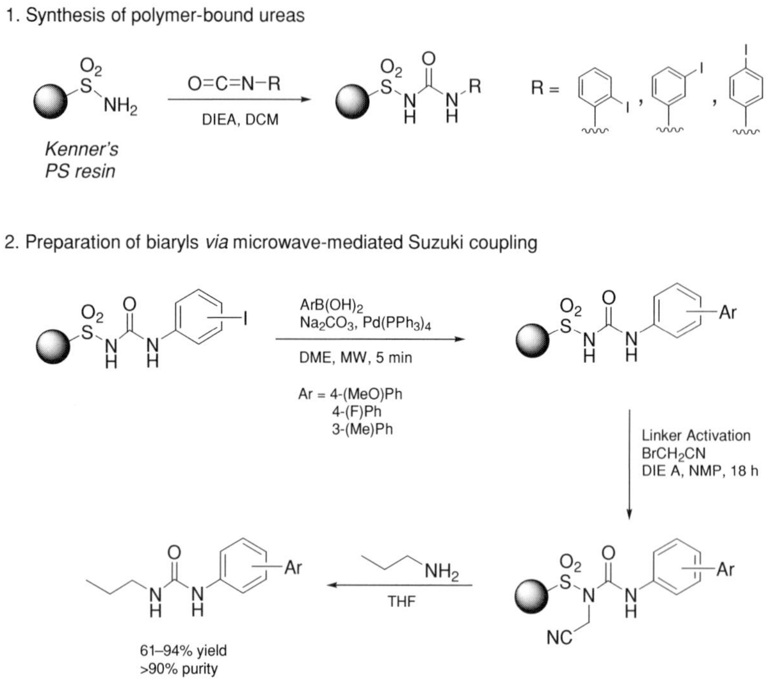

1. Synthesis of polymer-bound ureas

Kenner's
PS resin

2. Preparation of biaryls *via* microwave-mediated Suzuki coupling

Ar = 4-(MeO)Ph
4-(F)Ph
3-(Me)Ph

Linker Activation
BrCH₂CN
DIE A, NMP, 18 h

61–94% yield
>90% purity

Scheme 7.34　Microwave-promoted synthesis of urea-derivatives on solid-support (purity based on crude ¹H NMR; purified yields on initial loading (0.73 mmol g⁻¹) of alkyl safety-catch linker).

　　The aforementioned procedure enabled the generation of large biaryl urea compound libraries employing just a simple domestic microwave oven. The final cleaved products were obtained in high purity as determined by a combination of TLC and ¹H NMR.

　　In a more recent study, multi-directional cyclative cleavage leading to bicyclic di-hydropyrimidinones has been employed[17]. This approach required the synthesis of 4-chloroacetoacetate resin as the key starting material, which was prepared by microwave-assisted acetoacetylation of commercial available polystyrene Wang resin (Scheme 7.35) under open vessel conditions[17]. This resin precursor was subsequently treated with urea and various aldehydes in a Biginelli-type multi-component reaction,

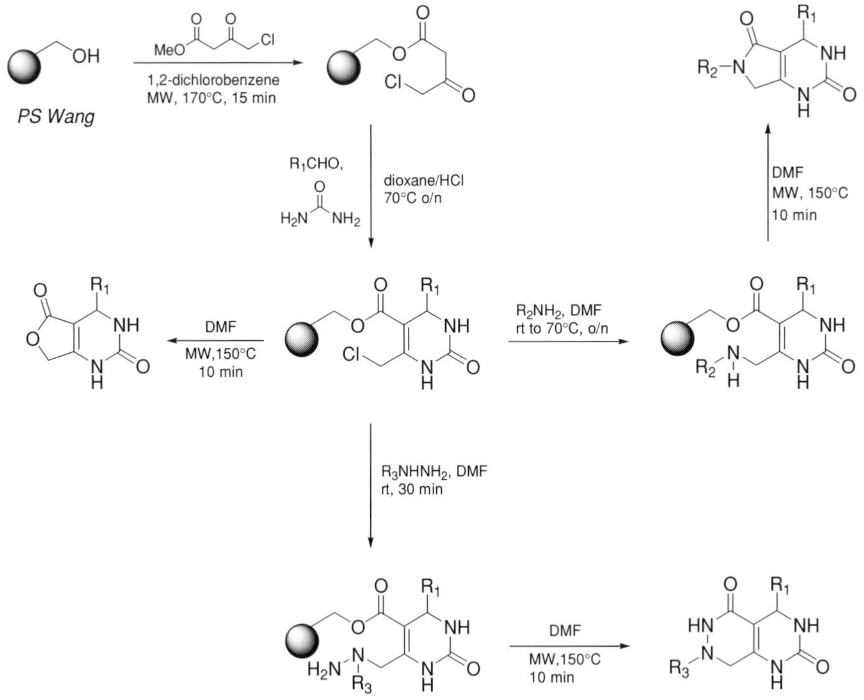

Scheme 7.35   Preparation of various bicyclic dihydropyrimidinones employing cyclative cleavage.

leading to the corresponding resin-bound dihydropyrimidinones. The progress of this transformation could be monitored by resin bead FTIR, which involved verification of the disappearance of the α-chloroketone C=O absorption (1748 cm^{-1}).

The desired furo[3,4-*d*]pyrimidine-2,5-dione scaffold was obtained by a novel protocol for cyclative release under microwave irradiation. The resin-bound monocyclic pyrimidine intermediate was pre-swollen in DMF in a sealed microwave vial and was irradiated in a dedicated single-mode microwave cavity at 150°C for 10 min. After cooling to ambient temperature, the resin was filtered off and washed with DMF. The filtrate and washings were combined and evaporated to dryness to afford the corresponding furo[3,4-*d*]pyrimidine-2,5-dione in high purity (as determined by HPLC–UV/MS measurement) (Scheme 7.35). Alternatively, pyrrolo[3,4-*d*]pyrimidine-2,5-diones were synthesised using the same pyrimidine resin precursor, which was first treated with a representative set of primary amines to substitute the chlorine. Subsequent cyclative cleavage was carried out as described previously, leading to the corresponding pyrrolopyrimidine-2,5-dione products in high purity but moderate yield (Scheme 7.35). The synthesis of pyrimido[4,5-*d*] pyridazine-2,5-diones was carried out in a similar manner employing several hydrazines (R$_3$ = H, Me, Ph) for the nucleophilic substitution, prior to cyclative cleavage (Scheme 7.35). Because of the high nucleophilicity of the hydrazines, reaction times for the substitution step could be reduced to 30 min. In the case of phenylhydrazine, concomitant cyclisation could not be avoided, which led to very low overall yields of the isolated products. Overall, this novel protocol enables the synthesis of a number of heterobicyclic scaffolds under microwave conditions. Further decoration of the pyrimidine moieties of these heterobicyclic templates may potentially lead to a larger number of derivatives for high-throughput evaluation of their biological properties.

In an earlier report, the microwave-mediated intramolecular carbanilide cyclisation to hydantoins was described[44]. Since the hydantoin moiety imparts a broad range of biological activities, several protocols involving both reactions in solution and on solid-phase have been investigated. Within this report, the first microwave-assisted synthetic approach to hydantoins is described (Scheme 7.36).

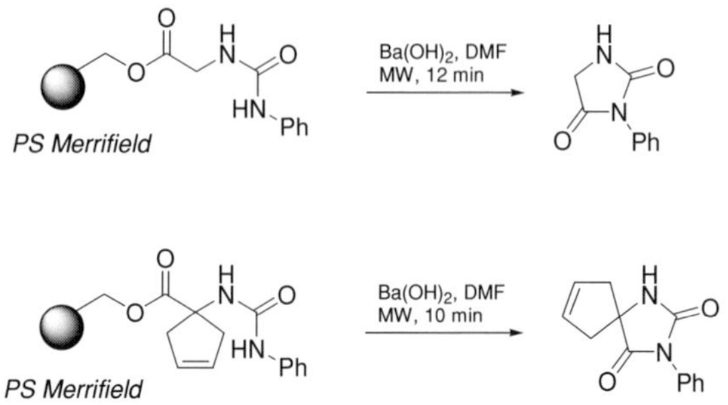

Scheme 7.36   Microwave-mediated intramolecular carbanilide cyclisation on solid support.

From preliminary studies, it was known that the carbanilide cyclisation required rather long reaction times under conventional thermal heating (8–48 h). Therefore, it was obvious to investigate the effect of microwave irradiation upon this reaction. Reaction studies were carried out employing a single-mode microwave cavity with a variety of several base and solvent combinations; barium hydroxide in DMF proved to be the best combination for this cyclisation reaction. Under these solution-phase conditions, carbanilides could be converted in high yields to the corresponding hydantoins within 2–7.5 min. With the appropriate solid-supported protocol, the carbanilide cyclisation would act as a method for resin release of the hydantoins; reaction times could be drastically reduced to several minutes compared to 48 h under thermal heating. For this solid-phase approach, conventional i-PrOCH$_2$ functionalised polystyrene resin (Merrifield Linker) was employed. After attachment of the corresponding substrate, the resin was pre-swollen in a solution of barium(II) hydroxide in DMF within an appropriate sealed microwave vial (Scheme 7.36). The vial was heated in the microwave cavity for 5 × 2 min cycles (overall 10 min) with the reaction mixture being allowed to cool to room temperature in between irradiation cycles. The resin was then filtered off and washed with DMF. The filtrate and washings were combined, quenched with aqueous citric acid solution and extracted with ethyl acetate. However, only poor isolated yields could be obtained with this method, but nonetheless the advantage of microwave irradiation for this application probably warrants further investigation.

7.3.9. Miscellaneous

Several other reaction types on solid support have also been investigated utilising microwave heating. For instance, Yu and co-workers monitored the addition of resin-bound amines with isocyanates employing resin bead FTIR measurement[45]. Using this method, the differences in reaction progress under microwave heating and thermal conditions were investigated (Scheme 7.37).

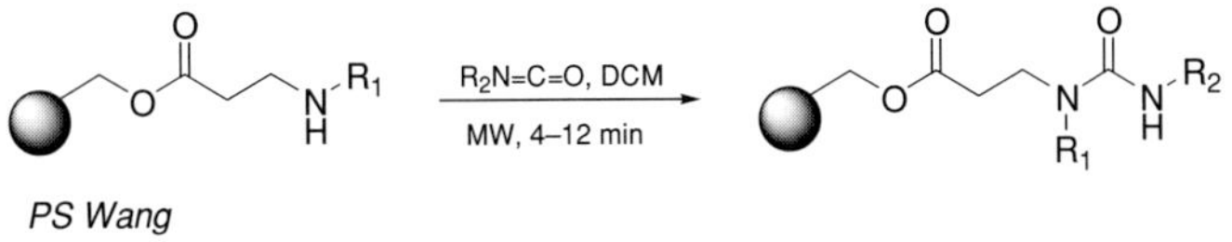

Scheme 7.37 Microwave-assisted addition of isocyanates.

The corresponding isocyanates were each added to the respective resin-bound amine suspended in DCM within an open glass tube. The resulting reaction mixtures were each irradiated in a single-mode microwave cavity for 2 min intervals. After each step, samples were collected for resin bead FTIR analysis. Reaction progress was indicated by increasing intensity of the new 1670 cm^{-1} carbonyl stretch. Table 7.2 illustrates the rate-enhancement effect for five examples; microwave-assisted reactions required a maximum of 12 min in contrast to more than 2 h required under classical room temperature conditions.

A very interesting approach towards solid-supported synthesis under microwave heating was introduced by Chandrasekhar and co-workers[46]. These authors developed a synthesis of N-alkyl imides on solid-phase under solvent-free conditions employing $TaCl_5$-doped silica gel as a reaction mediator. Surprisingly, in this rather unusual method, dry and unswollen polystyrene resin is involved. However, the reaction proceeded within 5–7 min with the N-alkyl imide product being obtained in good yield, after subsequent cleavage from the polystyrene resin-silica gel mixture employing TFA/DCM (Scheme 7.38).

Employing two resin-bound amines and three different anhydrides in the above solid-phase protocol, a set of six cyclic imides (see Fig. 7.15) were synthesised in good yields.

In a typical experiment, 1 mmol of resin bound amine, 25% excess of anhydride were mixed with 1 g of activated silica gel and 10 mol% of TiO_2-doped silica gel. This mixture was placed in a beaker in the centre of a domestic microwave oven and irradiated for 1 min intervals (5–7 min overall) with thorough agitation after each step. The resin–silica gel mixture was allowed to cool to room temperature prior to cleavage. The product was cleaved from the resin–silica gel mixture by treatment with 2 × 10 ml of 33% TFA in DCM. After filtration and evaporation of the solvent, the corresponding imides were obtained in moderate to good yields.

Table 7.2 Comparison of reaction times for resin bound addition reactions (Scheme 7.37)

Entry	R_1	R_2	Reaction time (MW)	Reaction time (rt)
1	2-MeO-benzyl	Ph	4 min	120 min
2	4-Cl-benzyl	Ph	6 min	n.d.
3	α-Me-benzyl	Ph	4 min	n.d.
4	Benzyl	Ph	12 min	210 min
5	Benzyl	4-MeOPh	6 min	150 min

HO
NH2
Cl
PS Merrifield
SiO2/TaCl5
MW, 5 min
no solvent
TFA/DCM
65%

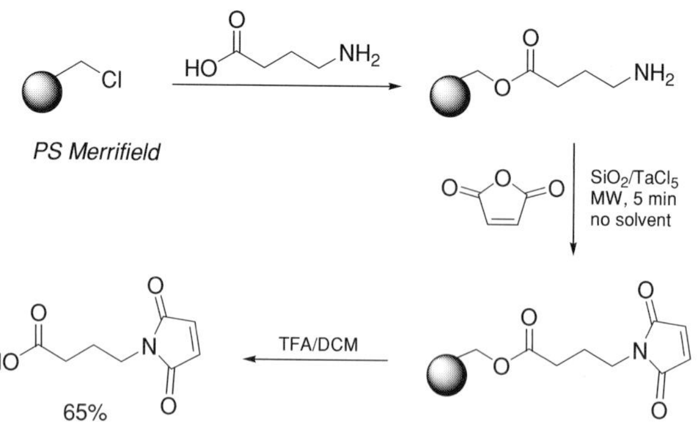

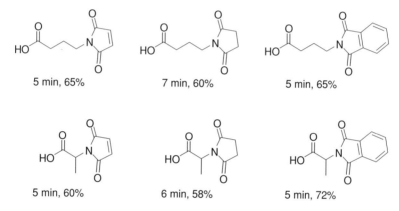

Scheme 7.38 Solvent-free preparation of 4-(1,3-2,3-dihydro-1*H*-2-isoindolyl)butanoic acid.

5 min, 65%
7 min, 60%
5 min, 65%
5 min, 60%
6 min, 58%
5 min, 72%

Figure 7.15 Structural representation of imides obtained by synthesis on solid support in solid state (microwave reaction time for imide formation and isolated yield of imide products after resin cleavage quoted).

A recent study describes the application of solid-supported cyclohexane-1,3-dione as a so-called 'capture and release' reagent for amide synthesis, as well as a novel scavenger resin[47]. Within this report, a three step synthesis of polymer bound cyclohexane-1,3-dione (CHD resin, Scheme 7.39) from cheap and readily available starting materials is described. The key step in this reaction is the microwave-assisted complete hydrolysis of the 3-methoxy cyclohexen-1-one resin to the desired CHD resin, employing a dedicated single-mode microwave cavity.

In order to show its potential for resin 'capture and release' methodology, the CHD resin was employed for the generation of a 16-member amide library. Resin 'capture and release' methodology would enable impurity removal and product purification. Five different acid chlorides were coupled to CHD resin under microwave heating conditions. Treatment of the resulting resin with various primary and secondary amines released the desired amides in moderate to high yields with generally good purities (Scheme 7.40). It was also demonstrated that the CHD resin could be recycled in this procedure, as products were afforded in good yields and comparable purity in a second run using the recycled CHD resin.

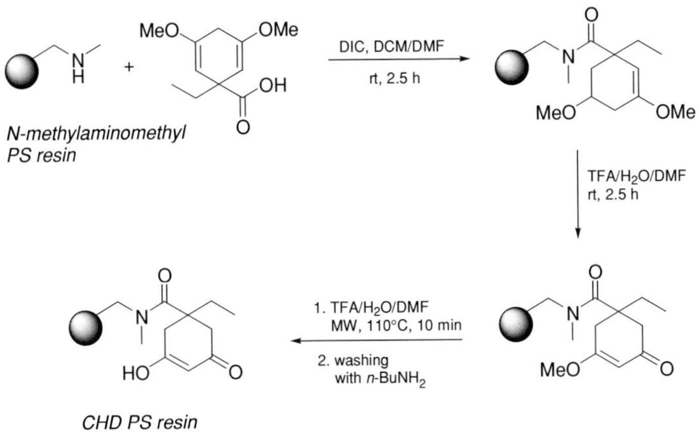

Scheme 7.39   Preparation of resin bound cyclohexan-1,3-dione (CHD resin).

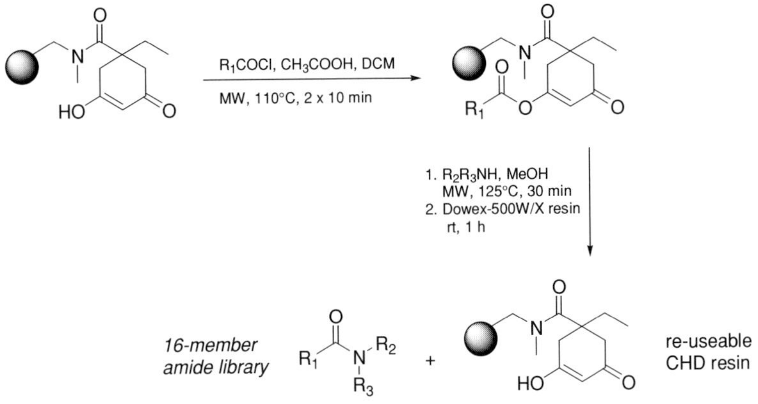

Scheme 7.40   Microwave-assisted generation of amide library using CHD resin.

In order to explore the potential of CHD resins as scavenger materials, 1-ethyl-3,5-dimethoxycyclohexa-2,5-dienecarboxylic acid was anchored to commercially available trisamine resin (Scheme 7.41), yielding high-loading cyclohexane-1,3-dione scavenger resin (CHD-SR).

The scavenging ability of this novel resin as an allyl cation scavenger was demonstrated in the palladium catalysed O-Alloc deprotection of the O-Alloc benzyl alcohol (Scheme 7.42). Benzyl alcohol was obtained in high yield with only small traces of by-product, thereby eliminating the need for further purification. The relevant C-allylation of the resin was observed by the presence of C-allyl signals in the corresponding MAS-probe ^{1}H NMR.

The versatility of cyclohexane-1,3-dione functionalised resins has been illustrated by their synthetic application as both capture and release reagents and resin scavengers. In addition, CHD resins show considerable potential for further use as linkers in several other solid-phase applications.

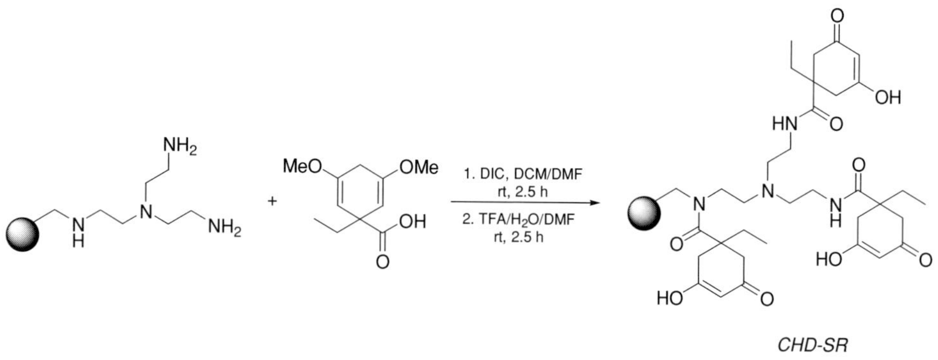

Scheme 7.41 Preparation of high-loading cyclohexane-1,3-dione scavenger resin (CHD-SR).

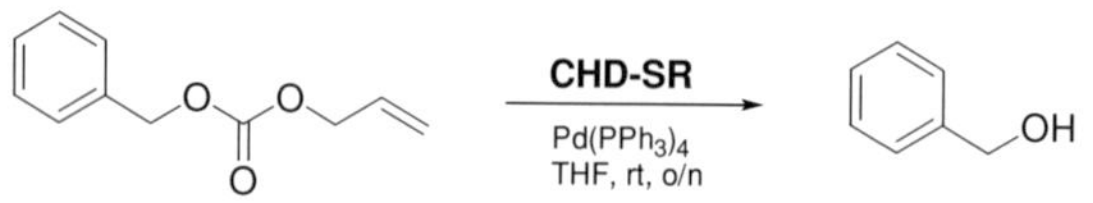

Scheme 7.42 Palladium-catalysed deprotection of O-Alloc benzyl alcohol employing the scavenging resin CHD-SR.

7.3.10. Case study: pyrazinone Diels–Alder chemistry

Finally, to highlight the advantages of both microwave-mediated protocols and solid-phase synthesis, a recently presented study[48] is discussed in this chapter. This is the microwave-assisted solid-phase Diels–Alder cycloaddition reaction of $2(1H)$-pyrazinones with dienophiles (Scheme 7.43).

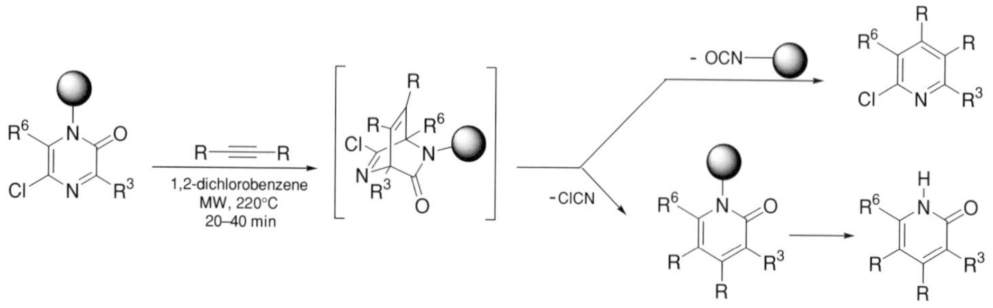

Scheme 7.43 General reaction sequence for microwave assisted traceless-linking concept for $2(1H)$-pyrazinone Diels–Alder cycloadditions with acetylenic dienophiles.

After the breakdown of the resin-bound adduct formed from Diels–Alder reaction of the $2(1H)$-pyrazinone with an acetylenic dienophile, separation of the resulting pyridines from the pyridinone by-products was achieved by applying a traceless-linking concept, whereby the pyridinones remained on the solid support with concomitant release of the pyridine products into solution (Scheme 7.43). This indicates the first successful solid-phase chemistry (linking, Diels–Alder reaction and cleavage) involving the 2(1H)-pyrazinone scaffold. Furthermore, a novel, tailor-made and readily available linker derived from inexpensive syringaldehyde (Scheme 7.44) was developed.

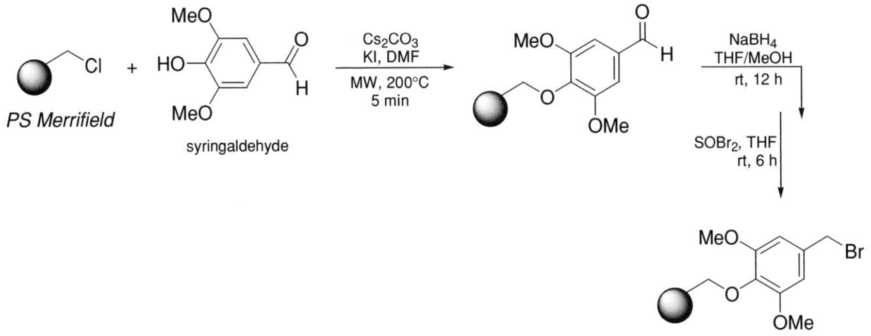

Scheme 7.44 Preparation of brominated syringaldehyde-resin.

The novel linker was produced by cesium carbonate activated coupling of commercially available syringaldehyde to Merrifield resin, under microwave heating conditions. Subsequently, the aldehyde moiety was reduced at room temperature within 12 h and the benzylic position was finally brominated by treatment with a large excess of thionyl bromide (10 equiv) leading to the desired polymeric support (Scheme 7.44).

For the development of an appropriate cleavage strategy, a preliminary elaborated solution-phase model study was adapted for a solid-phase approach, using various acid labile linkers such as Wang resin, HMPB-AM resin, and the above mentioned syringaldehyde-based resin. It should be highlighted that the syringaldehyde-based resin has been proven to be superior to both other linkers, especially for cleavage purposes (Scheme 7.45).

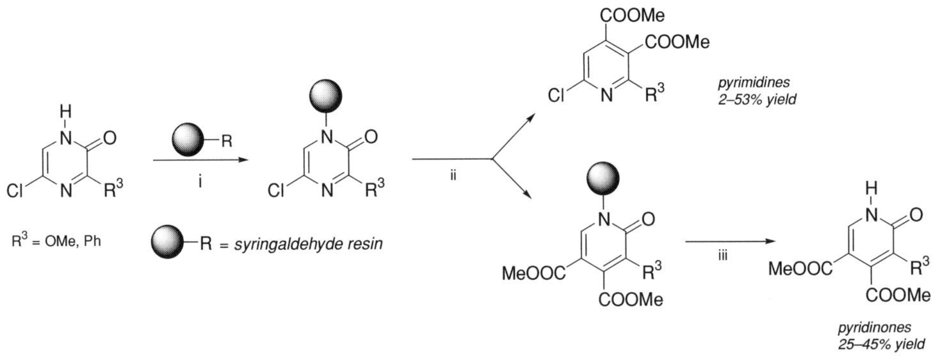

Scheme 7.45 Intermolecular 2(1H)-pyrazinone Diels–Alder reactions on solid support. Reagents and conditions: (i) Cs$_2$CO$_3$, DMF, MW, 70°C, 5 min; (ii) dimethyl acetylene dicarboxylate (DMAD), 1,2-dichlorobenzene, MW, 220°C, 20–40 min; (iii) TFA–CH$_2$Cl$_2$ (1:4 or 1:9), MW, 120°C, 10–40 min.

The resulting pyridines could be easily separated from the polymer-bound by-products by employing a simple filtration step and subsequent evaporation of the solvent. The remaining resins were each washed and dried. After drying, the resins were each treated with TFA/DCM cleavage solutions to obtain the corresponding pyridinones. Interestingly, the expected pyridinone products could not be released from Wang resin using the standard cleavage conditions of 95% TFA in DCM at room temperature. Instead, the corresponding *N*-*p*-hydroxybenzylated pyridinones were obtained. Attempts to cleave the polymer-bound pyridinones from HMBP-AM resin using different mixtures of trifluoracetic acid in DCM also resulted in the formation of several by-products. In contrast, the microwave-assisted cleavage of the pyridinone products at 120°C for 10 min proceeded well and required significantly less acidic conditions, as in the earlier Wang resin example. Utilising the novel syringaldehyde-resin, a smooth release from the support could be performed upon microwave heating of a suspension of the resin-bound pyridinones in TFA/DCM (5:95) at 120°C for only 10 min. The very mild cleavage conditions for this new linker as well as its stability towards different reaction conditions and its easy accessibility makes this a highly suited linker for ongoing pyrazinone chemistry.

Additionally, for comparison purposes, all steps in the solid-phase protocol (linking, cycloaddition, cleavage) were carried out both under thermal and controlled microwave heating conditions. In general, significant rate enhancements were found for each step and it has been demonstrated that microwave mediation could be combined with the efforts of solid-supported chemistry to enable the development of novel pathways in organic synthesis.

7.4. Other types of supports

Apart from the traditional solid supports (see above), several publications also report the successful use of microwave enhancement for supported transformations involving soluble polymers[49–54], fluorous phase conditions[55,56], and ionic liquids grafted onto polymeric supports[57,58].

An interesting and recently published study presented the functionalisation of insoluble polystyrene resin with soluble polyethylenglycol (PEG) (Scheme 7.46) to produce hybrid polymers that combine the advantages of both PEG and polystyrene as polymeric supports[52].

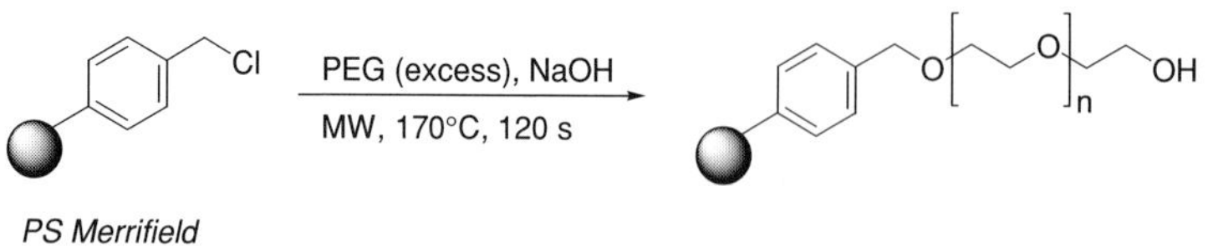

PS Merrifield

Scheme 7.46   Microwave-assisted synthesis of PEGylated Merrifield resin.

It should be highlighted that in some cases the distinction between a traditional solid-phase synthesis and a process involving solid-supported reagents is often difficult. The reader is advised to refer also to the chapter on the integration of microwave-assisted synthesis with solid supported reagents and scavengers (Chapter 6).

7.5. Conclusion

The combination of microwave-assisted chemistry and solid-phase synthesis applications is a logical consequence of the increased speed and effectiveness offered by microwave dielectric heating. While this technology is heavily used in the pharmaceutical and agrochemical research laboratories already, a further increase in the use of microwave-assisted solid-phase synthesis both in industry and in academic laboratories can be expected. This will depend also on the availability of modern microwave instrumentation specifically designed for solid-phase chemistry, involving for example dedicated vessels for bottom filtration techniques.

7.6. References

1. Nicolaou, K.C., Hanko, R. and Hartwig, W. (Eds.), *Handbook of Combinatorial Chemistry*, Wiley-VCH, Weinheim, 2002.
2. Dörwald, F.Z., *Organic Synthesis on Solid Phase*, Wiley-VCH, Weinheim, 2002.
3. Merrifield, R.B., Solid phase peptide synthesis. The synthesis of a tetrapeptide, *J. Am. Chem. Soc.*, 1963, **85**, 2149–2154.
4. Frank, R., Heikens, W., Heisterberg-Moutsis, G. and Blöcker, H., A new general approach for the simultaneous chemical synthesis of large numbers of oligonucleotides: segmental solid supports, *Nucl. Acids Res.*, 1983, **11**, 4365–4377.
5. Geysen, H.M., Meloen, R.H. and Barteling, S.J., Use of peptide synthesis to probe viral antigens for epitopes to a resolution of a single amino acid, *Proc. Natl. Acad. Sci. USA*, 1984, **81**, 3998–4002.
6. Houghten, R.A., General method for the rapid solid phase synthesis of large numbers of peptides: specificity of antigen-antibody interaction at the level of individual amino acids, *Proc. Natl. Acad. Sci. USA*, 1985, **82**, 5131–5135.
7. (a) Furka, Á., Sebestyén, F., Asgedom, M. and Dibó, G., Highlights of modern biochemistry, In: *Proceedings of the 14th International Congress of Biochemistry*, Prague, CZ, 1988, Vol. 13, p.14, VSP, Utrecht; (b) Furka, Á., Sebestyén, F., Asgedom, M. and Dibó, G., General method for rapid synthesis of multi-component peptide mixtures, *Int. J. Peptide Prot. Res.*, 1991, **37**, 487–493.
8. Hayes, B., *Microwave Synthesis. Chemistry at the Speed of Light*, CEM Publishing, Matthews, NC, 2002.
9. Loupy, A. (Ed.), *Microwaves in Organic Synthesis*, Wiley-VCH, Weinheim, 2002.
10. Lew, A., Krutzik, P.O., Hart, M.E. and Chamberlin, A.R., Increasing rates of reaction: microwave-assisted organic synthesis for combinatorial chemistry, *J. Comb. Chem.*, 2002, **4**, 95–105.
11. Kappe, C.O., High-speed combinatorial synthesis utilizing microwave irradiation, *Curr. Opin. Chem. Biol.*, 2002, **6**, 314–320.
12. Lidström, P., Westman, J. and Lewis, A., Enhancement of combinatorial chemistry by microwave-assisted organic synthesis, *Comb. Chem. High Throughput Screen*, 2002, **5**, 441–458.
13. Larhed, M. and Hallberg, A., Microwave-assisted high-speed chemistry: a new technique in drug discovery, *Drug Discovery Today*, 2001, **6**, 406–416.
14. Kappe, C.O. and Stadler, A., Microwave-assisted combinatorial chemistry, In: Loupy, A. (Ed.), *Microwaves in Organic Synthesis*, Wiley-VCH, Weinheim, 2002, pp. 405–433.
15. Gabriel, C., Gabriel, S., Grant, E.H., Halstead, B.S.J. and Mingos, D.M.P., Dielectric parameters relevant to microwave dielectric heating, *Chem. Soc. Rev.*, 1998, **27**, 213–223.
16. (a) Terret, N.K., *Combinatorial Chemistry*, Oxford University Press, New York, 1998; (b) Winter, M. and Warrass, R., Resins and anchors for solid-phase organic synthesis, In: Fenniri, H. (Ed.), *Combinatorial Chemistry*, Oxford University Press, New York, 2000, pp.117–138.
17. Pérez, R., Beryozkina, T., Zbruyev, O.I., Haas, W. and Kappe, C.O., Traceless solid phase synthesis of bicyclic dihydropyrimidones using multidirectional cyclization cleavage, *J. Comb. Chem.*, 2002, **4**, 501–510 and references cited therein.
18. Combs, A.P., Saubern, S., Rafalski, M. and Lam, P.Y.S., Solid supported aryl/heteroaryl C—N cross-coupling reactions, *Tetrahedron Lett.*, 1999, **40**, 1623–1626.
19. Glass, B.M. and Combs, A.P., Case study 4–6: rapid parallel synthesis utilizing microwave irradiation, In: Sucholeiki, I. (Ed.), *High-Throughput Synthesis. Principles and Practices*, Marcel Dekker, New York, 2001, Ch. 4.6, pp. 123–128,

20. Glass, B.M. and Combs, A.P., Rapid parallel synthesis utilizing microwave irradiation (Article E0027), In: Kappe, C.O., Merino, P., Marzinzik, A., Wennemers, H., Wirth, T., Vanden Eynde, J.J. and Lin, S.K. (Eds.), *Fifth International Electronic Conference on Synthetic Organic Chemistry*, MDPI, Basel, Switzerland, 2001, CD-ROM edition.

21. Coleman, C.M., MacElroy, J.M.D., Gallagher, J.F. and O'Shea, D.F., Microwave parallel library generation: comparison of a conventional- and microwave-generated substituted 4(5)-sulfanyl-1H-imidazole library, *J. Comb. Chem.*, 2002, **4**, 87–93.

22. Yu, H.-M., Chen, S.-T. and Wang, K.-T., Enhanced coupling efficiency in solid phase peptide synthesis by microwave irradiation, *J. Org. Chem.*, 1992, **57**, 4781–4784.

23. Erdelyi, M. and Gogoll, A., Rapid microwave-assisted solid phase peptide synthesis, *Synthesis*, 2002, 1592–1596.

24. Olivos, H.J., Alluri, P.G., Reddy, M.M., Salony D. and Kodadek, T., Microwave-assisted solid phase synthesis of peptoids, *Org. Lett.*, 2002, **4**, 4057–4059.

25. Finaru, A., Berthault, A., Besson, T., Guillaumet G. and Berteina-Raboin, S., Microwave-assisted solid phase synthesis of 5-carboxamido-N-acetyltryptamine derivatives, *Org. Lett.*, 2002, **4**, 2613–2615.

26. Daga, M.C., Tadde, M. and Varchi, G., Rapid microwave-assisted deprotection of N-Cbz and N-Bn derivatives, *Tetrahedron Lett.*, 2001, **42**, 5191–5194.

27. Stadler, A. and Kappe, C.O., High-speed couplings and cleavages in microwave-heated, solid-phase reactions at high temperatures, *Eur. J. Org. Chem.*, 2001, 919–925.

28. Stadler, A. and Kappe, C.O., The effect of microwave irradiation on carbodiimide-mediated esterifications on solid support, *Tetrahedron*, 2001, **57**, 3915–3920.

29. Yang, H., Peng, Y., Song, G. and Qian, X., Microwave-assisted preparation of functionalized resins for combinatorial synthesis, *Tetrahedron Lett.*, 2001, **42**, 9043–9046.

30. Westman, J. and Lundin, R., Solid-phase synthesis of aminopropenones and aminopropenoates; efficient and versatile synthons for combinatorial synthesis of heterocycles, *Synthesis*, 2003, **7**, 1025–1030.

31. Strohmeier, G.A. and Kappe, C.O., Rapid parallel synthesis of polymer-bound enones utilizing microwave-assisted solid phase chemistry, *J. Comb. Chem.*, 2002, **4**, 154–161.

32. Larhed, M., Lindeberg, G. and Hallberg, A., Rapid microwave-assisted Suzuki coupling on solid phase, *Tetrahedron Lett.*, 1996, **37**, 8219–8222.

33. Kaiser, N.-F., Bremberg, U., Larhed, M., Moberg, C. and Hallberg, A. Fast, convenient, and efficient molybdenum-catalyzed asymmetric allylic alkylation under noninert conditions: an example of microwave-promoted fast chemistry, *Angew. Chem.*, 2000, **112**, 3742–3744.

34. Alterman, M. and Hallberg, A., Fast microwave-assisted preparation of aryl and vinyl nitriles and the corresponding tetrazoles from organo-halides, *J. Org. Chem.*, 2000, **65**, 7984–7989.

35. Bengtsson, A., Hallberg, A. and Larhed, M., Fast synthesis of aryl triflates with controlled microwave heating, *Org. Lett.*, 2002, **4**, 1231–1233.

36. Scharn, D., Wenschuh, H., Reineke, U., Schneider-Mergener J. and Germeroth, L., Spatially addressed synthesis of amino- and amino-oxy-substituted 1,3,5-triazine arrays on polymeric membranes, *J. Comb. Chem.*, 2000, **2**, 361–369.

37. Scharn, D., Germeroth, L., Schneider-Mergener J. and Wenschuh, H., Sequential nucleophilic substitution on halogenated triazines, pyrimidines, and purines: a novel approach to cyclic peptidomimetics, *J. Org. Chem.*, 2001, **66**, 507–513.

38. Austin, R.E., Okonya, J.F., Bond, D.R.S. and Al-Obeidi, F., Microwave-assisted solid-phase synthesis (MASS) of 2,6,9-trisubstituted purines, *Tetrahedron Lett.*, 2002, **43**, 6169–6171.

39. Hoel, A.M.L. and Nielsen, J., Microwave-assisted solid-phase ugi four-component condensations, *Tetrahedron Lett.*, 1999, **40**, 3941–3944.

40. Launay, D., Booth, S., Clemens, I., Merritt, A. and Bradley, M., Solid-phase-mediated synthesis of isonitriles, *Tetrahedron Lett.*, 2002, **43**, 7201–7203.

41. Frutos Hoener, A.P., Henkel, B. and Gauvin, J.-C., Novel one-pot microwave-assisted gewald synthesis of 2-acyl amino thiophenes on solid support, *Synlett*, 2003, 63–66.

42. Kuster, G.J. and Scheeren, H.W., The preparation of resin-bound nitroalkanes and some applications in high pressure promoted cycloadditions, *Tetrahedron Lett.*, 2000, **41**, 515–519.

43. Kumar, H.M.S., Anjaneyulu, S., Reddy, B.V.S. and Yadav, J.S., Microwave-assisted rapid claisen rearrangements on solid phase, *Synlett*, 2000, 1129–1130.

44. Gong, Y.-D., Sohn, H.-Y. and Kurth, M.J., Microwave-mediated intramolecular carbanilide cyclization to hydantoins employing barium hydroxide catalysis, *J. Org. Chem.*, 1998, **63**, 4854–4856.

45. Yu, A.-M., Zhang, Z.-P., Yang, H.-Z., Zhang, C.-X. and Liu, Z., Wang resin bound addition reactions under microwave irradiation, *Synth. Commun.*, 1999, **29**, 1595–1599.

46. Chandrasekhar, S., Padmaja, M.B. and Raza, A., Solid phase-solid state synthesis of N-alkyl imides from anhydrides, *Synlett*, 1999, 1597–1599.

47. Humphrey, C.E., Easson, M.A.M., Tierney, J.P. and Turner, N.J., Solid-supported cyclohexane-1,3-dione (CHD): a "capture and release" reagent for the synthesis of amides and novel scavenger resin, *Org. Lett.*, 2003, **5**, 849–852.
48. Kaval, N., Van der Eycken, J., Caroen, J.,Dehaen, W.,Strohmeier, G.A., Kappe, C.O. and Van der Eycken, E., An exploratory study on microwave-assisted solid-phase diels-alder reactions of 2(1*H*)-pyrazinones: the elaboration of a new tailor-made acid-labile linker, *J. Comb. Chem.*, 2003, **5**, 560–568.
49. Blettner, C.G., König, W.A., Stenzel, W. and Schotten, T., Microwave-assisted aqueous suzuki cross-coupling reactions, *J. Org. Chem.*, 1999, **64**, 3885–3890.
50. Vanden Eynde, J.J. and Rutot, D., Microwave-mediated derivatization of poly(styrene-*co*-allyl alcohol), a key step for the soluble polymer-assisted synthesis of heterocycles, *Tetrahedron*, 1999, **55**, 2687–2694.
51. Sauvagnat, B., Lamaty, F., Lazaro, R. and Martinez, J., Poly(ethylene glycol) as solvent and polymer support in the microwave assisted parallel synthesis of aminoacid derivatives, *Tetrahedron Lett.*, 2000, **41**, 6371–6375.
52. Yaylayan, V.A., Siu, M., Bélanger, J.M.R. and Paré, J.R.J., Microwave-assisted PEGylation of merrifield resins, *Tetrahedron Lett.*, 2002, **43**, 9023–9025.
53. Bendale, P.M. and Sun, C.-M., Rapid microwave-assisted liquid-phase combinatorial synthesis of 2-(arylamino)benzimidazoles, *J. Comb. Chem.*, 2002, **4**, 359–361.
54. Porcheddu, A., Ruda, G.F., Sega, A. and Taddei, M., A new, rapid, general procedure for the synthesis of organic molecules supported on methoxy-polyethylene glycol (MeOPEG) under microwave irradiation conditions, *Eur. J. Org. Chem.*, 2003, 907–912.
55. Larhed, M., Hoshino, M., Hadida, S., Curran, D.P. and Hallberg, A., Rapid fluorous stille coupling reactions conducted under microwave irradiation, *J. Org. Chem.*, 1997, **62**, 5583–5587.
56. Olofsson, K., Kim, S.-Y., Larhed, M., Curran, D.P. and Hallberg, A., High-speed, highly fluorous organic reactions, *J. Org. Chem.*, 1999, **64**, 4539–4541.
57. Fraga-Dubreuil, J. and Bazureau, J.P., Grafted ionic liquid-phase-supported synthesis of small organic molecules, *Tetrahedron Lett.*, 2001, **42**, 6097–6100.
58. Fraga-Dubreuil, J., Famelart, M.-H. and Bazureau, J.P., Ecofriendly fast synthesis of hydrophylic poly(ethyleneglycol)-ionic liquid matrices for liquid-phase organic synthesis, *Org. Proc. Res. Dev.*, 2002, **6**, 374–378.

8 Timesavings associated with microwave-assisted synthesis: a quantitative approach

CHRISTOPHER R. SARKO

8.1. Introduction

The use of microwave-assisted organic synthesis has been claimed in a number of publications to accelerate the production of novel chemical entities[1–4]. However, little effort has been put forth to quantitate these timesavings and most claims have few experimental controls. This chapter will focus on direct comparisons between microwave-assisted synthesis and traditional thermal heating. The primary objective is to discuss the examples our group has presented, which contain direct comparisons, and to highlight controlled direct comparisons from the literature[5–8]. The secondary objective is to focus on overall productivity increases associated with microwave-assisted synthesis.

There have been significant advances in robotics and automation that have had a profound effect upon the ability to perform high throughput organic synthesis[9]. However, these advances have been primarily limited to the area of combinatorial library production. In general, very little work has been performed in the area of increasing throughput in the library design and development stages. In this chapter, we provide examples that show the overall timesavings associated with microwave-assisted synthesis in chemistry development.

Our final objective is to show new developments in microwave-technology hardware, specifically to illustrate examples of chemistry and new tools, such as plate-based systems, for improving the throughput of microwave-assisted library production.

8.2. Timesavings associated with microwave-assisted synthesis

Since the first reported usages of microwave-assisted synthesis, the timesavings associated with this tool have been as widely accepted as remarkable[10–12]. However, few if any of these reports show controlled reactions comparing conditions between thermal and microwave methods. The key with any experimental design is to have the proper controls and a concise testing method to rule out as many variables as possible from the experiment. Our initial efforts in this field were rather rudimentary, such as comparing refluxing reaction conditions to those generated in the microwave field, but as we began to probe the effect of microwave heating it became obvious that these were not accurate controls. The next logical stage was to compare temperatures of a sealed vessel in a temperature-controlled heating bath with internal temperature measurement of the sample via a thermocouple and the microwave vessel with infrared temperature readings. It should be pointed out that the external temperature measurement in the microwave device had been earlier calibrated towards an internal thermocouple. The results from these experiments along with others in the literature are the focus of this subsection.

Our initial foray into microwave chemistry was with a reaction that had proven inaccessible using traditional thermal techniques. The reaction was a simple [3 + 2] cycloaddition reaction between a di-substituted maleimide and a simple azomethine ylide generated *in situ* (Scheme 8.1)[13]. The reaction with the unsubstituted maleimide had yielded excellent results for a variety of dipoles; however, even simple methyl substitution had dramatic reductions in product yield (Table 8.1). The interest in our group was in generating novel three-dimensional scaffolds for library generation and the di-substituted maleimides would be a key entry point into these compounds, so the need for the products had us attempt these reactions in sealed pressure vessels. While we were pleased to obtain some of the desired product, the extremely low yield had eliminated this class of compounds from consideration.

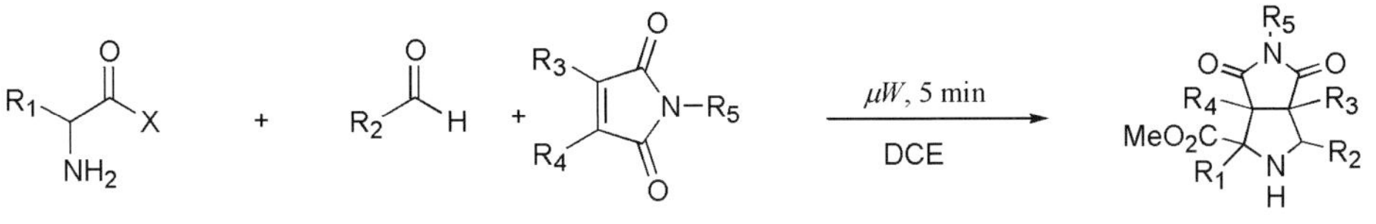

Scheme 8.1

At this time, we had access to a microwave system from Personal Chemistry called the Smith Synthesizer (Personal Chemistry AB, Uppsala, Sweden) and so we attempted this difficult cycloaddition reaction[13]. As evident in Table 8.1, the results with the microwave were remarkably improved compared to the conventionally heated counterpart. The product yield and purity was substantially higher than what was observed in the pressure tubes. With this first positive example, we were encouraged to try systems that had not been able to produce an observable product in the pressure tubes. The condensation of a fused cyclohexyl maleimide had not produced any product in our previous efforts, but with microwave heating for a short 5 min reaction time at 180°C we were able to isolate a satisfactory amount of the desired product.

Table 8.1 [3 + 2] Cycloaddition reaction results[a]

R1	X	R2	R3, R4	R5	Crude purity[b] (%)	Isolated yield[c] (%)
H	OCH_3	C_6H_5	H,H	$C_6H_5CH_2$	62	87
CH_3	OCH_3	$4\text{-}NO_2C_6H_4$	H,H	$4\text{-}ClC_6H_4$	56	82
$C_6H_5CH_2$	OCH_3	$3\text{-}BrC_6H_4$	CH_3, CH_3	C_6H_5	68	84
$(CH_3)_2CH$	OCH_3	$2\text{-}MeC_6H_4$	$-C_4H_8-$	$4\text{-}MeO_2CC_6H_4$	67	75
$CH_3SCH_2CH_2$	OCH_3	$4\text{-}FC_6H_4$	H,H	$C_6H_5CH_2$	72	89

[a] Microwave irradiation performed in a Personal Chemistry Smith Synthesizer™.
[b] Purity determined by LC-MS analysis of crude products (integration area at 254 nm) unless otherwise noted.
[c] Isolated compound after treatment with PS-TsNHNH₂.

8.3. Acceleration of combinatorial library design and development stages

Before one can quantify the timesavings associated with microwave-assisted synthesis in respect to library production, we must first provide a background for the production of a typical compound library. After analyzing several historical libraries produced at Boehringer Ingelheim, we were able to produce a range of timelines, Fig. 8.1. The typical process ranges from 15 to 22 weeks depending on molecular complexity and the method of synthesis. The entire process can furthermore be broken into several subunits consistent throughout all library projects.

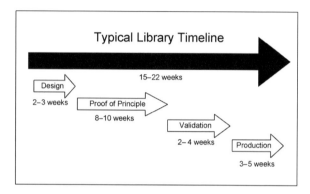

Figure 8.1   Typical library production time based upon historical data at Boehringer Inghelheim, Ridgefield.

The first stage in library production is the design stage wherein we arrive at the concept, evaluate literature precedent, determine the novelty of the proposed scaffolds in regards to our internal compound collection and external patent sources, and viability of the proposed synthetic route.

The second stage is the proof of principle: In this phase, we take the initial theoretical library idea and begin to apply chemistry experiments to validate experimental designs and potential library schemes; at this stage, one also evaluates the method of library production (solid/solution/hybrid phases). In this phase, which is usually the longest phase in any library production process, we will perform the initial experiments, optimize the chemical yields and purities, modify the experiments to generate easily removable by-products, which can be removed by traditional parallel purification methods (i.e. SPE, Resin capture), and determine the most feasible route to the final product.

After the proof of principle stage we perform library validation; here we determine which reagents will be compatible with others and which reagents will produce the cleanest and purest compounds as well as allow for maximal diversity in the overall library. Analytical data will be generated from this validation example providing key information for analytical method development for the 'full-blown' library production.

The final stage is obviously production, which entails actual synthesis of the library and placement into a format for submission into the compound collection. The improvements in automated synthesis have had a profound effect on the production phase. Accordingly, the longest period of time in library development and production is spent in the proof of principle and library validation stages, as seen in Fig. 8.1.

Several concepts for the acceleration of library production have been proposed, one could streamline the process by placing people in areas of expertise (e.g., have a team responsible for validation, another responsible for developing reaction schemes and performing preliminary experiments); however, the drawback to this approach is employee burnout, the repetitiveness of the daily routine can have a negative impact on the creative process. You could also use literature procedures as the sole source of libraries. However, you sacrifice the novelty of your scaffolds and thus your competitive advantage. Furthermore, it is commonplace to perform reaction optimizations in parallel or combinatorially giving you the increase in speed from multi-dimensional optimization; however, in principle this is difficult to perform for all libraries. For example, if you were looking to optimize a Suzuki coupling, you could easily generate all combinations of catalyst, ligand and base. However, considering you will most likely choose only one optimal condition a large amount of time is spent in organizing, performing, working up and analyzing the data from very many highly unlikely combinations to find that one optimal method. Thus, the best way to increase productivity would simply be to have all your reactions just go faster. In this subsection we focus on the relative timesavings associated with microwave-assisted synthesis and going from anecdotal to actual timesavings calculated under controlled conditions.

The initial attempts in our group to quantify the timesavings associated with microwave-assisted chemistry began as we started comparing reactions of interest in both a traditional thermal and microwave-accelerated setting. We had previously generated a library of 2-aminoquinolines in our group[14]. The key multi-component condensation reaction required heating to 110°C for 24 h to proceed at an acceptable

Table 8.2   Direct thermal and microwave comparisons – 2-aminoquinoline synthesis

Entry	R1, R2	R3, R4	R4	Temp (°C)	Solvent	Time	% Completion[a]
1	5-CI	morpholine	H	reflux	p-xylene	4 h	52
2	5-CI	morpholine	H	110	Toluene	6 h	39[b]
3	5-CI	morpholine	H	110	Toluene	24 h	50[b]
4	5-NO$_2$,H	cyclohexyl, Me	H	110	Toluene	6 h	10
5	5-NO$_2$,H	cyclohexyl, Me	H	110	Toluene	24 h	31
6	5-NO$_2$,H	cyclohexyl, Me	H	85	DCE	24 h	11
7	5-NO$_2$,H	cyclohexyl, Me	H	150	DMF	24 h	16
8	5-NO$_2$,H	N-methylpiperzine	H	180[c]	DCE	8 min[d]	63
9	5-NO$_2$,H	N-methylpiperzine	H	180[c]	DCE	10 min[d]	73

[a] Purity determined by LC-MS analysis of crude products (integration area at 254 nm) unless otherwise noted.
[b] Purity determined by ELSD-MS analysis of crude products (integration area).
[c] Microwave irradiation done in a Personal Chemistry Smith Synthesizer™.
[d] Time includes "3 min required for enamine formation.

Table 8.3 Direct comparison of thermal and microwave results – indole library

Entry	R1	Amide yield (%)	Cyclization conditions	Indole yield (%)
1	Phenyl	91	TiCl$_3$-THF TMSC1, Zn MeCN, 80°C, 8 h	32
2	Phenyl	91	TiCl$_3$-THF TMSC1, Zn MeCN, 170°C, 5 min	79
3	3,4-Dimethoxy phenyl	94	TiCl$_3$-THF TMSC1, Zn MeCN, 170°C, 5 min	67
4	3,4-Dichloro phenyl	90	TiCl$_3$-THF TMSC1, Zn MeCN, 170°C, 5 min	71
5	p-Anisyl	87	TiCl$_3$-THF TMSC1, Zn MeCN, 170°C, 5 min	82
6	Pyridyl	86	TiCl$_3$-THF TMSC1, Zn MeCN, 170°C, 5 min	66
7	Furyl	72	TiCl$_3$-THF TMSC1, Zn MeCN, 170°C, 5 min	73

percent completion (see Table 8.2). The usage of microwave irradiation greatly accelerated this reaction and produced the desired compounds in higher yield and improved purity. Furthermore, the use of more volatile solvent (1,2-dichloroethane) allowed for an easier work-up of the reaction products.

Our second attempt at using the microwave to accelerate reaction rates was on a library of indoles generated by a McMurray coupling reaction. Heating the reaction mixture at 80°C for 8 h produced poor yields of the desired product (entry 1, Table 8.3). However, when the microwave was utilized the isolated yield was raised to 79% and the reaction was completed in only 5 min. This reaction acceleration allowed for the production of a 300-membered library, which produced compounds of high quality in a significantly reduced time.

8.3.1. The contest

In an attempt to define the exact amount of timesavings associated with the use of microwave-assisted library synthesis, our group devised a method by which the actual control thermal reactions would be performed side-by-side with the microwave reactions by two independent researchers. Furthermore, it was decided that the two scientists would not share experimental results or methodologies so as to not affect each other's research. This series of experiments was labeled the 'Contest'[15].

In this library example, the two scientists were each given the same scaffold to synthesize. The 'contestants' could use any route of synthesis; however, only one of the chemists had access to the microwave system. The two scientists would keep accurate measurement of the time spent on the synthesis, including both performing reactions and literature searches. At the end of the experiment, the scientists would compare the time spent on the synthesis and the routes taken to the final products. The project would be considered completed, when both chemists had produced five distinct validation samples using the final library production conditions.

The initial step was to approach the literature and identify previous syntheses of the scaffold shown in Scheme 8.2. The literature survey provided two routes to the desired intermediates[16]; however, no single publication had generated the desired final compounds, so some level of reaction evaluation and optimization would be required for the synthesis.

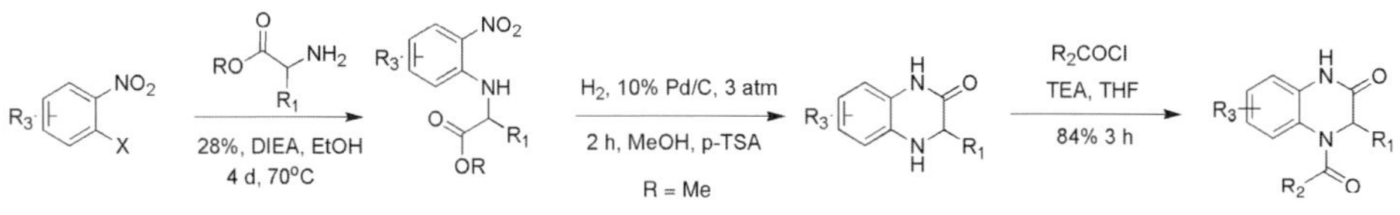

Scheme 8.2

8.3.2. The thermal approach

After reviewing the literature, the scientist set out on optimization of the first key step, a S_NAr reaction. Typical S_NAr reaction conditions were initially attempted, examples of which can be seen in Table 8.4. Typical reflux heating in solvents such as DMF and DMSO provided poor yields of the desired products even after prolonged heating. Only with the use of a pressurized vessel were the satisfactory levels of the desired product obtained. However, it should be pointed out that under typical library production conditions, sealed tubes could not be used to produce the desired compounds, and in all actuality the use of refluxing DMSO would also be problematic. After examining

Table 8.4 Thermal route synthetic Step 1

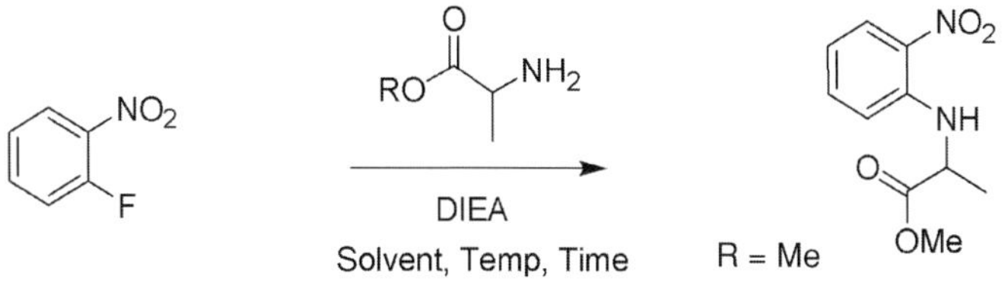

Solvent	Temp (°C)	Time (h)	Yield (%)
DCE	80	48	3
DMF	110	48	0[a]
DMA	120	48	9
DMA	120	72	14
NMP	135	72	15
DMSO	150	24	45[b]
DMSO (sealed tube)	170	24	72

[a] Only product-like material observed was 15% of the carboxylic acid.
[b] An additional 17% was observed as the carboxylic acid.

the possible methods to generate this intermediate, it was determined that the sealed vessel was the only method for adequate compound production. The optimization of this stage required 15 working days from concept through to optimization.

The second step was the reduction of the nitro group and subsequent cyclization to form the quinazalone core. Reduction under standard conditions using tin(II) chloride provided only a small amount of the desired product (13%), which was contaminated with a large amount of the uncyclized intermediate (see Table 8.5). It was possible to cyclize the material with prolonged heating and the use of a catalytic amount of protic acid. Other standard reduction conditions, such as dithionite and transfer hydrogenation reductions provided none of the desired product. Reduction using hydrogen at elevated temperatures provided the desired product in good (56%) yield after 24 h. However, it should once again be pointed out that the use of hydrogen gas in a block-based synthesis can be a potential safety issue and heating the reaction to 40°C is considered to increase the danger of this reaction. These safety issues were a concern during optimization of the nonmicrowave pathway. Total time for optimization of this reaction step was 10 days.

The last reaction step was the acylation of the 4-quinazalone nitrogen. There was little literature precedent for this reaction step; however, our group was very experienced in the field since we had performed many acylation reactions over the years. Initial efforts using triethylamine and THF/DMF had afforded poor yields of the desired products (see Table 8.6). Performing the reactions in neat pyridine slightly increased the yields, and the use of DMAP at 80°C in DMF produced the best yields obtained to date (a mere 14%). Additional reaction conditions were attempted with no increase in reaction yields. Total time for optimization of this reaction step was 12 days with only a 14% optimized yield.

Overall the total time spent on the nonmicrowave method was 37 working days with an overall yield of 4%. At this point, the chemist had determined that a synthetic route for this scaffold could not be completed that would be amenable to parallel synthesis and had concluded that further optimization of the methodology would not be successful in producing a library of compounds.

Table 8.5 Thermal route synthetic Step 2

Conditions	Temp (°C)	Time (h)	Yield (%)
SnCl$_2$, DMF	60	48	13[a]
Na$_2$S$_2$O$_4$, AcCN	60	48	0
HCO$_2$NH$_4$, Pd/C	80	48	0
H$_2$, Pd/C, 1 atm	25	24	47
H$_2$, Pd/C, 1 atm	40	24	56

[a] Majority of the material was the uncyclised aminoester.

Table 8.6 Thermal route synthetic Step 3

Conditions	Temp (°C)	Time (h)	Yield (%)
TEA, THF	25	48	3
TEA, DMF	25	48	7
Pyridine	60	48	12
DMAP, DMF	25	48	5
DMAP, DMF	80	48	14

8.3.3. *The microwave approach*

Let us turn our attention to the scientist having access to microwave technology. It should be illustrated that this scientist was not restricted to only using the microwave for synthesis, but could use any technology available, including microwave synthesis. A striking similarity in the synthetic route is immediately obvious: both scientist had identified the same references and had chosen the same synthetic routes independent of each other. This fact helps to illustrate the timesavings directly associated with microwave technology.

Once again the scientist chose to use the $S_N Ar$ reaction to gain access to the key *o*-nitroaniline intermediate; however, it is quite apparent from Table 8.7 that the use of microwave synthesis resulted in rapid optimization of this key reaction and significantly improved yields. The use of *n*-butanol as solvent for this substitution reaction gave high yields (88%) in only 15 min. It should also be pointed out that while in the thermal example DMSO was the optimal solvent, in the microwave example this solvent resulted in decomposed starting materials.

The second step was also similar to the thermal method. The scientist chose to reduce the nitro group by transfer hydrogenation. While this reaction resulted in poor yields and mostly uncyclized materials in the thermal approach, in the microwave example the yields were typically high with the desired cyclized intermediate predominating in the reaction mixture (see Table 8.8).

The final step is perhaps the most illustrative of the power of using microwave technology. While there were many unsuccessful attempts to acylate the nitrogen under thermal conditions, the results were quite different utilizing the microwave (see Table 8.9). Yields were typically high for a variety of acylating reagents and all reactions were completed in less than 10 min producing high yields of the desired final products.

While the scientist in the nonmicrowave approach had concluded that this library could not be generated using his synthetic strategy, the microwave approach had shown the ability to rapidly optimize the reaction conditions to produce the final product. It should also be highlighted that while the thermal approach required 37 working days

Table 8.7 Microwave route synthetic Step 1

Solvent	Equiv. AA	Base (equiv)	Time (min)	Temp (°C)	Yield (%)
DCE	1.5	DIEA(3)	5	180	25
DCE	1.5	DIEA(3)	10	180	30
DCE	1.5	DIEA(3)	15	180	33
DMF	1.5	DIEA(3)	10	180	47
NMP	1.5	DIEA(3)	10	180	63
DMSO	1.5	DIEA(3)	5	180	decomposed
AcCN	1.5	DIEA(3)	10	140	0
—	2	—	5	180	30
—	2	—	10	180	10
—	2	—	5	220	7
DCM	2	—	15	120	13
EtOH	2	—	15	150	54
i-PrOH	2	DIEA(4)	15	180	58
n-BuOH	2	DIEA(4)	15	180	88

Table 8.8 Microwave route synthetic Step 2

R_1	Time (min)	Yield (%)
H	5	100
Me	5	100
Bn	5	98
$CH_2CH(CH_3)_2$	5	97

to reach a conclusion, the scientist using the microwave-assisted approach had reached a positive endpoint to the experiment in only 2 working days (Fig. 8.2).

This example is unique in that the two groups were unaware of what methods and techniques were applied and since both groups had no previous knowledge of the methods of generating this scaffold, it provides a unique insight into the timesavings directly associated with microwave-assisted synthesis in library development and optimization.

8.4. New advances in microwave technology

The advent of microwave-assisted synthesis has allowed chemists to rapidly optimize reaction conditions for combinatorial library development[17–19]. But how can one efficiently produce libraries of compounds employing chemistry developed and optimized

Table 8.9 Microwave route synthetic Step 3

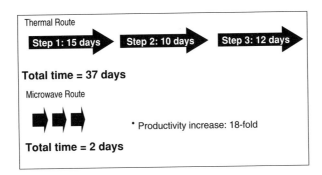

Entry	R_1	R_2	Yield (%)
1	H	p-ClPh	100
2	H	m-MeOPh	96
3	H	n-Propyl	92
4	Me	p-ClPh	100
5	Me	m-MeOPh	99
6	Me	n-Propyl	91
7	Bn	p-ClPh	87
8	Bn	m-MeOPh	89
9	Bn	n-Propyl	90

Thermal Route

Step 1: 15 days   Step 2: 10 days   Step 3: 12 days

Total time = 37 days

Microwave Route

• Productivity increase: 18-fold

Total time = 2 days

Figure 8.2 Microwave library optimization timesavings.

using microwave-assisted synthesis? A major issue with most commercially available microwave systems is the low throughput; typically up to 20 samples per hour can be processed. In an effort to increase throughput our group developed a novel 96 well plate-based system directed towards high-throughput microwave-assisted synthesis[20]. This system is a closed-vessel system allowing reactions to be performed under significant pressures and temperatures.

Similar plate-based systems have been reported in the literature[21]. The final goal of these efforts was to be able to synthesize several hundred compounds per day using microwave optimized reaction conditions.

Our initial attempts at using a plate-based systems for microwave synthesis relied upon commercially available systems. A simple plate turntable was available for the CEM Mars 5 system (CEM Corporation, Matthews, North Carolina, U.S.). Heating DMF to 130°C in the plates illustrated that the thermal stability of the plastic was an issue. The wells when heated above 90°C with solvent tended to melt and deform resulting in loss of solvent. In our investigation into higher temperature plates, we

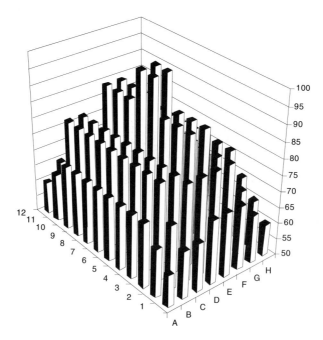

Figure 8.3   Plate heating profile for DMF, reference set to 100°C.

came across two viable alternatives: PTFE (Teflon) plates and HTPE (high-temperature polyethylene). Due to the high cost of the Teflon plates, we settled upon HTPE plates that were available from Whatman (Whatman PLC, Kent, U.K.) at a low cost and could thus be used as disposables.

One significant issue that was immediately recognized was the temperature differential across the plate in a multi-mode microwave system. Thus, the heating profile over a typical polypropylene deepwell (2 ml) microtitre plate was determined by using the available temperature measurement device (a thermocouple) inserted into a reference well located near the center of the microtitre plate. When heating DMF to 100°C, the temperature of the reference cell quickly rose to the desired set point. However, the outside of the plate never reached the desired temperature and displayed a temperature differential greater than 40°C (Fig. 8.3). This phenomenon could be illustrated further in the formation of 5-aminopyrazoles, which has been shown in our group to be a temperature-sensitive reaction (Scheme 8.3). Temperatures exceeding 120°C results in the decomposition of the starting materials and temperatures below 80°C do not

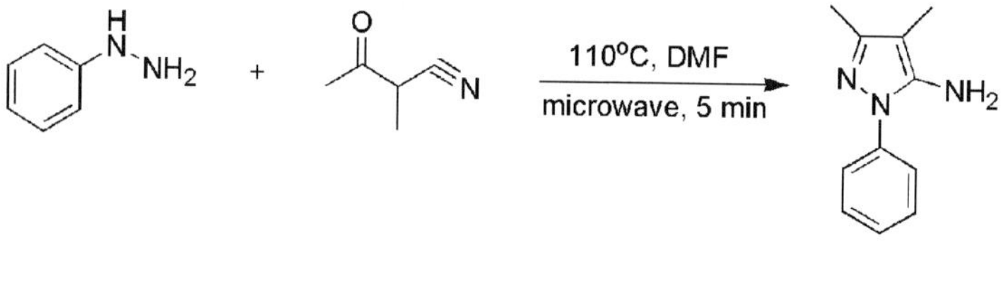

Scheme 8.3

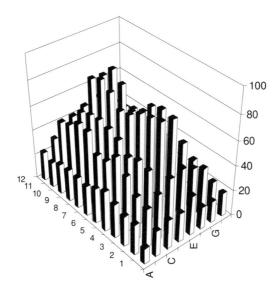

Figure 8.4   Percent purity using open microwave plate.

produce the product. Therefore, this reaction system would be an excellent probe for how the obtained temperature gradient would affect the outcome of a temperature-sensitive reaction. The plate was charged with the same stock reagents to all wells and the thermocouple was inserted into well E3. As can be seen in Fig. 8.4, the purity of the compound in the wells at both the periphery and in the middle were significantly re-duced relative to the thermocouple control well. We quickly ascertained that the lower temperatures of the exterior wells were the result of both radiative heat loss and a lower microwave coupling. The heat loss could be controlled to some extent by placing a cover over the plate to retain some of the heat. However, the low-efficiency coupling to the microwave field was a more pressing issue. The reason for the poor coupling is the lack of neighboring wells containing solvent. The surrounding wells act as a load to improve the efficiency of the microwave dielectric heating, and wells on the exterior of the mi-crotitre plate that do not have these surrounding wells thus lack the ability to efficiently couple to the field. Our solution to this problem came from literature on materials science, where graphite or carbon has been reported as efficient loads for microwave heating[22]. Placing graphite pellets in the exterior allowed for improved coupling to the microwave field and thus increased the effective temperature in these wells. Figure 8.5 illustrates the same reaction as in Fig. 8.4 performed with graphite pellets in the outer two rows and columns of the microtiter plate. The purity profile shows that not only are the yields on the periphery of the microtiter plate increased, but the yields of product in the center of the plate are also improved, due to the increased energy being applied to the plate edge.

Our next effort was to more accurately mimic the conditions that we were using for library optimization in the single-mode microwave, namely pressurized reaction vessels. It has been shown that the key to higher yields in microwave-assisted synthesis is not solely the direct heating but also the pressure effect—the ability to run reactions at higher

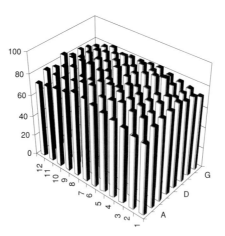

Figure 8.5   Purity profile of microwave plate with graphite added.

pressures and accordingly higher temperatures. A simple approach was first attempted in using capmats to seal the plates; however, the capmats were not as robust as the plates and tended to melt into the reaction vessels. We then used Teflon-lined capmats (Arctic White Corporation, Bethlehem, PA, USA), which allowed for higher temperatures, but keeping the plates sealed under pressure required a mechanical clamping system. Seeing how the entire plate-based system would have to be placed into the microwave field, all material used would have to be microwave transparent. For this purpose no commercial alternatives were found and our efforts resulted in a pressurized-plate-based system composed of in-house generated components. The microwave plates were capped with a Teflon-lined capmat and then held in place with a HTPE lid (Fig. 8.6). The

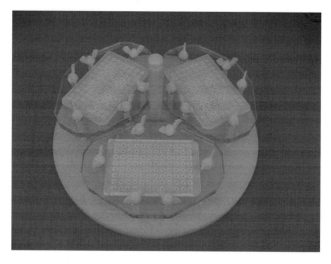

Figure 8.6   Sealed microwave pressure plates.

Table 8.10 Library validation with plate-based microwave reactor

ratio: 1/1.2

R$_1$	R$_2$	R$_3$	Purity[a] (%)
Ph	2-Furyl	4-Et-Ph	90
Me	3-Me-Ph	Et$_2$OCC$_2$H$_4$	100
2-Pyridyl	3-Cl-Ph	i-Pr	100
2-CF$_3$-Ph	t-Bu	Cyclohexyl	93
2,5-di-Ci-Ph	Ph	2-Thiophenyl-CH$_2$	98

[a] By ELSD.

lid was anchored into position by four nylon-threaded bolts capped with nylon wing-nuts. A thermocouple was introduced into the plates *via* a pressurized glass needle; this needle would seat under the HTPE plate and allow the thermocouple to accurately read the temperature of the reference vessel. To date, this system has been pressure rated to 10 bar and over 100 plate-based reactions have been performed without failures. A typical example of the types of chemistry performed in the plate-based system can be seen in Table 8.10. This simple acylation reaction can now be easily performed on several hundred compounds per day. The results in Table 8.10 highlight that the reactions performed in the plate are comparable in yield and purity to those found during optimization on the single-mode microwave system.

8.5. References

1. Lew, A., Krutzik, P.O., Hart, M.E. and Chamberlin, A.R., Increasing rates of reaction: microwave-assisted organic synthesis for combinatorial chemistry, *J. Combi. Chem.*, 2002, **4**(2), 95–105.
2. Frere, S., Thiery, V. and Besson, T., Microwave acceleration of the Pechmann reaction on graphite/montmorillonite K10: application to the preparation of 4-substituted 7-aminocoumarins, *Tetrahedron Lett.*, 2001, **42**(15), 2791–2794.
3. Sridar, V., Rate acceleration of Fischer-indole cyclization by microwave irradiation, *Indian J. Chem., Sect. B*, 1996, **35B**(7), 737–738.
4. Grigg, R., Martin, W., Morris, J. and Sridharan, V., Synthesis of Δ3-pyrrolines and Δ3-tetrahydropyridines via microwave-accelerated ring-closing metathesis, *Tetrahedron Lett.*, 2003, **44**(26), 4899–4901.
5. Frattini, S., Quai, M. and Cereda, E., Kinetic study of microwave-assisted Wittig reaction of stabilised ylides with aromatic aldehydes, *Tetrahedron Lett.*, 2001, **42**(39), 6827–6829.
6. Stadler, A. and Kappe, C.O., Solid phase coupling of benzoic acid to Wang resin: a comparison of thermal versus microwave heating. In: *Proceedings of ECSOC-3*, Sept. 1–30, 1999. Also in Proceedings of ECSOC-4, Sept. 1–30, 2000, pp. 1234–1239.
7. Stadler, A. and Kappe, C.O., Microwave-mediated Biginelli reactions revisited. On the nature of rate and yield enhancements, *J. Chem Soc., Perkin Trans. 2*, 2000, (7), 1363–1368.
8. Raner, K.D., Strauss, C.R., Vyskoc, F. and Mokbel, L., A comparison of reaction kinetics observed under microwave irradiation and conventional heating, *J. Org. Chem.*, 1993, **58**(4), 950–953.
9. Calvert, S., Stewart, F.P., Swarna, K. and Wiseman, J.S., The use of informatics and automation to remove bottlenecks in drug discovery, *Curr. Opin. Drug Discovery Develop.*, 1999, **2**(3), 234–238.
10. Gedye, R.N., Smith, F.E. and Westaway, K.C., The rapid synthesis of organic compounds in microwave ovens, *Can. J. Chem.*, 1988, **66**(1), 17–26.

11. Giguere, R.J., Bray, T.L., Duncan, S.M. and Majetich, G., Application of commercial microwave ovens to organic synthesis, *Tetrahedron Lett.*, 1986, **27**(41), 4945–4948.
12. Gedye, R., Smith, F., Westaway, K., Ali, H., Baldisera, L., Laberge, L. and Rousell, J., The use of microwave ovens for rapid organic synthesis, *Tetrahedron Lett.*, 1986, **27**(3), 279–282.
13. Wilson, N S., Sarko, C.R. and Roth, G.P., Microwave-assisted synthesis of a [3+2] cycloaddition library, *Tetrahedron Lett.*, 2001, **42**(51), 8939–8941.
14. Wilson, N.S., Sarko, C.R. and Roth, G.P., Microwave-assisted synthesis of 2-aminoquinolines, *Tetrahedron Lett.*, 2002, **43**(4), 581–583.
15. Sarko, C.R., *Microwave-Accelerated Combinatorial Library Design and Development*, ACS Prospectives: Proceedings in Combinatorial Chemistry 2002, Leesburg, VA.
16. TenBrink, R.E., Im, W.B., Sethy, V.H., Tang, A.H. and Carter, D.B., Antagonist, partial agonist, and full agonist imidazo[1,5-a]quinoxaline amides and carbamates acting through the GABAA/benzodiazepine receptor, *J. Med. Chem.*, 1994, **37**(6), 758–768.
17. Giacomelli, G., Porcheddu, A., Salaris, M. and Taddei, M., Microwave-assisted solution-phase synthesis of 1,4,5-trisubstituted pyrazoles, *Eur. J. Org. Chem.*, 2003, (3), 537–541.
18. Westman, J., Lundin, R., Stalberg, J., Ostbye, M., Franzen, A. and Hurynowicz, A., Alkylaminopropenones and alkylamino-propenoates as efficient and versatile synthons in microwave-assisted combinatorial synthesis, *Comb. Chem. High Throughput Screen.*, 2002, **5**(7), 565–570.
19. Kappe, C.O., High-speed combinatorial synthesis utilizing microwave irradiation, *Curr. Opin. Chem. Bio.*, 2002, **6**(3), 314–320.
20. Lord. J., Routberg, M. and Sarko, C., High-throughput Microwave-Accelerated Combinatorial Chemistry: A Plate-Based Approach, *J. Combi. Chem.* (submitted).
21. Coleman, C.M., MacElroy, J.M.D., Gallagher, J.F. and O'Shea, D.l F., Microwave parallel library generation: comparison of a conventional- and microwave-generated substituted 4(5)-sulfanyl-1H-imidazole library, *J. Combi. Chem.*, 2002, **4**(1), 87–93.
22. Gorshenev, V.N., Bibikov, S.B. and Spector, V.N., Simulation, synthesis and investigation of microwave absorbing composite materials, *Synthetic Metals*, 1997, **86**(1–3), 2255–2256.

9 Scale-up of microwave-assisted organic synthesis

BRETT A. ROBERTS and CHRISTOPHER R. STRAUSS

9.1. Introduction

In 1986 it was first reported that organic reactions could be conducted by heating in sealed containers in domestic microwave ovens[1,2]. Rate enhancements of up to three orders of magnitude were disclosed[3]. However, temperature and pressure measurement were technically difficult to achieve and in some instances the vessels deformed or exploded[1–3].

From these publications, workers interested in exploring the microwave technique perceived it to be simultaneously beneficial through increased rates, yet hazardous in the presence of flammable organic solvents. Subsequently, a vast body of work was carried out with domestic microwave ovens, but under solvent-free conditions and without recourse to sample mixing or temperature measurement. This continued across a broadening front on the laboratory scale. These and other developments in microwave chemistry have been reviewed extensively in journals, book chapters[4–20] and in a recent monograph[21].

Although relatively inexpensive in terms of purchase price, domestic microwave ovens are not designed to contain chemical explosions or toxic fumes and are incompatible with corrosive and inflammable compounds[13]. Flux densities within the microwave cavity can vary considerably, resulting in 'hot' and 'cold' spots. Mode stirrers and circulation of the vessel within the cavity are sometimes incorporated to assist in overcoming the non-uniformity of the energy distribution, but such measures are usually inadequate. Domestic ovens operate on duty cycles and intervals between zero and full power can be many seconds. This is not always ideal for heating common foods and beverages, let alone for performing delicate chemical manipulations. Intermittent bursts of power afford poor temperature control in chemical reactions. When these matters are considered in conjunction with the initial lack of dedicated pressure vessels for microwave-assisted organic chemistry, it is not surprising that explosions occurred[1–3] and that reactions are not necessarily reproducible in domestic microwave systems[6].

During the decade after 1986, disadvantages of domestic microwave ovens for organic chemistry became apparent to many researchers. Because of the unavailability of specifically designed commercial equipment, however, workers with volatile organic solvents, still found it necessary to adapt domestic units for reflux conditions or for operations under pressure. For reactions at reflux[22–30], domestic microwave ovens were modified by fitting with a shielded opening to prevent microwave leakage and through which the reaction vessel within the microwave cavity could be connected to an external condenser[22,23,30]. Commercial systems adopting this approach have since appeared. When microwave-transparent coolants (chilled hydrocarbons, ice or solid carbon dioxide) are used, along with the reaction vessel, the condenser can also be located within the microwave cavity, thereby avoiding potential problems with microwave leakage[26,29,31].

The first reported examples of microwave-assisted organic synthesis involved pressurised conditions[1-3]. Giguere and his collaborators[2] used small sealed tubes or screw-cap vials, which were placed in a microwave-transparent casing packed with vermiculite to absorb the contents in the event of an explosion. Heating times of up to 15 min were employed with a domestic microwave oven operating on full power. Because of the technical difficulties involved with direct measurement, reaction temperatures were estimated retrospectively. As the reactants and packing absorbed microwave energy, heat was transferred to the reaction mixture essentially by a combination of direct microwave and normal conduction heating. Giguere's approach was later adopted for transformations of long chain fatty acid esters[32-34], racemisation[35], intramolecular Diels–Alder reactions[36,37], decarboxylation[38], reactions of ethers[39], Ferrier rearrangement[40], Baylis–Hillman reaction[41] and for isoflavone synthesis[42].

In their pioneering studies, Gedye and his collaborators[1,3] used a domestic oven, and commercially available screw cap pressure vessels made from either polytetrafluoroethylene (PTFE) or perfluoroalkoxy (PFA) Teflon (registered trade name of DuPont). The pressure was measured and the final temperature was estimated with an infrared sensor directed at the vessel immediately after the heating was completed. For safety, maxima for solvent volumes, heating times and power settings were stipulated. By 1995, Gedye's methodology[1,3] had been applied to organometallic chemistry[43-46], as well as to Fischer cyclisations[47], additions of halocompounds to styrene[48] and Tipson–Cohen reactions[49].

In the early 1970s, more than a decade before the advent of microwave-assisted organic chemistry, microwave heating with pressure vessels had been introduced into analytical laboratories to speed the rate of digestion and dissolution of solid samples, such as ores, hair and foodstuffs[50]. Digestion typically involves treatment of a small sample with an excess of a strong oxidant, usually an acid such asnitric, perchloric or sulphuric. The aim is to degrade a material to produce a clear solution, usually for quantitative analysis without charring or physical loss of elements. Although it is essential that digestions are standardised to obtain precise results, process parameters such as temperature, reaction time, sample stirring and cooling rate are not essential for reproducible outcomes and typically they are not actively controlled[50].

As technologies for microwave-assisted digestion gained favour over slower, traditional heating methods, equipment manufacturers emerged in the 1970s and 1980s to satisfy the market for these analytical applications. In the 1990s, some attempted to adapt their digestion technologies for synthesis, but unfortunately, without fully appreciating the essential differences between the applications. In contrast with digestion, synthesis is a constructive process and apart from aspects of combinatorial chemistry involving parallel transformations, it is usually product specific. Reactants that are not particularly stable to strong acids and bases may be required and in much larger amounts than those used for digestion. They may be expensive, toxic and highly reactive, even on the laboratory scale. The ability to define and control conditions, including temperature, time, sample stirring, addition or withdrawal of materials and post-reaction cooling, frequently is vital for consistent, satisfactory outcomes and reproducibility. It is not always essential, however, as Majetich and Hicks[51] demonstrated through the performance of 45 different reactions with a system that did not possess these facilities. Their preparations were conducted on the gram scale in organic solvents, while a barostat indirectly maintained the temperature below 200°C. It could be argued that such examples retarded the commercial development of more appropriate equipment that could carry the field forward faster. Equipment manufacturers and chemists alike now appreciate that safe, dedicated reactors are required for microwave-assisted organic chemistry. Such systems have been available on the laboratory scale since the late 1990s.

With speed and convenience as major attractions, little wonder is it that the rate of increase of the field (based on number of publications) continues to grow exponentially. Well over 1000 papers have been published on microwave-assisted organic chemistry and they probably disclose more than 10 000 examples of microwave synthesis. Few of these procedures, however, have been further developed into large-scale industrial processes. In that context, it is important to appreciate that synthetic chemists merely desire defined reaction conditions, but process chemists and chemical engineers require and demand them. Thus, consideration of scale-up of the microwave technique and the associated issues is opportune. This requires an appreciation of key factors including the mechanisms of microwave heating, advantages and disadvantages, as well as approaches to microwave chemistry. This chapter summarises these aspects.

9.2. Mechanisms and effects of microwave heating

With conventional heating, energy transfer occurs mainly through conduction and convection. With microwaves, the primary mechanism is dielectric loss[4,52]. The dielectric loss factor (loss factor, ε'') and the dielectric constant (ε') of a material are two determinants of the efficiency of heat transfer to the sample. Their quotient ($\varepsilon''/\varepsilon'$) is the dissipation factor (tan δ), high values of which indicate ready susceptibility to microwave energy.

Materials dissipate microwave energy mainly by dipole rotation and ionic conduction. Dipole rotation refers to the alignment of molecules that have permanent or induced dipoles, with the electric field component of the radiation. At 2450 MHz (2.45 GHz), the most commonly used frequency (see later), the field oscillates 4.9×10^9 times per second, and sympathetic agitation of the molecules generates heat. The efficacy of heat production through dipole rotation depends upon the characteristic dielectric relaxation time of the sample, which in turn is dependent on temperature and viscosity. Ionic conduction occurs through migration of dissolved ions with the oscillating electric field. In this case, heat generation is due to frictional losses that depend on the size, charge and conductivity of the ions, and their interactions with the solvent.

Provided that there is adequate mixing, temperature gradients within the sample will be minimised by microwave bulk heating and the extent of conduction and convection will be low. The differences in means of energy transfer between conventional and microwave heating have numerous, diverse implications for the performance of thermal chemical reactions. They are reflected in the methodologies, monitoring, types of equipment and vessels employed. Significant influences in microwave heating are introduced by the samples themselves. The dimensions, volume, shape, composition, physical and chemical properties can be important as can the reactions being performed, the media employed and a range of safety issues. Many of these factors become increasingly important with scale.

Owing to the potential for disruption to communications and radar bands, the entire microwave spectrum is not readily available for heating applications[52]. By international convention, the frequencies 915 ± 25 MHz, 2450 ± 13 MHz, 5800 ± 75 MHz and 22125 ± 125 MHz have been allocated for industrial and scientific microwave heating and drying. For synthetic chemical applications, equipment operating at 2450 MHz, corresponding to a wavelength of 12.2 cm, is most commonly available.

The depth of penetration of electromagnetic radiation[52] increases with larger wavelengths (or decreasing frequencies) and can vary with temperature. With pure water, the depth of penetration at 2450 MHz increases from 6 mm at 3°C to about 40 mm at 60°C[52], indicating that the capacity to absorb microwaves bears an inverse relationship with temperature. This trend holds for a range of organic solvents, suggesting that variations in microwave susceptibility occur concomitantly when the dielectric properties of a specific medium alter with temperature changes[13]. This variability can be attenuated (in either direction) if, as reactions proceed, the products have dielectric properties that differ substantially from those of the starting materials.

Potentially sudden and sometimes unpredictable factors resulting from changes in the physical properties of the samples, including the influence of size alone, are major considerations when scaling up reactions. Although high rates of microwave heating are usually advantageous, a difficulty can arise when the dielectric loss exhibited by a material increases with temperature. In such circumstances, the material will absorb microwave energy with increasing efficiency at higher temperatures. The rate of temperature rise will also increase with temperature and a thermal runaway may ensue. This phenomenon can offer considerable benefit in some circumstances, particularly for selective heating of catalysts. In most cases, however, microwave-driven thermal runaways need to be prevented. This problem can be avoided in most cases by careful monitoring and control of temperature and power input[13].

Conversely, since solvents show a general decrease in dielectric constant with temperature, efficiency of microwave absorption will diminish with temperature rise and can lead to poor balancing ('matching') of the input and absorbed microwave energy. This becomes marked as liquids approach the supercritical fluid state. Solvents and reaction temperatures should be chosen with these considerations in mind, particularly as excess input microwave energy can lead to the build up of electric charge within the sample, followed by discharge through arcing.

Differences in sample size and composition can also affect heating rates. In the latter case, this particularly applies when ionic conduction becomes possible through the addition or formation of salts. For compounds of low-molecular weight, the dielectric loss contributed by dipole rotation decreases with rising temperature, but that due to ionic conduction increases. Therefore, as an ionic sample is microwave irradiated, the heating results predominantly from dielectric loss by dipole rotation initially, but the contribution from ionic conduction becomes more significant with temperature rise.

As a mixture gains complexity owing to the conversion of starting materials during a reaction, an increasing tendency for microwave absorption would be expected, unless polar components (e.g., ethanol and formic acid) were undergoing condensation to form a considerably less polar product (ethyl formate). In that event, the reaction temperature may not be sustainable without increasing the power, and arcing could occur within the sample unless the microwave load is matched.

Several workers have claimed that under the influence of microwaves, some reactions proceed faster than under conventional conditions at the same temperature because of various non-thermal 'microwave effects'[48,53–56]. Other investigators have rejected the theory of specific activation at a controlled temperature in homogeneous media[57–62]. A study by Stadler et al.[63] on the rate enhancements observed in solid-phase reactions revealed that the significant rate enhancements were a result of direct, rapid 'in-core' heating of the solvent by microwave energy and not a specific non-thermal 'microwave effect'. The existence or otherwise of non-thermal 'microwave effects' continues to be a source of great debate and if proven would have serious potential consequences for scale-up, particularly if such effects were unpredictable.

Although some of the speculation about non-thermal 'microwave effects' appears to emanate from a misconception that microwave radiation can excite rotational transitions, the frequencies at which these occur are much higher than 2.45 GHz. For example, the first absorption lines of OCS, CO, HF and MeF occur at 12.2, 115, 1230 and 51 GHz, respectively. Internal bond rotations (torsional vibrations) also require higher frequencies, in the order of 100–400 cm^{-1} or 3000–12 000 GHz, for excitation[61,62].

These values indicate that when a compound absorbs microwaves at 2.45 GHz, the radiation does not directly excite the molecule to higher vibrational or rotational energy levels and that dielectric heating accounts for the temperature rise. Regardless of the heating method, the internal energy of the material will be increased and partitioned among translational, vibrational and rotational energy, and it follows that there should be no kinetic differences between microwave-irradiated and conventionally heated reactions when the temperature is known and the reaction solution is thermally homogeneous. Kinetics data obtained for several reactions, including some for which previous workers had claimed a non-thermal 'microwave effect,' were in agreement with this conclusion in all cases[62].

From the foregoing discussion, the propensity of a sample to undergo microwave heating is related to its dielectric and physical properties. Compounds with high dielectric constants (e.g. ethanol and dimethylformamide) tend to absorb microwave irradiation readily, while less polar substances (such as aromatic and aliphatic hydrocarbons) or compounds with no nett dipole moment (e.g. carbon dioxide, dioxane and tetrachloromethane) and highly ordered crystalline materials are poorly absorbing. In the case of H_2O, the solid form, ice, has dielectric properties (ε', 3.2; tan δ, 0.0009; ε'', 0.0029)[64] that render it essentially microwave transparent and that differ significantly from those of liquid water at 25°C (ε', 78; tan δ, 0.16; ε'', 12.48). Such properties would facilitate thermal microwave reactions in liquid water (which absorbs microwaves readily) or by microwave heating of a reaction in a vessel made of or encased in ice. Bose et al.[55] carried out a reaction under the latter conditions. Although the workers concluded that the chemistry was 'not due to thermolysis', the result was later reinterpreted[13].

In another example of differential heating, a two-phase water/chloroform system (1:1 by volume) was heated in a microwave batch reactor (MBR)[65]. About 40 s after commencement, the temperatures of the aqueous and organic phases were 105 and 48°C, respectively, because of the differences in the dielectric properties of each solvent. A sizeable differential could be maintained for several minutes before cooling was begun. Differential heating is particularly advantageous for Hofmann eliminations. In a typical example, a mixture of N-(4'-ethoxy-2-benzoylethyl)-N,N,N,-trimethylammonium iodide, water and chloroform was heated with stirring for 1 min at 110°C (temperature of aqueous phase). During the reaction the product, 4'-ethoxyphenyl vinyl ketone, was extracted and diluted into the poorly microwave absorbing, cooler, organic phase. The ketonic monomer was obtained in 97% yield, a result not achieved by conventional heating of the starting material under vacuum[65].

Differential heating has also been used to good effect for large-scale preparations in which a catalyst is held at high temperature, while cooler organic substrates are passed over it. An interesting application involved passage of CO_2 and water over a microwave-heated nickel catalyst to afford low-molecular weight hydrocarbons, alcohols and acetone[66]. The DuPont HCN industrial process constitutes another excellent example, as it involves the condensation of methane and ammonia over alumina, maintained at 1200°C by microwave heating (Scheme 9.1)[67–69].

$$CH_4 + NH_3 \xrightarrow[\text{Catalyst}]{1200°C} HCN + 3H_2$$

Scheme 9.1   The DuPont HCN industrial process[68].

Although non-thermal 'microwave effects' were not involved in the above examples of differential heating, comparable conditions would have been difficult, if not well nigh impossible to obtain by traditional heating methods. Cundy has suggested that differential heating also could account for outcomes of some chemical reactions involving the adsorption of organic starting materials onto inorganic supports such as clays, alumina and silica gel[11].

Superheating also can account for acceleration of reactions under microwave conditions[70,71]. Mingos has estimated that this can lead to 10–50-fold reductions in reaction times in comparison with conventional reflux conditions, but that rate enhancements of 100–1000-fold at atmospheric pressure would be required before specific 'microwave effects' could be invoked[72].

9.3. Approaches to microwave-assisted organic chemistry

Microwaves offer general advantages, some that are more specific and others that are highly specific for particular methods of synthetic chemistry[14]. The main general advantages include the following:

(1)   Microwave energy can be introduced remotely, without contact between the source and the chemicals.
(2)   Energy input to the sample starts and stops immediately when the power is turned on or off.
(3)   Heating rates are higher than can be achieved conventionally, if at least one of the components can couple strongly with microwaves.

A major difference among the respective methods for microwave-assisted organic chemistry is the presence or absence of solvents. Solvent-free conditions have been pursued using neat reactants only, reactants adsorbed onto solid supports or reactants in the presence of phase-transfer catalysts. Methods employing solvents use either pressurised systems or open vessels for reflux and superheating. Although these are presented in a little more detail in the following subsections, the reader is suggested to consult comprehensive reviews dealing with these aspects[7–21].

9.3.1. *Solvent-free methods*

Applications under solvent-free conditions have been the subject of the greatest activity and have been reviewed extensively[7,10,16,17]. Besides apparent potential benefits in minimising solvent usage, reactions can be conducted conveniently and rapidly, without temperature measurement in domestic microwave ovens. A dedicated commercial reactor for applications at atmospheric pressure has also been used[7].

The simplest solvent-free method involves irradiation of neat reactants in an open container. In the absence of reagents or supports, the scope for such processes appears to be limited to relatively straightforward condensations that can be conducted without added catalysts, or to intramolecular thermolytic processes such as rearrangement or elimination.

As mentioned earlier, outcomes can be highly dependent upon sample size. Small samples tend to have larger surface to volume ratios than their larger counterparts. Thus they can suffer relatively greater heat losses from the surface. In extreme cases, the losses of thermal energy can be comparable with the microwave energy absorbed and the sample does not heat up. Increasing the sample size in such circumstances can lead to dramatic increases in the amount of energy retained by the sample. In an early literature example, a small sample (about 1 g) of starch appeared to be unaffected by microwave irradiation, but a larger sample (about 10 g) heated readily and rapidly decomposed[73].

Loupy *et al.* have argued that if the mixture of neat starting materials is heterogeneous (comprising a solid and a liquid) reaction could occur by dissolution of the solid in the liquid or by adsorption of the liquid onto the solid and in either case a diluting solvent would slow the reaction[7].

For solvent-free 'dry media' reactions, the organic reactants are adsorbed onto acidic or basic supports such as alumina, silica, bentonite, montmorillonite K10 or KSF clays and zeolites and subjected to irradiation, often in domestic microwave ovens without temperature measurement. The supports also can be doped with inorganic reagents. If a strong base is required, KF on alumina can ionise carbon-containing acids up to pK_a 35, while clays like montmorillonite K10 offer acidities comparable with those of nitric or sulphuric acids. As there is no solvent present during the microwave irradiation, safety concerns with hazards involving fire and explosion are minimised and reactions can be performed in open vessels. 'Dry media' seem well suited to transformations involving a single organic species, for example, as in deprotection, rearrangement, oxidation and dehydration. Condensations including alkylation of carboxylates and acetalisation have also been reported.

Solid–liquid solvent-free phase-transfer catalysis (PTC) is specific for anionic reactions including base-catalysed isomerisation[7]. Usually, a catalyst (typically a tetraalkylammonium salt or a cationic complexing agent) is added to an equimolar mixture of an electrophile and a nucleophile, one of which serves as both a reactant and the organic phase. A support of polyethylene glycol (MW 3400) has also been employed as the organic phase[74].

9.3.2. Scale-up of solvent-free methods

The former French company Prolabo developed two microwave systems for synthesis[7]. The machines were employed in several research laboratories mainly for solvent-free organic chemistry. They had monomodal rectangular waveguide sections that also served as microwave cavities. Cylindrical tubes could be inserted and rotated to increase thermal homogeneity and if required condensers could be fitted. Temperature measurement was by infrared pyrometry. Computer control enabled reaction monitoring with respect to temperature or power.

Loupy et al.[7] demonstrated that the smaller of the two microwave systems could process sample sizes of 30–40 g or 70 ml. For scale-up of microwave assisted 'dry media' organic chemistry, a larger reactor was developed. Examples of reactions conducted on this instrument, some of which are presented in Table 9.1, included esterification[75],

Table 9.1 Examples of scaled-up of solvent-free organic chemistry[79]

$$CH_3COOK \quad + nC_8H_{17}Br \quad \xrightarrow{PTC} \quad CH_3COOnC_8H_{17} \quad (1)$$

a   b

Scale-up factor	a, b (mol)	Final temperature (°C)	Yield (%)
1	0.05, 0.05	172	98
40	2.0, 2.0	174	98

(2)

Scale-up factor	a, b (mmol)	Final temperature (°C)	Yield (%)
1	6.0, 5.0	140	92
90	540, 450	140	89

(3)

Table 9.1 *Continued*

Scale-up factor	a, b, c (mmol)	Final temperature (°C)	Yield (%)
1	5.0, 10, 1	120	90
49	245, 490, 49	140	82

b Ac$_2$O

c ZnCl$_2$ (4)

Scale-up factor	a, b, c (mmol)	Final temperature (°C)	Yield (%)
1	11.1, 100, 1.47	125	98
18	200, 1800, 26	125	86

b ZnCl$_2$

c $^nC_{10}H_{21}OH$ (5)

Scale-up factor	a, b, c (mmol)	Final temperature (°C)	Yield (%)
1	5.0, 5.0, 7.5	110	74
50	250, 250, 375	110	72

b KOH / Al$_2$O$_3$ (6)

Scale-up factor	a, b (mmol)	Final temperature (°C)	Yield (%)
1	2.24, 11.2	100	96
50	112, 560	100	85

b KOH/ Al$_2$O$_3$ (7)

Scale-up factor	a, b (mmol)	Final temperature (°C)	Yield (%)
1	2.3, 9.2	100	99
20	46, 184	100	95

synthesis of dioxolanes, dithiolanes, oxathiolanes[76], carbonitriles, thiocarbamates[77], dithiazolium compounds[78], alkylation, acetylation, glycosylation and dealkylation[79].

The examples in Table 9.1 demonstrate scale-up by two orders of magnitude for solvent-free reactions. Hundreds of grams of product were obtained on the larger scale, with little difference in the final percentage yield or reaction conditions. The phenyla-cylation of 1,2,4-triazole (entry 2) was not only amenable to scale-up by a factor of 90, but enhanced selectivity to the $N - 1$ product occurred under microwave irradiation. Classical heating affords a mixture of $N - 1$ and $N - 4$ alkylated products as well as quaternary salts[80].

Esveld et al.[81,82] developed a continuous dry media reactor (CDMR) for pilot-scale applications. It consisted of a multi-modal tunnel microwave cavity operating at a frequency of 2.45 GHz with a power range from 0 to 6 kW irradiated on a surface of 0.6 m^2. Temperatures of up to 250°C were achieved. A web conveyor travelling at 17 cm min^{-1} transported the solid-phase reaction mixture to the oven in low, open Pyrex supports closely packed on a polytetrafluoroethylene (PTFE)-coated glass fibre. An open flat bed process was employed to facilitate easy evaporation.

Solvent-free reactions were taken to pilot-scale without any need to alter the major conditions. In 1 day the workers produced 100 kg of a wax ester in 95% purity[82]. They concluded that integration of microwave heating with chemical processing did not demand the development of new technology, but that the chemical and microwave engineering requirements needed to be met.

Recently, Deetlefs and Seddon[83] reported the solvent-free synthesis and scale-up of ionic liquids under microwave irradiation. Using a commercial microwave reactor they prepared ionic liquids based upon the 1-alkylpyridinium, 1-alkyl-3-methylimida-zolium, 1-alkyl-2-methylpyrazolium and 3-alkyl-4-methylthiazolium cations, on scales from 50 mmol to 2 mol (Scheme 9.2). Under microwave irradiation, because of efficient

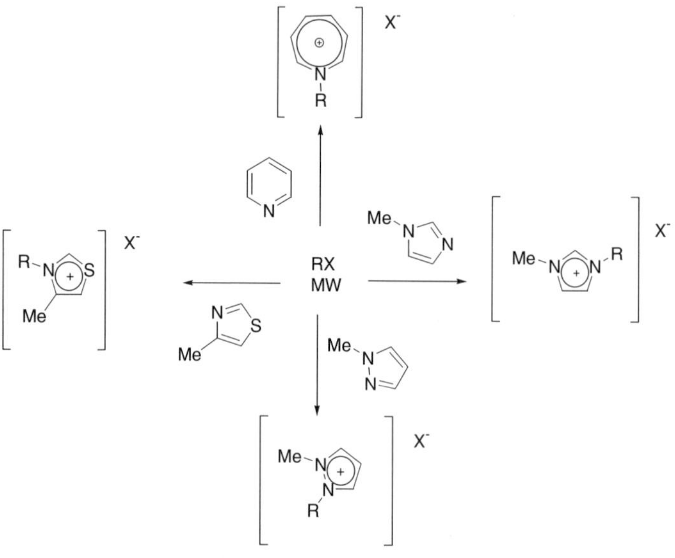

Scheme 9.2   Examples of solvent free, microwave-assisted synthesis of ionic liquids[83].

energy absorption through ionic conduction, the reaction time was drastically reduced, organic waste was minimised and the ionic liquid was synthesised in high purity and yield. These aspects combined to afford a much more environmentally friendly process.

In the area of drug discovery, the requisite scale-up may be relatively low, from a few milligrams to several grams. In these circumstances, large cavity microwave ovens are not required and smaller, high throughput machines can represent a convenient means for scaling up. Parallel synthesis (i.e. the same reaction performed simultaneously in many wells of a multiwell plate) and automation involving several consecutive replicate syntheses to produce greater quantities of the same material can represent sound alternative approaches to address this issue. In addition, scale-out can afford increased throughputs. In this case it would involve the use of multiple small instruments for the one reaction on the same small scale, rather than a single larger instrument for the same reaction on a greater scale. Thus, automated microwave operated reactors for synthesis on the gram scale or below can be employed in a number of different ways to scale up.

9.3.3. *Advantages and disadvantages of solvent-free methods*

The following benefits have been reported for microwave-heated reactions under solvent-free conditions[7,10]:

(1) Avoidance of large volumes of solvent reduces emissions and the need for redistillation.
(2) Work-up is simple, by extraction, distillation or sublimation.
(3) Recyclable solid supports can be used instead of polluting mineral acids and oxidants.
(4) The absence of solvent facilitates scale-up.
(5) Safety is enhanced by reducing risks of overpressure and explosions.
(6) Reactions are quite often cleaner, faster and higher yielding than conventional synthesis.
(7) Synthesis on a scale of multi-hundreds of grams has been achieved.

Deficiencies include a low ratio of organic reactants to solid support and typically a lack of facilities for mixing reactions and for measuring temperature. If the reaction temperature is not known and/or not uniform throughout the sample, reactions may not be reproducible between microwave systems. With variability in operation and performance of domestic microwave ovens, few specific, literature syntheses have been reproduced by others and occasionally such attempts have resulted in alternative outcomes or failure[84,85]. Some of these deficiencies have been recognised and addressed through the development of specific equipment.

It is not always acknowledged that solvents are used for 'dry media' reactions, both to load the reactants onto the support and to elute the products after reaction. If the supports are polar materials such as alumina or silica gel, which are commonly used in liquid chromatography, substantial quantities of solvent may be required to remove the organics. For clean processing, recycling of the solvent and the support would be essential. The latter does not appear to have been demonstrated and may prove difficult if residual organic reactants and/or products are retained strongly. The development and establishment of generally applicable protocols for recycling spent supports would minimise waste and significantly advance the scope of the technique.

Issues for scale-up concern the depth of penetration of the microwaves and the need for uniformity in preparation of the sample and in the heating. With large batches of material, without adequate mixing, temperature gradients could be high. Efficient mixing and temperature measurement within the sample are necessary. Thus far, post-reaction cooling has not been rate limiting for throughput apparently mainly because of the relatively low reaction scales employed. Methods for rapid cooling of large-scale batches do not appear to have been developed.

Potential hazards, particularly for scale-up, concern toxic effects: firstly, of volatiles liberated from the supports during the reactive step, and secondly, of the supports themselves. Chemical composition and active surface states are critical determinants of biological response[86]. Minerals doped with inorganic or organic oxidants such as MnO_2, CrO_3, iodobenzene diacetate and sodium periodate, or reductants like $NaBH_4$ and catalysts including KF and CsF, which have been employed as 'dry media', could have severe biological side effects if inhaled.

9.3.4. *Methods employing solvents*

Apart from microwave plasma-mediated processes that are usually performed under reduced pressure and are beyond the present scope, in general, methods utilising solvents fall into two categories: Those that operate at atmospheric pressure and those that operate at elevated pressure. The former, include 'microwave-induced organic reaction enhancement' (MORE) chemistry[87–89], superheating and reflux conditions and are briefly described hereafter. Although they may offer convenience on the laboratory scale, extending in some cases to the preparation of hundreds of gram (examples of which are presented later), they present disadvantages or appear to offer few tangible benefits for further scale-up.

MORE chemistry[19,87–89] employs polar, high-boiling solvents with open vessels in unmodified domestic microwave ovens. The solvents have dielectric properties suitable for efficient coupling of microwave energy and rapid heating to temperatures, which are high, but typically 20–30°C below boiling point. With this technique, reactions (on the milligram to several hundred gram scale) have been performed within minutes. Disadvantages include the limitation to high-boiling polar solvents such as dimethylsulphoxide, ethylene glycol, diglyme, triglyme, N-methylmorpholine, N,N-dimethylformamide and 1,2-dichlorobenzene that have relatively similar boiling points and that can present difficulties for recycling and for isolation of products.

Solvents with relaxation times greater than 65 ps (corresponding to 2450 MHz) will have loss tangents that increase with temperature and will be prone to superheating under microwave irradiation[90]. Baghurst and Mingos have reported temperatures up to 26°C above equilibrium boiling points at atmospheric pressure. An obvious disadvantage presenting a potential safety hazard (particularly when scaling up) would occur in the event of sudden nucleation, which would result in 'bumping' and boiling. When such events occur, reaction mixtures are difficult to contain within the vessel and commonly, vapour is released into the atmosphere.

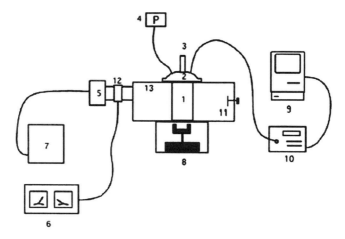

Figure 9.1   Schematic diagram of the MBR[65]. 1, Reaction vessel; 2, top flange; 3, cold-finger; 4, pressure meter; 5, magnetron; 6, forward/reverse power meters; 7, magnetron power supply; 8, magnetic stirrer; 9, computer; 10 , optic fibre thermometer; 11, load matching device; 12, waveguide; 13, multi-modal cavity (applicator).

Dedicated microwave reactors have been developed that are capable of reliable and safe operation with volatile organic solvents at elevated temperatures and pressures[13,65,91–95]. Independent investigations into optimal parameters for microwave chemistry support this approach[90,96].

A laboratory scale microwave batch reactor (MBR) was developed for synthesis or kinetics studies[65]. The MBR has a capacity of 25–200 ml and is capable of operating at up to 260°C and up to 10 MPa (100 atm), although in practice, pressures in excess of 5 MPa are seldom used. Aspects of the system are depicted schematically in Figs. 9.1 and 9.2 and an embodiment, fabricated in the CSIRO laboratories, is illustrated in

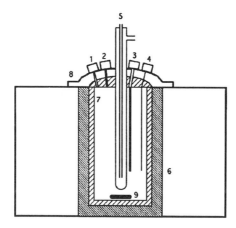

Figure 9.2   Schematic diagram of reaction vessel and associated components[65]. 1, Pressure transducer; 2, pressure relief valve; 3, sample addition/removal port; 4, optic fibre thermometer; 5, cold-finger; 6, retaining cylinder; 7, PTFE reaction vessel; 8, top flange; 9, magnetic stirrer bar.

Figure 9.3   Dr Brett Roberts at work with the MBR[65].

Fig. 9.3. The main features include rapid heating capability (1.2 kW microwave output), infinitely variable control of microwave power, measurement of absorbed and reflected microwave energy, a load matching device to maximise heating efficiency, direct measurement of the reaction temperature and pressure, a stirrer for mixing and to ensure uniform temperature within the sample, valving and plumbing to facilitate sample introduction and withdrawal during the heating period, chemically inert wettable surfaces and fittings, rapid cooling post-reaction and a facility for conducting reactions under an atmosphere of inert gas.

Microwave power input is computer controlled and enables heating to be conducted at high or low rates and designated temperatures to be retained for lengthy periods if desired. The safety features have been described and discussed elsewhere[65].

Robotically operated microwave batch reactors incorporating several of the design features of the MBR and an earlier prototype, but with a lower capacity (2–5 ml) have been developed commercially for rapid synthesis, primarily of candidates for drug discovery. These systems can operate under atmospheric or elevated pressure, the upper limits of which are dependent upon individual designs.

The continuous microwave reactor (CMR) was the first microwave system designed specifically for organic reactions (see Fig. 9.4)[91]. In that regard, as with the MBR, the

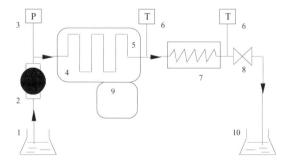

Figure 9.4   Schematic diagram of the CMR[91]. 1, Reactants for processing; 2, metering pump; 3, pressure transducer; 4, microwave cavity; 5, reaction coil; 6, temperature sensor; 7, heat exchanger; 8, pressure regulator; 9, microprocessor controller; 10, product vessel.

Figure 9.5   Commercial embodiment of the continuous microwave reactor (by Milestone MLS).

ability to heat a reaction mixture safely and rapidly, under controlled conditions, to a designated temperature and then to cool it with comparable efficiency afterwards were viewed as essential. To achieve this with the CMR, the microwave cavity was fitted with a vessel of microwave-transparent inert material. Plumbing in the microwave zone was attached to a metering pump and pressure gauge at the inlet end and a heat exchanger and pressure-regulating valve at the effluent end. The heat exchanger enabled rapid cooling of the effluent, under pressure, immediately after it exited the irradiation zone. Temperature was monitored immediately before and after cooling. Variables such as volume of the vessel within the microwave zone, flow rate and control of the applied microwave power allowed flexible operation. The plumbing was designed to withstand corrosive acids and bases and to minimise contact between metal surfaces and reaction mixtures. Feedback microprocessor control allowed setting of pump rates and temperatures for heating and cooling of reactions. Fail-safe measures ensured that the system would shut down if the temperature exceeded the maximum allowable by 10°C, or in the event of blockages or ruptures in the plumbing.

A commercially available embodiment (Fig. 9.5) produced under licence, had a volume of 120 ml within the microwave zone, 80 ml within the cooling zone and a pump that could produce flow rates of up to 100 ml/min. With this arrangement, residence times in the microwave zone (which for convenience are also defined as reaction times) were typically 2–10 min[97–99].

9.3.5. Scale-up of methods employing solvents

Similar to the solvent-free approaches discussed in Section 3.2, a combinatorial approach also has been employed to scale-up the synthesis of desired compounds[100]. The microwave-assisted reactions are performed on solvent-swollen polymeric beads and are classified herein as being carried out in the presence of a solvent. Examination of these supports after 20 min of microwave irradiation (700 W) revealed that neither the appearance nor swelling behaviour of the beads had altered[65].

Applications of this technique include peptide hydrolysis[101], peptide coupling[102], Suzuki couplings, Stille coupling[63,103,104], parallel synthesis of 1,3,5-triazines[105] and fluorous Stille coupling[106].

Jones and his co-workers have employed microwave-assisted organic chemistry for the preparation of labelled compounds containing radionuclides such as tritium or stable isotopes such as deuterium. Tritiation is usually performed on a much smaller scale (milligrams) than deuteration (grams) and attention must be paid to the radioactive waste produced. The widely used methods for incorporating tritium into organic compounds include hydrogen isotope exchange, hydrogenation, catalytic dehalogenation, reduction and methylations. These approaches all have disadvantages ranging from reduced specific activity and low solubility of T_2 gas to limited regio- and chemoselectivity.

The work of Jones and Lu has embraced a wide range of reactions including hydrogenations, borohydride reductions, aromatic dehalogenations, decarboxylations and hydrogen isotope exchange processes, (Scheme 9.3, Eqs. 1–3). In addition to the accelerated rates of reaction, new environmentally friendly routes have been developed, particularly solventless reactions that minimise waste production and facilitate containment[107–110].

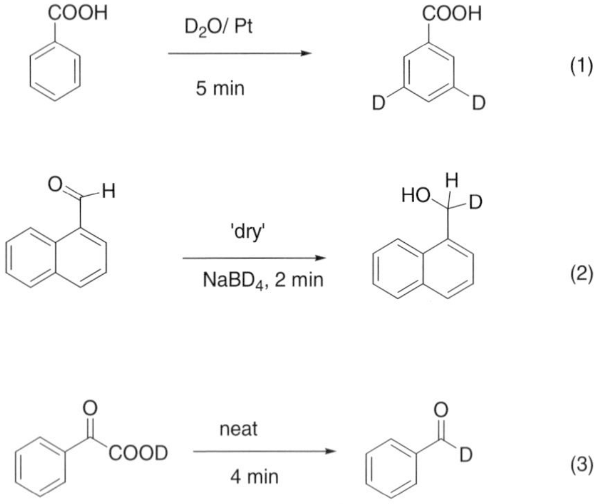

Scheme 9.3   Microwave-assisted deuterium exchange[107]

Several workers have employed monomodal cavities for microwave chemistry on the sub-gram scale. In some cases in which monomodal cavities have been used[7], special benefits of so-called 'focussed' microwaves have been claimed. As mentioned earlier, the dielectric properties of a sample can alter substantially with temperature and/or with changing chemical composition. Hence, regardless of whether multi-modal or unimodal cavities are employed, frequent tuning may be necessary if heating efficiency is to be retained. This aspect has often been overlooked by proponents of 'focussed' microwaves. The nett result is that transfer of microwave conditions between monomodal to multi-modal cavities is usually facile. With the MBR (which had a tunable multi-modal cavity), Cablewski *et al.* performed five reactions that had been conducted earlier on the gram scale or below with 'focussed' microwaves (T. Cablewski, B. Hellman, P. Pilotti, J. Thorn, and C.R. Strauss, personal communication; see also Ref. 117 for conference poster). These were scaled-up between 40- and 60-fold and reaction conditions and yields were comparable with those on the smaller scale. The reactions studied were diverse. They comprised Suzuki coupling, a Hantzsch reaction, benzoylation of diacetone glucose catalysed by polymer-bound dimethylaminopyridine, the synthesis of hydantoin from phenylisothiocyanate, and the Pd-catalysed asymmetric alkylation of 4-methoxyphenol with cyclohex-2-enyl carbonate in the presence of Trost's catalyst (Scheme 9.4, Eqs. 1–5).

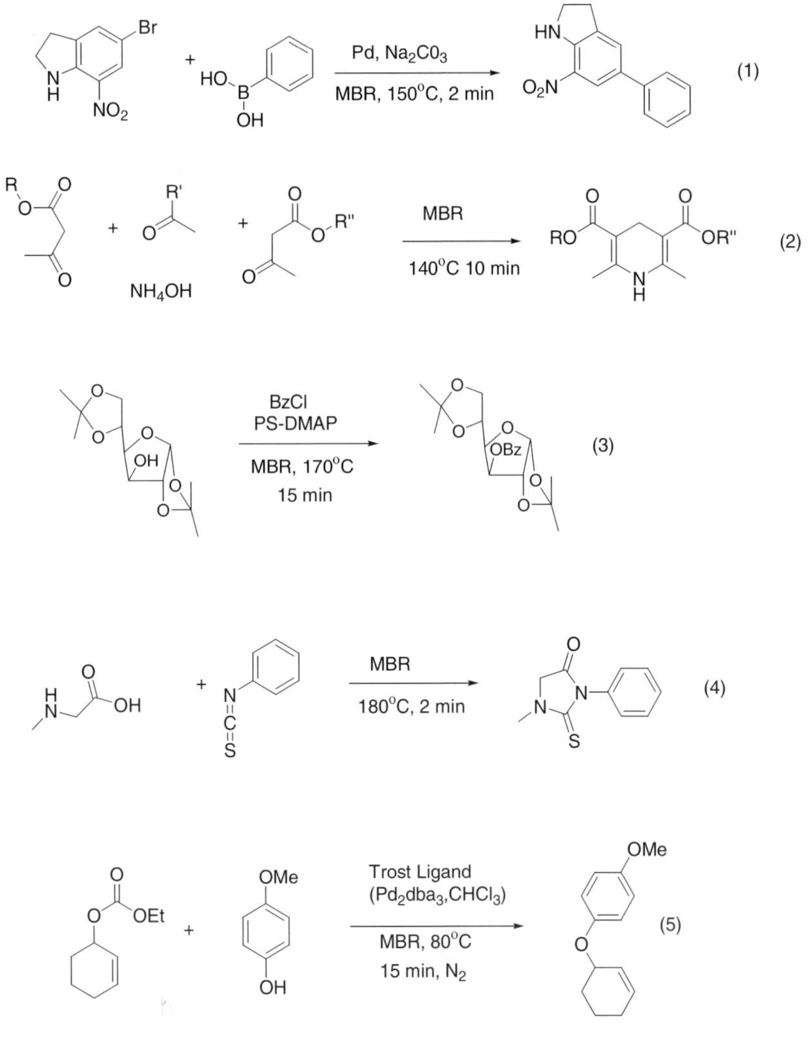

Scheme 9.4   Examples of scale-up in the MBR (T. Cablewski, B. Hellman, P. Pilotti, J. Thorn, and C.R. Strauss, personal communication; see also Ref. 117 for conference poster).

Results of Kappe *et al.* were consistent with these findings for scale-up of the hydroly-sis of benzamide by 100-fold from 50 mg (Scheme 9.5).

With the MBR, other examples of scale-up involving a triphenylphosphine-free one-pot Wittig olefination, a one-step three-component synthesis of imidazo annulated

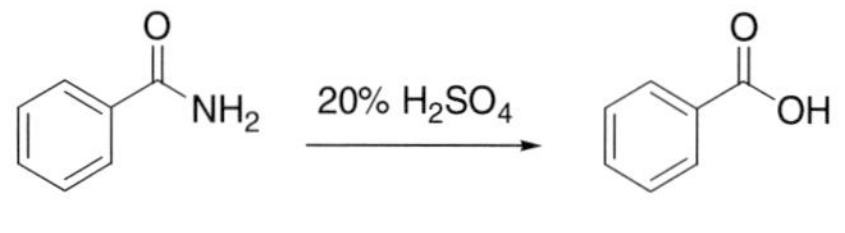

Scheme 9.5 Hydrolysis of benzamide[111].

pyridine and a metal-catalysed Suzuki coupling have been recently reported (See Personal Chemistry at www.Personalchemistry.com).

In a recent example Kappe and co-workers[112] scaled up a series of reactions from a single mode to a multimode parallel batch reactor. Typically reactions were scaled from 1 to 100 mmol. The transformations included

(1) multi-component reactions (Biginelli condensation and Kindler thioamide synthesis),
(2) Heck and Negishi reactions,
(3) solid-phase organic synthesis and
(4) Diels–Alder cycloaddition reactions using gaseous reagents.

Bose *et al*[113] have demonstrated the synthesis of aspirin from salicylic acid and acetic anhydride using commercial microwave reactors (Scheme 9.6). Without optimising the conditions, high purity aspirin was obtained in greater than 80% yield (500–800 g scale).

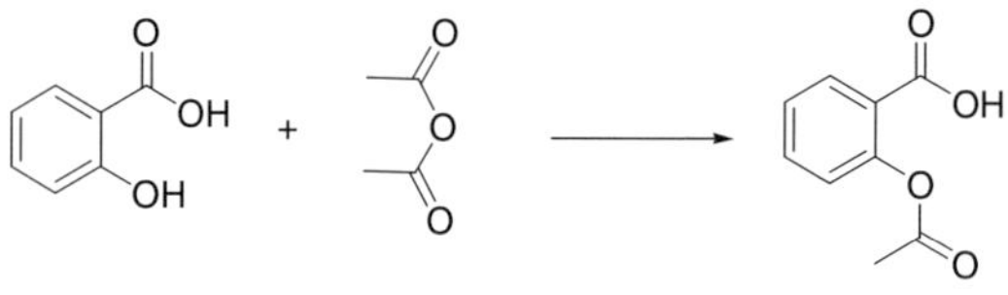

Scheme 9.6 Microwave-assisted synthesis of aspirin[113].

The same group also reported the *N*-acetylation of *p*-aminophenol (50 g) with acetic acid (200 ml) to produce paracetamol in 55% yield, unoptimised (Scheme 9.7)[113]. It is emphasised that the microwave methods were intended for the purposes of demonstration of scale only and not to represent alternative industrial processes. They have been included here for the same reason.

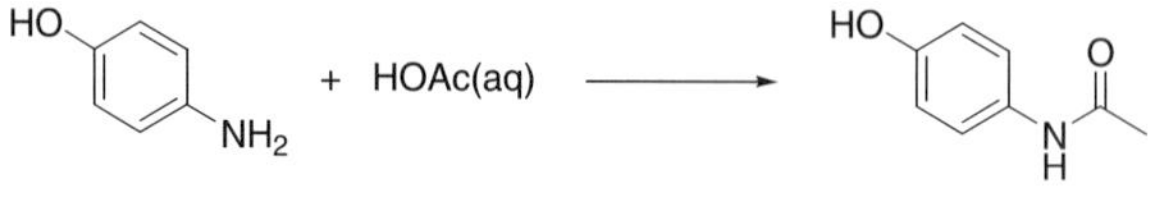

Scheme 9.7 Acetaminophen synthesis[113].

The protected amino acid tetrachlorophthalimidoacetic acid synthesised from tetrachlorophthalic anhydride and glycine using DMF as the solvent was conducted on a molar scale to yield 300 g of the product after 8 min (Scheme 9.8)[113].

Scheme 9.8 Tetrachlorophthalimidoacetic acid synthesis[113].

Under aqueous conditions, after 12 min of irradiation, 3-carboxycoumarin was obtained from salicylaldehyde in a 60% yield by a Knoevenagel reaction and lactonisation (Scheme 9.9)[113].

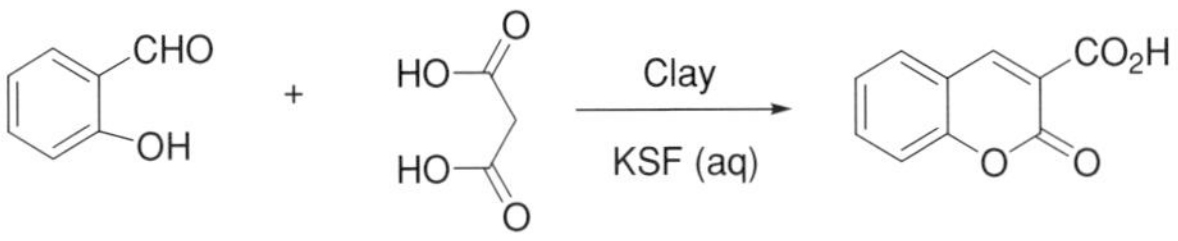

Scheme 9.9 3-Carboxycoumarin synthesis[113].

For each 10°C increase in reaction temperature, the required time for synthetic reactions is approximately halved. Provided that components are not degraded under the conditions used, a reaction taking 18 h at 80°C would be expected at 200°C to afford a comparable yield in only 16 s. This example indicates that not only do higher temperatures than normal offer opportunities for efficiencies in time and energy, but that some batch operations can be transformed into continuous processes merely by raising the temperature.

At 200°C or above, at atmospheric pressure, however, the range of solvents is limited. High-boiling solvents also are inconvenient to remove and to re-purify. These aspects militate against (but do not rule out) the use of MORE chemistry for scale-up beyond the hundreds of gram. On the other hand, temperatures in the order of 200°C can be attained at pressures of 2–3 MPa for solvents such as ethyl acetate, methanol, ethanol, acetonitrile, chloroform and acetone, all of which boil below 85°C at ambient pressure.

Since the pioneering work of Grieco and Breslow in the late 1980s, water has been investigated intensively, as a medium for non-enzymatic organic reactions[114–120]. Temperatures below boiling points have been employed mainly to exploit hydrophobic effects. At ambient temperatures, water is a poor solvent for most organic compounds. However, its ionic product increases 1000-fold between 25 and 240°C, so it becomes a stronger acid and base[13]. The dielectric constant decreases from 78 at 25°C to 20 at 300°C indicating that the polarity is lowered with temperature increase. Investigations into organic synthesis in high-temperature water, carried out with the MBR and CMR (as well as in conventionally heated autoclaves), have demonstrated the advantages of high-temperature water as a synthetic medium[121,122]. A few examples are given below.

As suggested by the lower dielectric constant, high-temperature water behaves like a pseudo-organic solvent, dissolving organic compounds that are much less soluble at ambient temperatures. This property was exploited for the hydrolysis of naturally occurring monoterpene alcohols without the need for prior derivatisation with typical water-solubilising groups such as phosphates or glycosidic units. Biomimetic reactions that normally would be acid catalysed proceeded on the underivatised compounds in the absence of added acidulant. Cooling of the mixtures rendered the products insoluble, readily isolable and the aqueous phase did not require neutralisation before work-up. These aspects hold particular attraction for Green Chemistry[123] and the manufacture of flavours and fragrances.

2,3-Dimethylindole was obtained from phenylhydrazine and butan-2-one, by a one-pot reaction in water (Scheme 9.10, Eq. 1). This is the first example of water as the reaction medium for Fischer indole synthesis, and significantly neither a preformed hydrazone nor addition of acid was required. Also, the decarboxylation of indole-2-carboxylic acid to afford indole is not trivial by traditional methods. However, in water at 255°C, it was quantitative within 20 min (Scheme 9.10, Eq. 2).

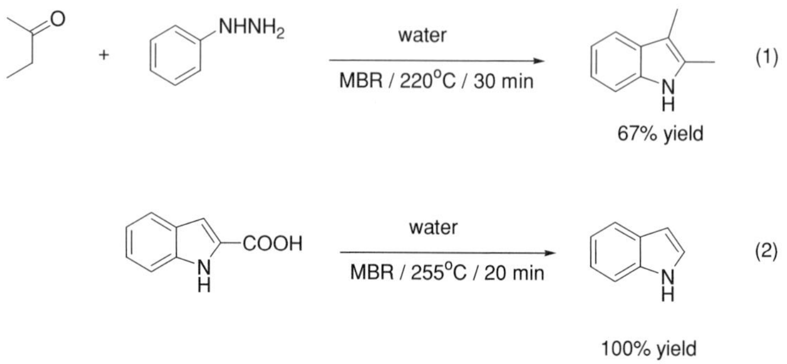

Scheme 9.10   Indole chemistry in water[121].

The diversity of applicable reactions, high selectivities often obtained with seemingly minor variations in the conditions, and demonstrated scale-up, including to continuous operation with the CMR, indicate that aqueous high-temperature media will become more important for the development of clean processes. Underpinning this are obvious additional advantages of water including low-waste work-up, low cost, negligible toxicity and safe handling and disposal.

Customarily the products would be recovered by extraction with organic solvent. This would saturate the aqueous phase with the solvent (and the solvent with water), thereby complicating disposal and offsetting environmental benefits gained through using water as the reaction medium in the first place. To avoid solvent extraction, hydrophobic resins can be employed for concentration and isolation of the products from aqueous media. Organics are retained on the resin and subsequently can be desorbed with a solvent such as ethanol, which is a useful solvent for Green Chemistry as it is readily recyclable and is both renewable and biodegradable. Advantages of non-extractive processes include convenience, high throughput and low waste owing to ready disposal of the spent water, recyclability of the resin and the solvent used for desorption.

At elevated temperature and pressure, shifts including reversal of equilibria can result in new and/or cleaner reactions. In that regard, a catalytic, thermal etherification has been demonstrated that can be carried out near neutrality and that produces minimal waste[124]. This represents a cleaner alternative to the traditional Williamson synthesis, in which the ether is produced through substitution of an alkyl halide (RX) by a strongly basic alkoxide or phenoxide. The major disadvantages of Williamson etherification are that a stoichiometric amount of waste salt is generated and in some cases base-catalysed elimination of hydrogen halide can compete.

The new process is suited to production by MBR or CMR and is shown in Scheme 9.11 for a symmetrical ether. An excess of alcohol (ROH) and a catalytic amount of RX are heated. A solvolytic displacement reaction between RX and ROH affords R_2O along with HX or its elements (hereafter referred to as HX; Scheme 9.11, Eq. 1). In contrast with the outcome at lower temperatures, where HX is employed to cleave ethers (i.e., the conditions favour the reverse reaction in Eq. 1), the liberated HX attacks another molecule of ROH to form water and to regenerate RX (Scheme 9.11, Eq. 2). If the forward rates of both reactions are comparable, the concentration of HX will be low throughout and that of RX will remain relatively constant. Although HX and RX are stoichiometric reactants or products in Eqs. 1 and 2, they do not appear in the sum, Eq. 3. The procedure involves condensation of two molecules of ROH to give R_2O plus water. The process has been demonstrated with primary and secondary alcohols, including compounds labile to base and acid. Advantages for this etherification are high atom economy, that salts are not formed, RX often is recoverable, the reaction does not require addition of strong acids or bases and that water is the major by-product.

$$RX + ROH \rightleftharpoons R_2O + HX \qquad \text{Equation 1}$$

$$HX + ROH \rightleftharpoons RX + H2O \qquad \text{Equation 2}$$

$$2ROH \rightleftharpoons R_2O + H_2O \qquad \text{Equation 3}$$

Scheme 9.11 Pathway for the etherification catalysed by RX[124].

The preparation of dibutyl ether is illustrative (Scheme 9.12). No reaction occurred with 1-butanol alone for 2 h at 200°C. However, in the presence of 10 mol%

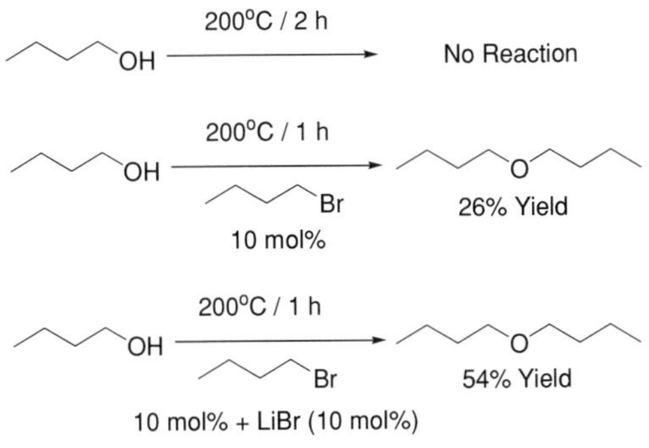

Scheme 9.12 Preparation of dibutyl ether[124].

1-bromobutane, 26% conversion of the alcohol to the ether was obtained after 1 h, without apparent depletion of the catalyst. Alkaline metal salts can accelerate solvolytic processes. Accordingly, the introduction of LiBr (10 mol%) along with the 1-bromobutane afforded a higher conversion within the same time.

Wang and co-workers also were early to recognise opportunities for continuous flow processes under microwave heating. They reported an open-tubular system that operated essentially at ambient pressure and employed a modified domestic oven. Although this approach has limitations, it was applied successfully to the following organic reactions[125] on scales exceeding 20 g:

(1) Esterification of *p*-hydroxybenzoic acid with butan-1-ol and methanol.
(2) Racemisation of optically pure amino acids in acetic acid.
(3) Acid hydrolysis of sucrose to glucose and fructose.
(4) S_N2 reaction of phenoxide with benzyl chloride to afford benzyl phenyl ether.
(5) Cyclisation of butane-1,4-diol and diethylene glycol.

Also, Marquié and co-workers have conducted Friedel–Crafts reactions on a large laboratory scale via a continuous flow process[126]. They reported the acylation of aromatic ethers and sulphonylation of mesitylene, isolating up to 300 g and 250 g of product, respectively (Schemes 9.13 and 9.14).

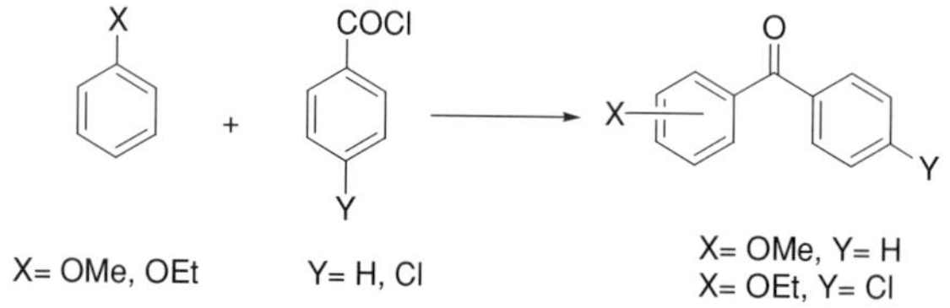

X= OMe, OEt   Y= H, Cl

X= OMe, Y= H
X= OEt, Y= Cl

Scheme 9.13 Acylation of aromatic ethers[126].

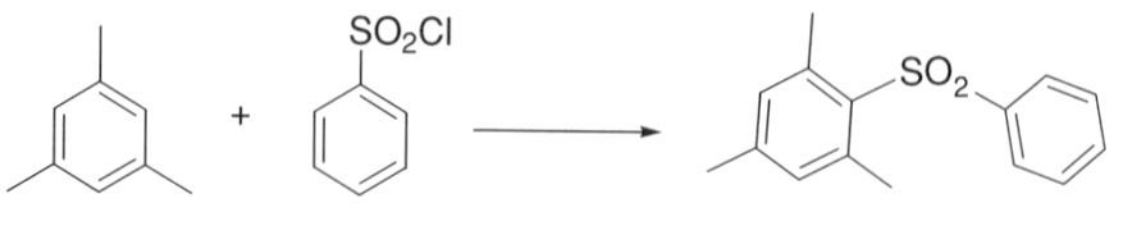

Scheme 9.14 Sulphonylation of mesitylene[126].

With a commercially available CMR, Shieh and co-workers methylated phenols, indoles and benzamidazoles with dimethyl carbonate and DBU as catalyst[99] (Scheme 9.15). In a general procedure, 5 g of substrate was passed through the reactor every 6 min and the process was monitored by HPLC until the reactions were complete. Reactions that normally required days under conventional heating were completed within minutes.

Chemat and co-workers have developed a continuous flow microwave reactor that is suitable for homogeneous and heterogeneous reactions[127]. Although their system had only a small cavity (66 ml), it was possible to treat significant quantities of reagents (30–330 ml/min). This underscored the feasibility of scaling up the manufacture of fine

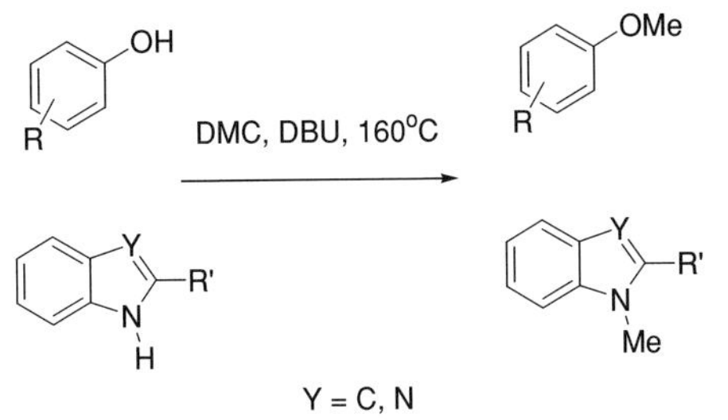

Scheme 9.15 Methylation of phenols, indoles and benzamidazoles[99].

chemicals by flowthrough processes. An important feature of the reactor was that only the catalyst was exposed to the microwave irradiation, resulting in selective heating. This is an extension of much of the excellent earlier work of Wan *et al.*[66,68] and constitutes another example of differential heating. Because the catalyst had significantly higher dielectric loss than the surrounding solvent, the chemical reactants were introduced to it at a low bulk temperature. Reaction occurred in the vicinity of the catalyst, at higher local temperatures and at increased reaction rates. The workers reported esterification under both homogeneous and heterogeneous conditions. Through selective heating of the catalyst an increase in reaction rate of up to 150% was achieved. When temperature of the catalytic surface was equivalent to the temperature of the bulk solution there was no difference in product conversion between classical and microwave heating.

9.3.6. *Advantages and disadvantages of methods utilising solvents*

Solvent-based methods have been claimed to offer the following advantages:

(1) The reaction conditions can be measured and reproduced.
(2) Solvents moderate the reaction and minimise the risk of pyrolysis.
(3) Reactions are cleaner, faster and higher yielding.
(4) Microwave energy is adsorbed directly by the sample and not via the vessel or a solid support.
(5) Thermally unstable products can be cooled rapidly, after the heating step.
(6) Minimising thermal gradients through stirring of the sample.
(7) The employable temperature range of many solvents is increased.
(8) Reactions that are known to require high temperatures and higher boiling solvents can be carried out under pressure at these temperatures, but in lower boiling solvents, facilitating work-up.
(9) Low-boiling reactants can be heated to high temperatures and rapidly cooled under pressure and thus losses of volatiles are minimised.
(10) Reactions can be easily sampled for analysis.
(11) Homogenous solvent-based methods can be scaled up through flowthrough technologies.
(12) Batch reactions using solvents carried out on a laboratory scale are directly amenable to scale up, since the conditions are defined.

(13)   Moderate to high temperature reactions can be carried out in vessels fabricated from inert materials such as PFA Teflon, PTFE or quartz.
(14)   In multiphase systems, differential (i.e., selective) heating is possible.

As indicated earlier, methods that employ solvents at atmospheric pressure appear to offer few benefits for scale-up beyond the multi-hundreds of grams and some even may be disadvantageous in that regard. With pressurised systems, however, the story is different. Such conditions can be obtained with the microwave reactors but not readily with typical glassware. Higher temperatures have led to reaction times up to three orders of magnitude shorter than those for the same preparations carried out conventionally.

Benefits of high temperature can also be gained with traditional autoclaves but the energy is usually applied to the reaction mixture conductively, by external heating. Consequently, the rate of temperature increase is usually low, thermal gradients develop and even by stirring batch reactions not all of the sample will be at the temperature of the applied heat. With microwaves, the whole sample can be irradiated and the energy input can be readily adjusted to match that required. Bulk heating, combined with efficient stirring diminishes temperature gradients.

Vessels for microwave-assisted chemistry are usually made from thermal insulators and the benefits of rapid heating can be diminished if the opportunity for work-up is delayed by slow cooling. Decomposition of thermally unstable products can also occur. In the CMR, rapid cooling takes place through an in-line heat exchanger adjacent to the microwave-heating zone. Mixtures can be cooled immediately, while still under pressure, to prevent losses of volatiles and to minimise decomposition of thermally labile products. The MBR has an incorporated cold-finger that contacts the reaction mixture directly. Cooling can be initiated at any time during operation and is efficient because it is not via the container. Temperature and pressure monitoring, as well as stirring, can be maintained during the cooling process, allowing access to the vessel at the earliest opportunity.

The continuous and batch microwave reactors have been particularly useful for heating reactions in which thermally labile products are formed. For example, alkyl 2-(hydroxymethyl)acrylates have considerable potential as functionalised monomers and synthons[128]. Published syntheses at ambient temperature, however, required several days and were not conducive to scale-up[129–133]. The microwave procedure involved a modified Baylis–Hillman reaction, in which the parent acrylate derivative was reacted with formalin in the presence of 1,4-diazabicyclo[2.2.2]octane (DABCO). Preparations from starting acrylates, including methyl, ethyl and n-butyl esters, were easily achieved within minutes with multiple passes through the CMR, at ca. 160–180°C (Scheme 9.16). Rapid cooling was required to limit hydrolysis, dimerisation and polymerisation. Yields

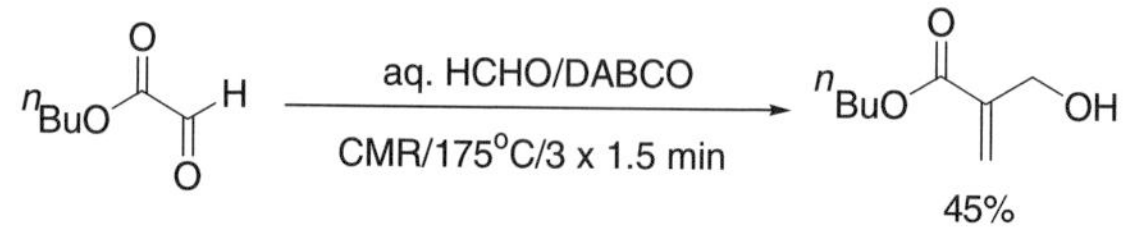

Scheme 9.16   Alkyl 2-(hydroxymethyl)acrylate synthesis[13].

were comparable with those obtained conventionally. Formalin required for the reaction also was obtained conveniently in the CMR, by acid-catalysed hydrolysis of an aqueous slurry of paraformaldehyde. Losses of formaldehyde were minimal because the system was sealed until after cooling.

Strategies for sustainability and reduction of CO_2 emissions include exploration of technologies based on renewable resources. Key challenges include the depolymerisation of cellulose to produce glucose or its oligomers from biomass, for use as feedstock for fermentation to ethanol, acetic acid and lactic acid[134]. Technically the difficulties arise because high temperatures and acidic conditions are required to cleave the glycosidic bonds, but the monosaccharide and oligosaccharides produced are more acid labile than the starting material. This demands a process employing rapid heating and cooling to capture what are in essence, the kinetic products. Microwave technology can be well suited to such applications. With 1% sulphuric acid, the developed method involved raising the temperature from ambient to 215°C within 2 min (in the MBR), maintaining this temperature for 30 s and cooling (Scheme 9.17). The entire operation was completed within 4 min and afforded glucose in nearly 40% yield, along with oligomeric materials.

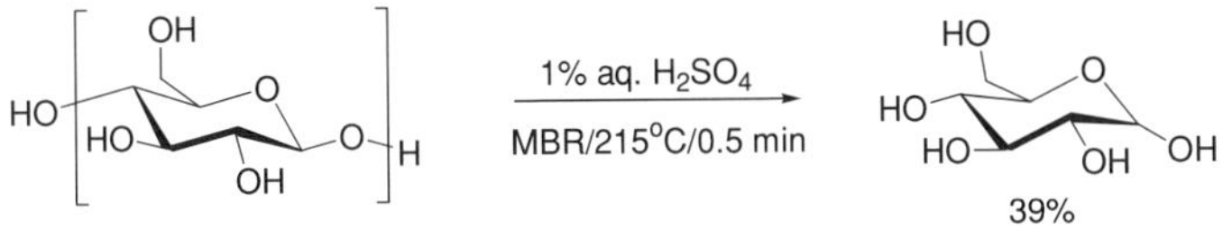

Scheme 9.17 Microwave-assisted depolymerisation of cellulose[13].

Since the vessels used are microwave-transparent, they will be no hotter than their contents. They are usually made from insulating polymeric materials like polytetrafluoroethylene (PTFE) and have inherent advantages for cleaner processing. In contrast to glass, PTFE is resistant to attack by strong bases or HF. In contrast to stainless steel, it is not corroded by halide ions. Also, conductive heat loss by PTFE is minimal and the low adhesivity can help to minimise detergent and organic solvent usage during cleaning operations that would otherwise generate considerable effluent.

In cases where metals or metal ions can contaminate the products, reaction vessels fabricated from inert polymeric materials restrict that possibility. A significant example involved the reaction of maltol with aqueous methylamine to give 1,2-dimethyl-3-hydroxypyrid-4-one. The product is a metal chelator employed for the oral treatment of iron overload. Consequently, it is an excellent metal scavenger but must be produced under stringent conditions that preclude metal complexation. Literature conditions involved heating maltol in aqueous methylamine at reflux for 6 h, the product was obtained in 50% yield, but required decolourisation with charcoal[135]. With the CMR, the optimal reaction time was 1.3 min, and the effluent was immediately diluted with acetone and the near colourless product crystallised from this solvent in 65% yield (Scheme 9.18). A microwave-based batch-wise preparation of 3-hydroxy-2-methylpyrid-4-one from maltol and aqueous ammonia was also developed.

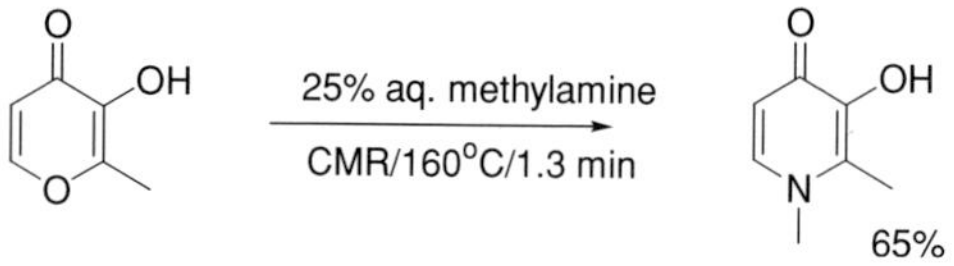

Scheme 9.18   Microwave-assisted 1,2-dimethyl-3-hydroxypyrid-4-one synthesis[13].

9.4. Safety

Microwave-assisted organic synthesis deals with high temperature, high-pressure re-actions and as a consequence of this safety will always be a major concern. The safety aspects increase in importance with the scale of the reaction. General safety procedures for microwave equipment, for example, measures to avoid and detect microwave leak-age, including those for laboratory use, have been documented[13,50,52]. There are also many other technical matters fundamental to microwave-assisted organic chemistry, some of which have been discussed in other sections, are summarised below.

Temperature measurement is paramount to both the performance of the chemistry and the safety when conducting microwave synthesis. Owing to the electromagnetic fields within a microwave cavity, temperature measurement is not trivial and in many applications it has been avoided, for example, for 'dry' media reactions and in the MORE approach. Thermocouples can be employed, but they must be shielded and/or grounded or induced currents and heating can eventuate, resulting in false readings or arcing. Modified thermocouples now have been developed. Fibre optic thermometers, not generally available until the 1990s, are also employed, along with infrared pyrometry, which records surface temperatures.

Excess input microwave energy can lead to arcing. For this to occur, the build up of an intense electrical field is necessary. The possibility exists when low-loss samples are being heated, when salt solutions suffer dielectric breakdown or when suspended metallic materials or sharp edges are present. To avoid the input of excess energy it is essential to measure forward and reflected power and to have infinitely variable microwave energy input (i.e., not a duty cycle) and/or a load matching device. The use of a 'dummy load', that is, a material that can absorb the excess energy, but that is held in the microwave zone independently of the reaction mixture, can also reduce the risk of arcing. In practice, water in a container or circulating through the cavity, is commonly used. Another frequently used technique involves the insertion of an isolator T-section into the waveguide. This section is designed to divert reflected power into the arm of the T, where it is absorbed by a dummy load.

The incorporation of turntables and mode mixers to the domestic oven have been employed to reduce non-uniformity within the cavity and consequently reduce the number of 'hot' and 'cold' spots. On scale-up it is important that homogeneity is maintained and the temperature within the bulk medium is consistent. With a single mode reactor, the sample can be located precisely within the cavity where the electric field strength will be maximal. This in turn allows tuning of the power input into the sample and very high internal heat transfer.

Tuning of multi-modal cavities also can be relatively simple, allowing a good match to the source and minimal reflected power. On a large scale, it is important that the tuning within the cavity be monitored in real time. With modern systems, computer control enables the efficiency of the microwave power to be monitored and adjusted accordingly. As mentioned above, this is especially important when the dielectric properties of a material change throughout the course of the reaction.

Also as discussed earlier, a difficulty can arise when a material being irradiated possesses a dissipation factor that *increases* with temperature. A microwave-driven thermal runaway can result unless the temperature is carefully monitored and the power controlled. On the other hand, solvents show a general *decrease* in dielectric constant with temperature. Efficiency of microwave absorption diminishes with temperature rise and can lead to poor matching of the microwave load, particularly as the fluid approaches the supercritical state. Solvents and reaction temperatures should be selected with these considerations in mind, as excess input microwave energy can lead to arcing.

Differences in sample size and composition can also affect heating rates. In the latter case, this particularly applies when ionic conduction becomes possible through the addition or formation of salts. For compounds of low-molecular weight, the dielectric loss contributed by dipole rotation decreases with rising temperature, but that due to ionic conduction increases. When working under pressure, it is essential to measure pressure. This can be used for reaction control. If pressures fall beyond acceptable upper and lower limits or the rate of pressure rise exceeds a tolerable value, operating software should automatically shut down the machine. In combination with efficient cooling this approach can avoid thermal runaways near their onset.

9.5. Tandem technologies involving microwaves

Recent research in microwave-assisted organic chemistry has been directed towards niche applications, with an increasing emphasis on environmentally benign methods including those utilising tandem processes and multi-disciplinary activities.

Reactors have been constructed that combine microwave heating with sonication[136–139], ultraviolet radiation[140–144] and electrochemistry[145–148]. The microwave ultrasound (MW-US) reactor propagates sound waves through a series of compression and rarefaction waves induced within the medium. At sufficiently high power, cavitation bubbles form and grow over a few cycles, accruing vapour or gas from the medium. The acoustic field associated with the bubbles is unstable, leading to sudden expansion and violent collapse that generates energy for chemical and mechanical effects[149].

By coupling an ultrasonic probe with a microwave reactor and propagating the ultrasound waves into the reactor via decalin introduced into their double jacket design, Chemat *et al.* studied the esterification of acetic acid with propanol and the pyrolysis of urea to afford a mixture of cyanuric acid, ameline and amelide (Scheme 9.19)[136]. Improved results were claimed compared to those obtained under conventional and microwave heating. The MW-US technique was also used to study the esterification of stearic acid with butanol and for sample preparation in chemical analysis[137,138].

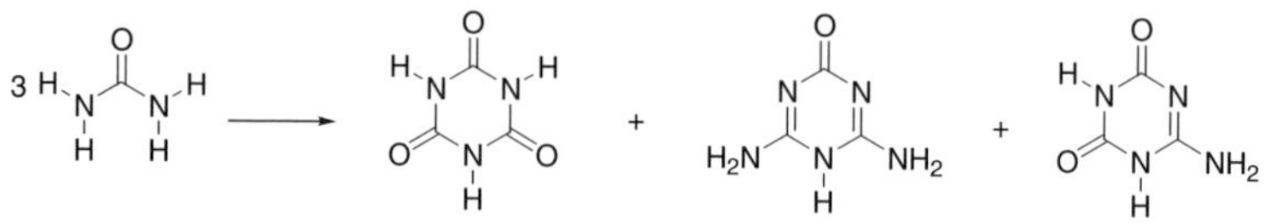

Scheme 9.19 Pyrolysis of urea[136].

More recently Peng and Song reported the rapid synthesis of a library of hydrazides in a MW-US combined reactor (Scheme 9.20)[139]. Unlike the aforementioned system that employed decalin as an energy transfer medium for the ultrasound irradiation, in their modified domestic oven, the horn was immersed directly into the reaction mixture.

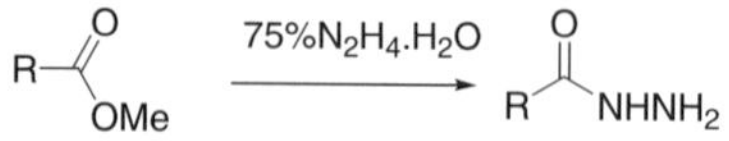

Scheme 9.20 MW-US synthesis of hydrazides[139].

The reported outcomes from these MW-US reactors are somewhat surprising as ultrasonic reactions are usually favoured at low temperatures and microwave heating could be expected to inhibit cavitation.

The combination of microwave and ultraviolet radiation (MW-UV) was first reported by Chemat and co-workers in 1999[140] and soon after by Církva and co-workers[141]. Chemat *et al.* designed a MW-UV combined reactor that was capable of handling solvent and dry media. The reactor had a capacity of 10–100 g at temperatures from 20 to 450°C at ambient pressure. An external UV mercury lamp was employed. The efficacy of the MW-UV reactor was demonstrated through the rearrangement of 2-benzoyloxyacetophenone to 1-(2′-hydroxyphenyl)-3-phenylpropan-1,3-dione. Compared with the microwave heating or UV radiation alone, the rate of reaction followed the sequence of MW-UV >UV >MW. Classical heating yielded only 5% rearrangement after several hours, but the combined radiation methodology yielded >90% in less than 2 h. Separately, within the same time frame, 70% rearrangement was obtained under UV radiation and <20% for microwave irradiation (Scheme 9.21).

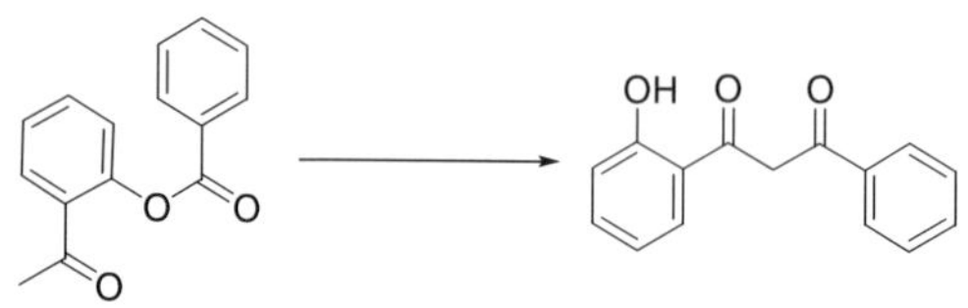

Scheme 9.21 Rearrangement of 2-benzoyloxyacetophenone[140].

Initially Církva and Hájek explored the simultaneous action of microwave and UV irradiation via the addition of tetrahydrofuran to perfluorohexylethene employing a microwave photoreactor with an electrodeless non-contact lamp. In this and their subsequent work[141–144], the lamp was placed into the reactor vessel inside the microwave cavity. This resulted in the generation of UV and microwave radiation concurrently. Several reactions were investigated with this reactor, including

(1) photolysis of phenacyl benzoate,
(2) photosubstitution reaction of chlorobenzene in methanol,
(3) photoreduction of acetophenone by 2-propanol and
(4) photo-Fries reaction of phenyl acetate.

The interaction of microwave radiation with electrochemistry was first reported in 1998[145]. The technique of microwave-assisted voltammetry involves focusing microwave energy at the electrode/solution (electrolyte) interface of an electrode immersed in a solution and placed in a microwave cavity. Either superheating or a stable high temperature of the solution near the electrode can be accommodated. Immediately after switching off the microwave power, the reaction system exhibited characteristics of that seen at room temperature, inferring that the processes involved high-intensity microwave conditions rather than bulk heating[148].·

Marken *et al.* concluded that microwave activation of electrochemical processes enables an increase in temperature at the electrode surface, a thermal gradient and a 'hot spot' zone within the diffusion layer to be achieved and a convective flow to be induced[146].

Microwave-activated voltammetry has been applied to the ferrocyanide/ferricyanide redox couple[145], reduction of $Ru(NH_3)_6^{3,146}$, enhanced PbO2 electro-deposition, stripping and electrocatalysis[147] and electrodehalogenation in non-aqueous media[148].

9.6. Concluding remarks

Since its inception in 1986, microwave-assisted organic chemistry has become an exciting and vibrant field for research and development. Although healthy debate continues regarding the existence or otherwise of specific, non-thermal 'microwave effects', the advantages and disadvantages of microwave techniques for organic synthesis are well documented. Early conclusions by some researchers that microwave energy was incompatible with organic solvents have been discounted. The output of publications on microwave-assisted organic chemistry continues to experience exponential growth, indicating that the microwave approach has met with broad acceptance. Systems developed for solvent-free reactions and other reactors for processes in the presence of solvents have been effective for speeding conventional reactions, particularly those involving sterically constrained components. They have also promoted thermal preparations of heat-labile compounds, for reactions that require high temperature and they have aided optimisation of established elevated-temperature reactions. Improved conditions obtained in comparison with literature methods, typically have involved a combination of increased convenience, savings in time, higher yields, greater selectivity, the need for less catalyst, or employment of a more environmentally benign solvent or reaction medium.

Economic and safety considerations have led to reductions in stockpiles of chemicals and decreasing transportation of hazardous substances. Industrially, reactor size is now important, with miniaturisation becoming an attribute. These factors suggest that in future, individual chemical reactors will be required for diverse tasks and may need to be readily relocatable. Microwave systems should fulfill those requirements.

As discussed in this chapter, the scope of demonstrated applications now extends from the sub-milligram level for radio-tracer work to the kilogram scale for preparative chemistry. Commercial microwave batch reactors have been introduced to accommodate such requirements. Continuous reactors have also been produced for use with 'dry media' or liquid-phase reactions and these allow higher throughputs.

Thus the term scale-up will have different meanings for different users and significantly, it does not only pertain to production of bulk or commodity chemicals. The engineering cost of microwave capacity is based on the installed kilowatt. For domestic and laboratory-scale microwave systems of 1200 watts or below, the magnetron (microwave generator) can be air-cooled. As the capacity increases, however, to 5 kW and above, more sophisticated oil-based or water-based cooling is required and this introduces extra size, complexity and cost to microwave systems. Another aspect that requires consideration is that the energy efficiency of the conversion of electricity into microwave power can be relatively low (in the order of 70%), which might make the microwave approach less attractive when the mass of material required is beyond the multi-tonne scale.

Microwave-assisted organic synthesis offers a very quick and direct route to intermediate quantities of material. When working on a large scale it is important to understand the mechanisms of microwave heating. When synthesis is designed correctly the microwave-assisted approach to scale is a safe and efficient tool for any chemist.

9.7. References

1. Gedye, R., Smith, F., Westaway, K., Ali, H., Baldisera, L., Laberge L. and Rousell, J., The use of microwave ovens for rapid organic synthesis, *Tetrahedron Lett.*, 1986, **27**, 279.
2. Giguere, R.J., Bray, T.L., Duncan, S.M. and Majetich, G., Application of commercial microwave ovens to organic synthesis, *Tetrahedron Lett.*, 1986, **27**, 4945.
3. Gedye, R.N., Smith, F.E. and Westaway, K.C., The rapid synthesis of organic compounds in microwave ovens, *Can. J. Chem.*, 1988, **66**, 17.
4. Mingos, D.M.P. and Baghurst, D.R., Applications of microwave dielectric heating effects to synthetic problems in chemistry, *Chem. Soc. Rev.*, 1991, **20**, 1.
5. Gedye, R., Smith, F. and Westaway, K., Microwaves in organic and organometallic synthesis, *J. Microwave Power Electromag. Energy*, 1991, **26**, 3.
6. Abramovitch, R.A., Applications of microwave energy in organic chemistry: a review, *Org. Prep. Proced. Int.*, 1991, **23**, 683.
7. Loupy, A., Petit, A., Hamelin, J., Texier-Boullet, F., Jacquault, P. and Mathé, D., New solvent-free organic synthesis using focused microwaves, *Synthesis*, 1998, 1213.
8. Wathey, B., Tierney, J., Lidström, P. and Westman, J., The impact of microwave-assisted organic chemistry on drug discovery, *Drug Discov. Today*, 2002, **6**, 373.
9. Lidström, P., Tierney, J., Wathey, B. and Westman, J., Microwave assisted organic synthesis: a review, *Tetrahedron*, 2001, **57**, 9225.
10. Diaz-Ortiz, A., De La Hoz, A. and Langa, F., Microwave irradiation in solvent-free conditions: an eco-friendly methodology to prepare indazoles, pyrazolopyridines and bipyrazoles by cycloaddition reactions, *Green Chem.*, 2000, **2**, 165.
11. Cundy, C.S., Microwave techniques in the synthesis and modification of zeolite catalysts: a review, *Collect. Czech. Chem. Commun.*, 1998, **63**, 1699.
12. Galema, S.A., Microwave chemistry, *Chem. Soc. Rev.*, 1997, **26**, 233.

13. Strauss, C.R. and Trainor, R.W., Developments in microwave-assisted organic chemistry, *Aust. J. Chem.*, 1995, **48**, 1665.
14. Strauss, C.R., Application of microwaves for environmentally benign organic chemistry, In: Clark J. and Macquarrie D. (Eds.), *Handbook of Green Chemistry & Technology*, Blackwell, London, 2002, p. 397.
15. Caddick, S., Microwave assisted organic reactions, *Tetrahedron*, 1995, **51**, 10403.
16. Langa, F., De la Cruz, P., De la Hoz, A., Diaz-Ortiz, A. and Diez-Barra, E., Microwave irradiation: more than just a method for accelerating reactions, *Contemp. Org. Synthesis*, 1997, **4**, 373.
17. Varma, R.S., Solvent-free organic syntheses, *Green Chem.*, 1999, **1**, 43.
18. Elander, N., Jones, J.R., Lu, S.Y. and Stone-Elander, S., Microwave-enhanced radiochemistry, *Chem. Soc. Rev.*, 2000, **29**, 239.
19. Bose, A.K., Manhas, M.S., Banik, B.K. and Robb, E.W., Microwave-induced organic reaction enhancement (MORE) chemistry: techniques for rapid, safe and inexpensive synthesis, *Res. Chem. Intermed.*, 1994, **20**, 1.
20. Perreux, L. and Loupy, A., A tentative rationalization of microwave effects in organic synthesis according to the reaction medium, and mechanistic considerations, *Tetrahedron*, 2001, **57**, 9199.
21. Loupy, A. (Ed.), *Microwaves in Organic Synthesis*, Wiley-VCH, Weinheim, 2002.
22. Linders, J.T.M., Kokje, J.P., Overhand, M., Lie, T.S. and Maat, L., Chemistry of opium alkaloids, Part XXV: Diels–Alder reaction of 6-demethoxy-β-dihydrothebaine with methyl vinyl ketone using microwave heating; preparation and pharmacology of 3-hydroxy-α, α, 17-trimethyl-6β, 14β-ethenomorphinan-7β-methanol, a novel deoxygenated diprenorphine analog, *Rec. des Trav. Chim. des Pays-Bas*, 1988, **107**, 449.
23. Baghurst, D.R. and Mingos, D.M.P., Design and application of a reflux modification for the synthesis of organometallic compounds using microwave dielectric loss heating effects, *J. Organomet. Chem.*, 1990, **384**, C57.
24. Woudenberg, R.H., Oosterhoff, B.E., Lie, T.S. and Maat, L., Chemistry of opium alkaloids, Part XXXVI: synthesis of 5β-alkylthebaines, 5β-alkyl-10α-ethylthebaines and 5β,10α,10β-triethylthebaine; Diels–Alder reactions to 5β-alkyl- and 5β-alkyl-10α-ethyl-6α,14α-ethenoisomorphinans, *Rec. des Trav. Chim. des Pays-Bas*, 1992, **111**, 119.
25. Linders, J.T.M., Briel, P., Fog, E., Lie, T.S. and Maat, L., Chemistry of opium alkaloids, Part XXVIII: preparation of 6-demethoxy-*N*-formyl-*N*-northebaine and its Diels–Alder reactions with methyl vinyl ketone and nitroethene; novel 8-nitro-substituted 6α,14α-ethenoisomorphinans and 6β,14β-ethenomorphinans, *Rec. des Trav. Chim. des Pays-Bas*, 1989, **108**, 268.
26. Pucíová, M. and Toma, Š., Synthesis of oximes in the microwave oven, *Coll. Czech. Chem. Commun.*, 1992, **57**, 2407.
27. Pucíová, M., Ertl, P. and Toma, Š., Synthesis of ferrocenyl-substituted heterocycles: the beneficial effect of the microwave irradiation, *Coll. Czech. Chem. Commun.*, 1994, **59**, 175.
28. Stuerga, D., Gonon, K. and Lallemant, M., Microwave heating as a new way to induce selectivity between competitive reactions: application to isomeric ratio control in sulfonation of naphthalene, *Tetrahedron*, 1993, **49**, 6229.
29. Zhang, Y.-W., Shen, Z.-X., Pan, B., Lu, X.-H. and Chen, M.-H., Research on the synthesis of 1,4-dihydropyridines under microwave, *Synthetic Commun.*, 1995, **25**, 857.
30. Matsumura-Inoue, T., Tanabe, M., Minami, T. and Ohashi, T., A remarkably rapid synthesis of ruthenium(II) polypyridine complexes by microwave irradiation, *Chem. Lett.*, 1994, 2443.
31. Dabirmanesh, Q. and Roberts, R.M.G., The synthesis of iron sandwich complexes by microwave dielectric heating using a simple solid CO_2-cooled apparatus in an unmodified commercial microwave oven, *J. Organomet. Chem.*, 1993, **460**, C28.
32. Lie Ken Jie, M.S.F. and Yan-Kit, C., The use of a microwave oven in the chemical transformation of long chain fatty acid esters, *Lipids*, 1988, **23**, 367.
33. Khan, M.U. and Williams, J.P., Microwave-mediated methanolysis of lipids and activation of thin-layer chromatographic plates, *Lipids*, 1993, **28**, 953.
34. Dasgupta, A., Banerjee, P. and Malik, S., Use of microwave irradiation for rapid transesterification of lipids and accelerated synthesis of fatty acyl pyrrolidides for analysis by gas chromatography-mass spectrometry: study of fatty acid profiles of olive oil, evening primrose oil, fish oils and phospholipids from mango pulp, *Chem. and Physics of Lipids*, 1992, **62**, 281.
35. Takano, S., Kijima, A., Sugihara, T., Satoh, S. and Ogasawa, K., Racemization of (–)-vincadifformine using a microwave oven, *Chem. Lett.*, 1989, 87.
36. Lei, B. and Fallis, A.G., Direct total synthesis of (+)-longifolene via an intramolecular Diels–Alder strategy, *J. Am. Chem. Soc.*, 1990, **112**, 4609.
37. Lu, Y.-F. and Fallis, A.G., An intramolecular Diels–Alder approach to tricyclic taxoid skeletons, *Tetrahedron Lett.*, 1993, **34**, 3367.
38. Jones, G.B. and Chapman, B.J., Decarboxylation of indole-2-carboxylic acids: improved procedures, *J. Org. Chem.*, 1993, **58**, 5558.

39. Jiang, Y. and Yuan, Y., The tribromolanthanoids (LnBr$_3$) catalysed reactions of benzyl alkyl ethers and carboxylic acids promoted by microwave irradiation, *Synthetic Commun.*, 1994, **24**, 1045.
40. Sowmya, S. and Balasubramanian, K.K., Microwave induced Ferrier rearrangement, *Synthetic Commun.*, 1994, **24**, 2097.
41. Kundu, M.K., Mukherjee, S.B., Balu, N., Padmakumar, R. and Bhat, S.V., Microwave mediated extensive rate enhancement of the Baylis–Hillman reaction, *Synlett*, 1994, 444.
42. Chang, Y.-C., Nair, M.G., Santell, R.C. and Helferich, Microwave-mediated synthesis of anticarcinogenic isoflavones from soybeans, *J. Agric. Food Chem.*, 1994, **42**, 1869.
43. Ali, M., Bond, S.P., Mbogo, S.A., McWhinnie, W.R. and Watts, P.M., Use of a domestic microwave oven in organometallic chemistry, *J. Organomet. Chem.*, 1989, **371**, 11.
44. Baghurst, D.R., Cooper, S.R., Greene, D.L., Mingos, D.M.P. and Reynolds, S.M., Application of microwave dielectric loss heating effects for the rapid and convenient synthesis of coordination compounds, *Polyhedron*, 1990, **9**, 893.
45. Baghurst, D.R., Mingos, D.M.P. and Watson, M.J., Application of microwave dielectric loss heating effects for the rapid and convenient synthesis of organometallic compounds, *J. Organomet. Chem.*, 1989, **368**, C43.
46. Greene, D.L. and Mingos, D.M.P., Application of microwave dielectric loss heating effects for the rapid and convenient synthesis of ruthenium(II) polypyridine, *Transition Met. Chem.*, 1991, **16**, 71.
47. Abramovitch, R.A. and Bulman, A., Fischer cyclizations by microwave heating, *Synlett*, 1992, 795.
48. Adámek, F. and Hájek, M., Microwave-assisted catalytic addition of halo compounds to alkenes, *Tetrahedron Lett.*, 1992, **33**, 2039.
49. Baptistella, L.H.B., Neto, A.Z., Onaga, H. and Godoi, E.A.M., An improved synthesis of 2,3- and 3,4-unsaturated pyranosides: the use of microwave energy, *Tetrahedron Lett.*, 1993, **34**, 8407.
50. Kingston, H.M. and Jassie, L.B. (Eds.), *Introduction to Microwave Sample Preparation*, American Chemical Society, Washington, DC, 1988.
51. Majetich, G. and Hicks, R., The use of microwave heating to promote organic reactions, *J. Microwave Power Electromag. Energy*, 1995, **30**, 27.
52. Metaxis, A.C. and Meredith, R.J., *Industrial Microwave Heating*, Peregrinus, London, 1988.
53. Sun, W.C., Guy, P.M., Jahngen, J.H., Rossomando, E.F. and Jahngen, E.G.E., Microwave-induced hydrolysis of phospho anhydride bonds in nucleotide triphosphates, *J. Org. Chem.*, 1988, **53**, 4414.
54. Lewis, D.A., Summers, J.S., Ward, T.C. and McGrath, J.E., Accelerated imidization reactions using microwave radiation, *J. Polym. Sci., Part A: Polym. Chem.*, 1992, **30**, 1647.
55. Bose, A.K., Manhas, M.S., Ghosh, M., Raju, V.S., Tabei, K. and Urbanczyk-Lipkowska, Z., Highly accelerated reactions in a microwave oven: synthesis of heterocycles, *Heterocycles*, 1990, **30**, 471.
56. Berlan, J., Giboreau, P., Lefeuvre, S. and Marchand, C., Organic synthesis with microwave: first example of specific activation in homogenous phase, *Tetrahedron Lett.*, 1991, **32**, 2363.
57. Pollington, S.D., Bond, G., Moyes, R.B., Whan, D.A., Candlin, J.P. and Jennings, J.R., The influence of microwaves on the rate of reaction of propan-1-ol with ethanoic acid, *J. Org. Chem.*, 1991, **56**, 1313.
58. Plazl, I., Esterification of benzoic acid with ethanol by conventional and microwave heating in stirred tank reactor, *Acta Chim. Slovenica*, 1994, **41**, 437.
59. Laurent, R., Laporterie, A., Dubac, J., Berlan, J., Lefeuvre, S. and Audhuy, M., Specific activation by microwaves: myth or reality? *J. Org. Chem.*, 1992, **57**, 7099.
60. Jahngen, E.G.E., Lentz, R.R., Pesheck, P.S. and Sackett, P.H., Hydrolysis of adenosine triphosphate by conventional or microwave heating, *J. Org. Chem.*, 1990, **55**, 3406.
61. Raner, K.D. and Strauss, C.R., Influence of microwaves on the rate of esterification of 2,4,6-trimethylbenzoic acid with 2-propanol, *J. Org. Chem.*, 1992, **57**, 6231.
62. Raner, K.D., Strauss, C.R., Vyskoc, F. and Mokbel, L., A comparison of reaction kinetics observed under microwave irradiation and conventional heating, *J. Org. Chem.*, 1993, **58**, 950.
63. Stadler, A. and Kappe, C.O., High-speed couplings and cleavages in microwave-heated, solid-phase reactions at high temperatures, *Eur. J. Org. Chem.*, 2001, 919.
64. Schiffmann, R.F., Microwave technology and applications, In: McKetta, J.J. and Cunningham, W.A. (Eds.), *Encyclopedia of Chemical Processing and Design, Vol. 30,* Marcel Dekker, New York, 1989, p. 202.
65. Raner, K.D., Strauss, C.R., Trainor, R.W. and Thorn, J.S., A new microwave reactor for batchwise organic synthesis, *J. Org. Chem.*, 1995, **60**, 2456.
66. Wan, J.K.S., Bamwenda G. and Depew, M.C., Microwave induced catalytic reactions of carbon dioxide and water: mimicry of photosynthesis, *Res. Chem. Intermed.*, 1991, **16**, 241.
67. Koch, T.A., Krause, K.R. and Mehdizadeh, M., Improved safety through distributed manufacturing of hazardous chemicals, *Process Safety Prog.* 1997, 16(1), 23–24.
68. Wan, J.K.S. and Koch, T.A., Application of microwave radiation for the synthesis of hydrogen cyanide, *Res. Chem. Intermed.*, 1994, **20**, 29.
69. Mehdizadeh, M., Engineering and scale-up considerations for microwave induced reactions, *Res. Chem. Intermed.*, 1994, **20**, 79.

70. Baghurst, D.R. and Mingos, D.M.P., Superheating effects associated with microwave dielectric heating, *J. Chem. Soc., Chem. Commun.*, 1992, 674.

71. Armstrong, B.F. and Neas, E.D., Development of a microwave distillation system for the analytical laboratory, *Separation Sci. Tech.*, 1990, **25**, 2007.

72. Mingos, D.M.P., The applications of microwaves in chemical syntheses, *Res. Chem. Intermed.*, 1994, **20**, 85.

73. Straathof, A.J.J., van Bekkum, H. and Kieboom, A.P.G., Preparation of 1,6-anhydroglucose from 1,4-glucans using microwave technology, *Rec. des Trav. Chim. des Pays-Bas*, 1988, **107**, 647.

74. Sauvagnat, B., Lamaty, F., Lazaro, R. and Martinez, J., Polyethylene glycol as solvent and polymer support in the microwave assisted parallel synthesis of amino acid derivatives, *Tetrahedron Lett.*, 2000, **41**, 6371.

75. Chemat, F., Poux, M., Di Martino, J.L. and Berlan, J., A new continuous-flow recycle microwave reactor for homogeneous and heterogeneous chemical reactions, *Chem. Eng. Tech.*, 1996, **19**(5), 420.

76. Perio, B., Dozias, M.-J. and Hamelin, J., Ecofriendly fast batch synthesis of dioxolanes, dithiolanes, and oxathiolanes without solvent under microwave irradiation, *Org. Process. Res. Dev.*, 1998, **2**, 428.

77. Besson, T., Dozias, M.-J., Guillard, J., Jacquault, P., Legoy, M.D. and Rees, C.W., Expeditious routes to 4-alkoxyquinazoline-2-carbonitriles and aryl thiocarbamates via (arylimino)-1,2,3-dithiazoles using microwave irradiation, *Tetrahedron*, 1998, **54**, 6475.

78. Guillard, J., Schmitt, V., Dozias, M.J. and Besson, T. The scale-up of focused microwave assisted organic synthesis: from the research to the pilot production scale, In: *Book of Abstracts, 216th ACS National Meeting*, Boston, August 23–27, 1998, ANYL-114.

79. Cléophax, J., Liagre, M., Loupy, A. and Petit, A., Application of focused microwaves to the scale-up of solvent-free organic reactions, *Org. Process. Res. Dev.*, 2000, **4**, 498.

80. Abenhaim, D., Diez-Barra, E., De la Hoz, A., Loupy, A. and Sanchez-Migallon, A., Selective alkylations of 1,2,4-triazole and benzotriazole in the absence of solvent, *Heterocycles*, 1994, **38**, 793.

81. Esveld, E., Chemat, F. and van Haveren, J., Pilot scale continuous microwave dry-media reactor, Part 1: design and modeling, *Chem. Eng. Technol.*, 2000, **23**, 279.

82. Esveld, E., Chemat, F. and van Haveren, J., Pilot scale continuous microwave dry-media reactor, Part II: application to waxy esters production, *Chem. Eng. Technol.*, 2000, **23**, 429.

83. Deetlefs, M. and Seddon, K., Improved preparations of ionic liquids using microwave irradiation, *Green Chem.*, 2003, **5**, 181.

84. Stadler, A. and Kappe, C.O., Microwave-mediated Biginelli reactions revisited: on the nature of rate and yield enhancements, *J. Chem. Soc., Perkin Trans. 2*, 2000, 1363.

85. Vidal, T., Petit, A., Loupy, A. and Gedye, R.N., Re-examination of microwave-induced synthesis of phthalimides, *Tetrahedron*, 2000, **56**, 5473.

86. Fubini, B. and Otero Arean, C., Chemical aspects of the toxicity of inhaled mineral dusts, *Chem. Soc. Rev.*, 1999, **28**, 373.

87. Banik, B.K., Manhas, M.S., Kaluza, Z., Barakat, K.J. and Bose, A.K., Microwave-induced organic reaction enhancement chemistry, 4: convenient synthesis of enantiopure α-hydroxy-β-lactams, *Tetrahedron Lett.*, 1992, **33**, 3603.

88. Bose, A.K., Manhas, M.S., Ghosh, M., Shah, M., Raju, V.S., Bari, S.S., Newaz, S.N., Banik, B.K., Chaudhary, A.G. and Baraket, K.J., Microwave-induced organic reaction enhancement chemistry, 2: simplified techniques, *J. Org. Chem.*, 1991, **56**, 6968.

89. Banik, B.K., Manhas, M.S., Newaz, S.N. and Bose, A.K., Studies on lactams, 92: facile preparation of carbapenem synthons via microwave-induced rapid reaction, *Bioorg. Med. Chem. Lett.*, 1993, **3**, 2363.

90. Gabriel, C., Gabriel, S., Grant, E.H., Halstead, B.S.J. and Mingos, D.M.P., Dielectric parameters relevant to microwave dielectric heating, *Chem. Soc. Rev.*, 1998, **27**, 213.

91. Cablewski, T., Faux, A.F. and Strauss, C.R., Development and application of a continuous microwave reactor for organic synthesis, *J. Org. Chem.*, 1994, **59**, 3408.

92. Constable, D., Raner, K., Somlo, P. and Strauss, C., A new microwave reactor suitable for organic synthesis and kinetic studies, *J. Microwave Power Electromag. Energy*, 1992, **26**, 195.

93. Strauss, C.R. and Trainor, R.W., Reactions of ethyl indole-2-carboxylate in aqueous media at high temperature, *Aust. J. Chem.*, 1998, 51, 703.

94. Strauss, C.R. and Trainor, R.W., In: Application of new microwave reactors for food and flavour research, Takeoka, G.R, Teranishi, R., Williams, P.J. and Kobayashi, A. (Eds.), *Biotechnology for Improved Foods and Flavors*, ACS Symposium Series 637, American Chemical Society, Washington, DC, 1996, Ch. 26, pp. 272–281.

95. Strauss, C.R., Microwave-assisted reactions in organic synthesis: are there any nonthermal microwave effects? Comments, *Angew. Chem. Int. Ed.*, 2002, **41**, 3589.

96. Whittaker, A.G. and Mingos, D.M.P., Arcing and other microwave characteristics of metal powders in liquid systems, *J. Chem. Soc., Dalton Trans.*, 2000, 1521.

97. Bagnell, L., Bliese, M., Cablewski, T., Strauss, C.R. and Tsanaktsidis, J., Environmentally benign procedures for the preparation and isolation of 3-methylcyclopentd-2-en-1-one, *Aust. J. Chem.*, 1997, **50**, 921.

98. Braun, I., Schulz-Ekloff, G., Wohrle, D. and Lautenschlager, W., Synthesis of AlPO4-5 in a microwave-heated, continuous-flow, high-pressure tube reactor, *Microporous Mesoporous Mat.*, 1998, **23**, 79.

99. Shieh, W.-C., Dell, S. and Repič, O., 1,8-Diazabicyclo[5.4.0]undec-7-ene (DBU) and microwave-accelerated green chemistry in methylation of phenols, indoles, and benzimidazoles with dimethyl carbonate, *Org. Lett.*, 2001, **3**, 4279.

100. Lew, A., Krutzik, P.O., Hart, M.E. and Chamberlin, A.R., Increasing rates of reaction: microwave-assisted organic synthesis for combinatorial chemistry, *J. Comb. Chem.*, 2002, **4**, 95 and references therein.

101. Yu, H.M., Chen, S.T., Chiou, S.H. and Wang, K.T., Determination of amino acids on Merrifield resin by microwave hydrolysis, *J. Chromatogr.*, 1988, **456**, 357.

102. Yu, H.M., Chen, S.T. and Wang, K.T., Enhanced coupling efficiency in solid-phase peptide synthesis by microwave irradiation, *J. Org. Chem.*, 1992, **57**, 4781.

103. Larhed, M., Lindeberg, G. and Hallberg, A., Rapid microwave-assisted Suzuki coupling on solid-phase, *Tetrahedron Lett.*, 1996, **37**, 8219.

104. Larhed, M., Moberg. C. and Hallberg, A., Microwave-accelerated homogeneous catalysis in organic chemistry, *Acc. Chem. Res.*, 2002, **35**, 717.

105. Scharn, D., Wenschuh, H., Reineke, U., Schneider-Mergener, J. and Germeroth, L., Spatially addressed synthesis of amino- and amino-oxy-substituted 1,3,5-triazine arrays on polymeric membranes, *J. Comb. Chem.*, 2000, **2**, 361.

106. Larhed, M., Hoshino, M., Hadida, S., Curran, D.P. and Hallberg, A., Rapid fluorous stille coupling reactions conducted under microwave irradiation, *J. Org. Chem.*, 1997, **62**, 5583.

107. Jones, J.R. and Lu, S.-Y., New microwave-enhanced deuteriation/tritiation studies, In: *Proc Int. Conf. on Microwave Chemistry*, 4–7 September, 2000, Institut National Polytechnique de Toulouse, p. 51.

108. Barthez, J.M., Filikov, A.V., Frederiksen, L.B., Huguet, M.L. Jones, J.R. and Lu, S.-Y., Microwave-enhanced metal- and acid-catalyzed hydrogen isotope exchange reactions, *Can. J. Chem.*, 1998, **76**, 726.

109. Erb, W. Th., Jones, J.R. and Lu, S.-Y., Microwave enhanced deuteriations in the solid state using alumina doped sodium borodeuteride, *J. Chem. Res. (S)*, 1999, 728.

110. Frederiksen, L.B., Grobosch, T.H., Jones, J. R, Lu, S.-Y. and Zhao, C.C., Microwave enhanced decarboxylation of aromatic carboxylic acids: improved deuteration/tritiation potential, *J. Chem. Res. (S)*, 2000, 42.

111. Horeis, G., Pichler, S., Stadler, A., Gössler, W. and Kappe, C.O., microwave assisted organic synthesis – back to the roots, In: *Fifth International Electronic Conference on Synthetic Organic Chemistry (ECSOC-5)* E0000.

112. Stadler, A., Yousefi, B.H., Dallinger, D., Walla, P., Van der Eycken, E., Kaval, N. and Kappe, C.O., Scalability of microwave-assisted organic synthesis: from single-mode to multimode parallel batch reactors, *Org. Process Res. Dev.*, 2003, **7**, 707.

113. Bose, A.K., Manhas, M.S., Ganguly, S.N., Sharma, A., Huarotte, M., Rumthao, S., Jayaraman, M. and Banik, B.K., MORE chemistry techniques for process development, In: *Fifth International Electronic Conference on Synthetic Organic Chemistry (ECSOC-5)* E0047.

114. Breslow, R., Hydrophobic effects on simple organic reactions in water, *Acc. Chem. Res.*, 1991, **24**, 159.

115. Grieco, P.A., Organic chemistry in unconventional solvents, *Aldrichim. Acta*, 1991, **24**, 59.

116. Lubineau, A., Augé, J. and Queneau, Y., Water-promoted organic reactions, *Synthesis*, 1994, 741.

117. Grieco, P.A., Brandes, E.B., McCann, S. and Clark, J.D., Water as a solvent for the Claisen rearrangement: practical implications for synthetic organic chemistry, *J. Org. Chem.*, 1989, **54**, 5849.

118. Li, C.-J., Organic reactions in aqueous media – with a focus on carbon–carbon bond formation, *Chem. Rev.*, 1993, **93**, 2023.

119. Blokzijl, W. and Engberts, J.B.F.N., Hydrophobic effects: opinion and fact, *Angew. Chem. Int. Ed. Engl.*, 1993, **32**, 1545.

120. Lubineau, A. and Augé, J., In: Water as solvent in organic synthesis, Knochel, P. (Ed.), *Topics in Current Chemistry, Vol. 206: Modern Solvents in Organic Synthesis*, Springer, Berlin, 1999, pp. 1–39.

121. An, J., Bagnell, L., Cablewski, T., Strauss, C.R. and Trainor, R.W., Applications of high-temperature aqueous media for synthetic organic reactions, *J. Org. Chem.*, 1997, **62**, 2505.

122. Bagnell, L., Cablewski, T., Strauss, C.R. and Trainor, R.W., Reactions of allyl phenyl ether in high-temperature water with conventional and microwave heating, *J. Org. Chem.*, 1996, **61**, 7355.

123. Strauss, C.R., A combinatorial approach to the development of environmentally benign organic chemical preparations, *Aust. J. Chem.*, 1999, **52**, 83.

124. Bagnell, L., Cablewski, T. and Strauss, C.R., A catalytic symmetrical etherification, *Chem. Commun.*, 1999, 283.

125. Chen, S.-T., Chiou, S.-H. and Wang, K.-T., Preparative scale organic synthesis using a kitchen microwave oven, *Chem. Comm.*, 1990, 807.

126. Marquié, J., Salmoria, G., Poux, M., Laporterie, A., Dubac, J. and Roques, N., Acylation and related reactions under microwaves, 5: development to large laboratory scale with a continuous-flow process, *Ind. Eng. Chem. Res.*, 2001, **40**, 4485.

127. Chemat, F., Esveld, D.C., Poux, M. and DiMartino, J.L., The role of selective heating in the microwave activation of heterogeneous catalysis reactions using a continuous microwave reactor, *J. Microwave Power Electromag. Energy*, 1998, **2**, 88.
128. Kress, A.O., Mathias, L.J. and Cei, G., Copolymers of styrene and methyl α-(hydroxymethyl)acrylate: reactivity ratios, physical behavior, and spectral properties, *Macromolecules*, 1989, **22**, 537.
129. Kū sefoğ lu, S.H., Kress, A.O. and Mathias, L.J., Functional methacrylate monomers: simple synthesis of alkyl α-(hydroxymethyl) acrylates, *Macromolecules*, 1987, **20**, 2326.
130. Ueda, M., Koyama, T., Mano, M. and Yazawa, M., Radical-initiated homopolymerization and copolymerization of ethyl α-hydroxymethylacrylate, *J. Polym. Sci., Part A: Polym. Chem.*, 1989, **27**, 751.
131. Mathias, L.J., Warren, R.M. and Huang, S., *tert*-Butyl α-(hydroxymethyl)acrylate and its ether dimer: multifunctional monomers giving polymers with easily cleaved ester groups, *Macromolecules*, 1991, **24**, 2036.
132. Avci, D. and Kū sefoğlu, S.H., Functionalization and crosslinking reactions of ethyl α-hydroxymethylacrylate, *J. Polym. Sci., Part A: Polym. Chem.*, 1993, **31**, 2941.
133. Avci, D., Kū sefoğ lu, S.H., Thompson, R.D. and Mathias, L.J., Ester derivatives of α-(hydroxymethyl) acrylates: itaconate isomers giving high molecular weight homopolymers, *Macromolecules*, 1994, **27**, 1981.
134. Danner, H. and Braun, R., Biotechnology for the production of commodity chemicals from biomass, *Chem. Soc. Rev.*, 1999, **28**, 395.
135. Kontoghiorghes, G.J. and Sheppard, L., Simple synthesis of the potent iron chelators 1-alkyl-3-hydroxy-2-methylpyrid-4-ones, *Inorg. Chim. Acta*, 1987, **136**, L11.
136. Chemat, F., Poux, M., Di Martino, J.-L. and Berlan, J., An original microwave-ultrasound combined reactor suitable for organic synthesis: application to pyrolysis and esterification, *J. Microwave Power Electromag. Energy*, 1996, **31**, 19.
137. Chemat, F., Poux, M. and Galema, S.A., Esterification of stearic acid by isomeric forms of butanol in a microwave oven under homogeneous and heterogeneous reaction conditions, *J. Chem. Soc., Perkin Trans. 2*, 1997, 2371.
138. Lagha, A., Chemat, S., Bartels, P.V. and Chemat, F., Microwave-ultrasound combined reactor suitable for atmospheric sample preparation procedure of biological and chemical products, *Analusis*, 1999, **27**, 452.
139. Peng, Y. and Song, G., Simultaneous microwave and ultrasound irradiation: a rapid synthesis of hydrazides, *Green Chem.*, 2001, **3**, 302.
140. Chemat, S., Aouabed, A., Bartels, P.V., Esveld, D.C. and Chemat, F., An original microwave-ultra violet combined reactor suitable for organic synthesis and degradation, *J. Microwave Power Electromag. Energy*, 1999, **34**, 55.
141. Církva, V. and Hájek M., Microwave photochemistry: photoinitiated radical addition of tetrahydrofuran to perfluorohexylethene under microwave irradiation, *J. Photochem. Photobiol., A: Chem.*, 1999, **123**, 21.
142. Klán, P., Literák, J. and Hájek, M., The electrodeless discharge lamp: a prospective tool for photochemistry, *J. Photochem. Photobiol., A: Chem.*, 1999, **128**, 145.
143. Literák, J. and Klán, P., The electrodeless discharge lamp: a prospective tool for photochemistry, Part 2: scope and limitation, *J. Photochem. Photobiol., A: Chem.*, 2000, **137**, 29.
144. Klán, P., Hájek, M. and Církva, V., The electrodeless discharge lamp: a prospective tool for photochemistry, Part 3: the microwave photochemistry reactor, *J. Photochem. Photobiol., A: Chem.*, 2001, **140**, 185.
145. Compton, R.G., Coles, B.A. and Marken, F., Microwave activation of electrochemical processes at microelectrodes, *Chem. Comm.*, 1998, 2595.
146. Marken, F., Tsai, Y.-C., Coles, B.A., Matthews, S.L. and Compton, R.G., Microwave activation of electrochemical processes: convection, thermal gradients and hot spot formation at the electrode·solution interface, *New J. Chem.*, 2000, **24**, 653.
147. Marken, F., Tsai, Y.-C., Saterlay, A.J., Coles, B.A., Tibbetts, D., Holt, K., Goeting, C.H., Foord, J.S. and Compton, R.G., Microwave activation of electrochemical processes: enhanced PbO_2 electrodeposition, stripping and electrocatalysis, *J. Solid State Electrochem.*, 2001, **5**, 313.
148. Tsai, Y.-C., Coles, B.A., Compton, R.G. and Marken, F., Microwave activation of electrochemical processes: enhanced electrodehalogenation in organic solvent media, *J. Am. Chem. Soc.*, 2002, **124**, 9784.
149. Mason, T.J., Ultrasound in synthetic organic chemistry, *Chem. Soc. Rev.*, 1997, **26**, 443.

Index

acid digestion, accelerated, 2, 15, 238
α-acylamino amides, 107–108
2-acylaminothiophenes, 203–204
acylation reactions, 147–8, 228, 254
4-acyloxypyrimidine, 147–8
alcohols
    by carbonyl reduction, 80–87
    cyclic phenylethylamino, 205–206
    deuterated, 82
    dielectric loss spectra, 8, 10, 11
    esterification, 152–3, 162–3
    loss tangents, 5
    nitro, 143
    relaxation times and dielectric
        properties, 5–6
    resolution of 1-phenol ethanol, 162
aldehydes *see* carbonyl compounds
alkaloids, 65, 69, 125
alkenes
    bromomethoxylation, 153
    hydroacylation, 80
    hydrogenation, 78, 155–6
3-(4-alkoxyphenyl)-3-methylbutan-3-
    ones, 157
alkylaminopropenones and propenoates,
    115, 116–17
alkylation
    allylic, 34–5, 194
    carboxylic acids, 144
    of phenolic compounds, 144–5, 253
alkynes, hydrogenation, 78–9
allylic alkylations, molybdenum-
    catalysed, 34–5, 194
O-allylsalicylic acids, 207–208
amination
    aryl bromides, 32–3
    5-bromoquinoline, 33–4
    reductive, 142
amines

acylation, 147–8
    aromatic, 32–3
    protection, 188
    reduction of imines, 87–90
    reduction of nitro compounds,
        90–93
    scavenging, 167–8
aminocarbonylation, 36–7
β-aminoketones, 111
α-amino phosphates, 126–7
aminopropenones and
        aminopropenoates, 192–3
aminopyrimidines, 29
aminoquinolines, 33, 225
ammonium formamide, 109
ammonium formate and hydrogenation,
        77, 78
aryl halides
    Buchwald–Hartwig couplings, 32–3
    and carbonylation reactions, 36–9
    cyanation of, 31–2
    Heck couplings, 29, 156–7
    Suzuki couplings, 27, 28, 28–9
N-arylimidazole, 33
aryl nitriles, 31
aryl triflates, 27
aza-Diels–Alder reaction, 114

Beckmann rearrangement, 153–5
benzamides, 38–9
benzamidazoles, 258–9
benzimidazo[1,2-*c*]quinazolines, 69
benzimidazoles, 49–50, 195–6
benzodiazepines, 63–5
benzofurans, 45
benzopyrans, 59–61
benzoxazoles, 52, 53
Biginelli reaction, 57, 61, 104, 107, 210
Bohlmann–Rahtz reaction, 57

273